The Composer Embalmed

NEW MATERIAL HISTORIES of MUSIC

a series edited by
James Q. Davies *and*
Nicholas Mathew

ALSO PUBLISHED IN THE SERIES:

Musical Vitalities: Ventures in a Biotic Aesthetics of Music
Holly Watkins

Sex, Death, and Minuets: Anna Magdalena Bach and Her Musical Notebooks
David Yearsley

The Voice as Something More: Essays toward Materiality
Edited by Martha Feldman and Judith T. Zeitlin

Listening to China: Sound and the Sino-Western Encounter, 1770–1839
Thomas Irvine

The Search for Medieval Music in Africa and Germany, 1891–1961: Scholars, Singers, Missionaries
Anna Maria Busse Berger

An Unnatural Attitude: Phenomenology in Weimar Musical Thought
Benjamin Steege

Mozart and the Mediation of Childhood
Adeline Mueller

Musical Migration and Imperial New York: Early Cold War Scenes
Brigid Cohen

The Haydn Economy: Music, Aesthetics, and Commerce in the Late Eighteenth Century
Nicholas Mathew

Tuning the World: The Rise of 440 Hertz in Music, Science, and Politics, 1859–1955
Fanny Gribenski

Music in the Flesh: An Early Modern Musical Physiology
Bettina Varwig

Creatures of the Air: Music, Atlantic Spirits, Breath, 1817–1913
J. Q. Davies

Sounding Human: Music and Machines, 1740/2020
Deirdre Loughridge

Format Friction: Perspectives on the Shellac Disc
Gavin Williams

The Composer Embalmed

Relic Culture from Piety to Kitsch

ABIGAIL FINE

The University of Chicago Press
Chicago and London

The University of Chicago Press, Chicago 60637
The University of Chicago Press, Ltd., London
© 2025 by The University of Chicago
All rights reserved. No part of this book may be used or reproduced in any manner whatsoever without written permission, except in the case of brief quotations in critical articles and reviews. For more information, contact the University of Chicago Press, 1427 E. 60th St., Chicago, IL 60637.
Published 2025

34 33 32 31 30 29 28 27 26 25   1 2 3 4 5

ISBN-13: 978-0-226-83605-8 (cloth)
ISBN-13: 978-0-226-84044-4 (paper)
ISBN-13: 978-0-226-84045-1 (e-book)
DOI: https://doi.org/10.7208/chicago/9780226840451.001.0001

This book has been supported by the AMS 75 PAYS Fund of the American Musicological Society.

Library of Congress Cataloging-in-Publication Data

Names: Fine, Abigail, author.
Title: The composer embalmed : relic culture from piety to kitsch / Abigail Fine.
Other titles: New material histories of music.
Description: Chicago ; London : The University of Chicago Press, 2025. | Series: New material histories of music | Includes bibliographical references and index.
Identifiers: LCCN 2024042411 | ISBN 9780226836058 (cloth) |
    ISBN 9780226840444 (paperback) | ISBN 9780226840451 (ebook)
Subjects: LCSH: Music fans—Germany—History—19th century. | Music fans—Austria—History—19th century. | Music fans—Germany—History—20th century. | Music fans—Austria—History—20th century. | Beethoven, Ludwig van, 1770–1827—Collectibles. | Beethoven, Ludwig van, 1770–1827—Monuments. | Beethoven, Ludwig van, 1770–1827—Relics. | Beethoven, Ludwig van, 1770–1827—Public opinion. | Mozart, Wolfgang Amadeus, 1756–1791—Collectibles. | Mozart, Wolfgang Amadeus, 1756–1791—Monuments. | Mozart, Wolfgang Amadeus, 1756–1791—Relics. | Composers—Monuments—Germany. | Composers—Monuments—Austria. | Music—Social aspects—Germany. | Music—Social aspects—Austria. | Music and tourism—Germany. | Music and tourism—Austria. | Composers—Homes and haunts.
Classification: LCC ML3916 F55 2025 | DDC 780.92/2—dc23/eng/20241104
LC record available at https://lccn.loc.gov/2024042411

Contents

List of Illustrations and Musical Example vii
Preface ix

Introduction 1
1 Beethoven's Nativity 20
2 Mozart on the Mountaintop 51
3 From Relic to Specimen 80
4 Beethoven's Masks and the Beautiful Death 105
5 Art-Religion *Verkitscht* 137
Coda 171

Acknowledgments 175
List of Archives and Abbreviations 179
Notes 181
Bibliography 219
Index 245

Illustrations and Musical Example

Illustrations

0.1   Stills from Mauricio Kagel's *Ludwig van*   x
1.1   Caricature of Joseph Joachim by Franz Stassen   29
1.2   Interior views of the Beethoven-Haus, Bonn   32
1.3   Friedrich Geselschap, *Beethoven's Birth*   33
1.4   Page from the Beethoven-Haus visitors' books   35
1.5   Atmospheric musical entries in the Beethoven-Haus visitors' books   37
1.6   Carl Berg's longest entry in the Beethoven-Haus visitors' books   38
1.7   Postcard with Egger-Lienz, *Ninth Symphony*   41
1.8   Front page of the *Illustrirtes Wiener Extrablatt*, 1903   47
2.1   The *Magic Flute* cottage on the Kapuzinerberg   52
2.2   Comparison of a design for Mozart's festival house with Wagner's festival house in Bayreuth   56
2.3   Interior of the *Magic Flute* cottage   59
2.4   Souvenir miniature of the *Magic Flute* cottage   59
2.5   Josef Hoffmann's design for an Egyptian temple to protect the *Magic Flute* cottage   61
2.6   Schutz's design for a protective cover for the *Magic Flute* cottage   62
2.7   Men's choral ceremony for the reopening of the *Magic Flute* cottage   64
2.8   Anton Romako, *Mozart at the Spinet*   73
2.9   Lindenschmit the Younger, *Hall of Heroes of German Music*   74
3.1   Haydn's skull in a decorative case   93

3.2   The comparative approach to composers' bones   99
3.3   Casts of Schubert's, Haydn's, and Beethoven's skulls   100
3.4   Ernst Klotz, "The sculptor kisses the scholarly table with devotion"   101
4.1   Lionello Balestrieri, *Beethoven*   106
4.2   Postcard with C. V. Muttich, *Beethoven's Sonate*   107
4.3   Comparison of Beethoven's life and death masks   108
4.4   Pages from product catalog advertising Beethoven's masks   108
4.5   C. W. Bergmüller, *Beethoven-Sonate*   123
4.6   Georg Wimmer, *Head of Beethoven over Nocturnal Waters*   124
4.7   Franz von Stuck, *Beethoven Mask with Laurel Wreath*   125
4.8   Beethoven's mask in still-life scenes   126
4.9   Etching by Walther Rath, "My songs will live on when I myself am gone"   127
4.10 Max Klinger, *Pietà*   131
5.1   Caricature of Liszt and Rákóczy rising from their graves   150
5.2   Comparison of Bruckner's tomb with that of Emperor Maximilian I of Mexico   164

Musical Example

1   Excerpt from Max von Weinzierl, "Des Künstlers Genius," 1877   67

Preface

Consider a poignant scene. Through point-of-view shots, we watch as Ludwig van Beethoven returns to Bonn and visits a twisted version of his birth-house museum. His convoluted tour brings him to a bathtub piled with corroded busts of lard and chocolate, an installation designed by the Swiss artist Dieter Roth, who specialized in dilapidated statuary made from edible materials. In a ritual that spans several minutes, Beethoven tenderly lifts the ruins of his own visage from the tub, grasping them at odd angles like archaeological artifacts (fig. 0.1). This scene would feel eerie, even horrific, if it weren't for the music that accompanies it: a string quartet arrangement of the last movement of Beethoven's Piano Sonata, op. 109, which intones longing and loss. The underscore transforms these busts into an homage to *Sehnsucht*, the Romantic yearning for that which is distant or absent, as reflected in the hollow core of the film's truncated title: *Ludwig van*, by Mauricio Kagel. In nineteenth-century funeral culture, families longed for the departed through relics and mementos that preserved a "beautiful death," a transfigured face fringed with flowers.[1] But there is little beauty in this ruin gazing. These busts are decidedly icky, made of schmutz and schmaltz that will not rinse clean. As foodstuffs, they signal mass consumption both literal and figurative. Across this long and repetitive scene, yearning crosses over into fixation.

Ludwig van was commissioned by the West German broadcasting station WDR as a provocation for Beethoven's bicentennial year of 1970. Those who have studied this film regard it as an artifact of reactionary politics, when a wave of West German artists energized by the student movements of 1968 critiqued bourgeois classicism through performance art and postmodern collage.[2] The more political scenes in Kagel's film reflect a tension with the old guard, such as the cameo appearance of media personality Werner Höfer, a

FIGURE 0.1. Stills from Mauricio Kagel, *Ludwig van* (1969). Kagel worked with artist Dieter Roth to design this scene. The tub of busts was later displayed as a freestanding installation.

former Nazi who had not yet retired from public life, and a desiccated caricature of the recently deceased pianist Elly Ney, a darling of the regime.[3] In an interview, Kagel insisted that his film took aim not at Beethoven, whom he appreciated deeply, but at the dry ersatz rituals of the concert hall that turn the composer into a surrogate Christ and his music into moral betterment.[4] Above all, the film critiques the inability to leave Beethoven well enough alone. In one scene, the composer's own art song speaks for him: *in questa tomba oscura*, "let me rest in my dark tomb."[5]

Throughout this film, the inability to let the dead rest amounts to a suffocating volume of *stuff*. In the oneiric birth-house museum, the reincarnated Beethoven encounters the superfluity of his visage on every surface. What begins as a normal assemblage of things transforms into a satire of what speakers of German call *Musealisierung*, the impulse to preserve and display: vitrines stuffed to the gills with medallions, bones, clothes, suitcases, and heaps of mundane and broken things relegated to the "wine cellar." Beethoven's death mask is cradled in a basket of straw like a baby in a manger, visible through a hole in the birth-room wall. Later, in a touristic spin on civic rituals, musicians process solemnly through a ferry along the Rhine as

PREFACE   xi

they play the funeral march from the Piano Sonata, op. 26. The film takes an anatomical turn. Model skeletons, larynxes, and ears demonstrate their function; a skull breaks open in an exploded view; a pianist plays the *Tempest* Sonata while wired into a pseudoscientific mechanism that measures a litany of odd ailments.

In its time, Kagel's satire might have alluded to the Cold War friction between West Germany and the Communist East, where Lenin's Mausoleum and other Soviet bodies turned the dead into propaganda, while the West (falsely) claimed to bury their dead with pious respect. But the belly laughs in the film arise from a much older suspicion of secular religion, with its forgery of the sacred that turns piety into kitsch. Throughout the film, Kagel conjures holy atmospheres only to deflate them. At the bathtub, Kagel bursts our bubble with a sudden pan to the alleyway below, where workers busily deface miniature busts of Beethoven and sweep them into oblivion. These desktop busts, still found today on living-room pianos, have become a symbol of the dilettante, an icon of art-religion *verkitscht*, or cheapened, through mass manufacture. Later, a shrine to musical works—that is, a room papered floor to ceiling with scores and populated by cardboard cutouts of composers— cuts abruptly to the "storage room," a closet packed with forgotten repertoire that topples onto the floor. By the end, a grandiose rendition of the *Waldstein* Sonata loops onto itself and collapses into a jaunty oom-pah, pulsating over footage of the local zoo. Kagel makes the devotion to German art music look alien and extreme, if affectionate and well intentioned.

Without its underscore, Dieter Roth's ruined pile of busts appears quite desolate. His installation suggests that composers survive not as relics, but as replicas, smashed and smeared as soon as we reach for them. But in Kagel's film, our attention is drawn equally to musical sound. Most of the film is accompanied by Kagel's arrangements of Beethoven's music, which scratch and wheeze as they mimic the missing partials of the composer's hearing. In contrast, the Sonata, op. 109 is remarkably whole, its black-and-white piano timbre now recast in sepia tones. As we confront the futility of the composer embalmed, music is the last surviving trace of the beautiful death.

Introduction

In 1906, three relics were donated to Vienna's Society for Friends of Music by the descendants of an Austrian diplomat: a lock of hair, a walking stick, and a spoon. All three had belonged to Beethoven, and their letter of provenance contains a facetious quote by Julius Schneller that is emblematic of the relic hunter: "Beethoven's cane is one of my greatest treasures; musical ladies who have given me the impression of artistic upbringing through their chaste and virginal conduct are allowed to kiss it. Thus far I have only permitted this honor three times."[1]

For those accustomed to seeing objects behind glass, it is startling to read about such tactile and erotic encounters. Prior to the age of museums, private collectors could grant the gift of touch to an inner circle like these ladies of artistic upbringing. In a comparable instance of gatekeeping, the mezzo-soprano Pauline Viardot enshrined the autograph manuscript of Mozart's *Don Giovanni* in a custom-built reliquary and invited famous guests like Tchaikovsky and Rossini to "genuflect" before it.[2] Even after relics went public in the vitrines of museums and music societies,[3] institutions recognized these habits of communion and became the new gatekeepers, couching themselves as pious sanctuaries. Their mission statements claimed that one could feel the phantom composer in rooms where he lived or died, and in some cases, the walls of a historic house became a surrogate body. Relics also populated the annals of medicine, as art-loving doctors displayed specimens of genius on plush pillows or in glass cases. Later, at the turn of the century, select relics were mass-produced, such as the plaster mask of Beethoven that found its way onto parlor walls. All these objects—from first-class relics such as body parts, to second-class or "contact" relics such as autographs, to ersatz relics like houses and replicas—constitute what we might broadly call "relic culture." With the late nineteenth-century decline of private collecting in favor of public display, every music lover, regardless of their connection to the dead, found a path into what felt like a prestigious circle of devotees.

The contradictory nature of that devotion, which took extreme forms in Germany and Austria between 1870 and 1930, is the subject of this book. Affection for composers, when expressed through contact with their material traces, could be earnest and tender, a form of friendship displaced onto the dead. Yet those same behaviors could also be competitive, transactional, and obsessive. Julius Schneller did not boast of how he cherished Beethoven's walking stick quietly by the fireside. Rather, he encouraged his guests to embrace the relic with erotic charge—to enact a relationship with Beethoven that exaggerates what friends in mourning might do, and that makes the collector a perverse voyeur. This book shows how fetishistic forms of violence can underlie the laudation that built the Western canon. To sanctify the remnants of composers, one had to tear those bodies asunder, or preserve them past the point of dignity.

If composers' relics and shrines resemble those of the saints, it is reasonable to ask whether this was an extension of German *Kunstreligion* (art-religion). This compound word originated with the writings of the Jena Romantics as they sought to enchant the listening experience in the wake of Enlightenment secularism.[4] In Germany, philosophical idealism was one of many secular domains, alongside art and scholarship, that were enmeshed with Christian theology.[5] By the later nineteenth century, art-religion combined aspects of Catholic revival, Lutheran Pietism, and German *Bildung*—defined as moral and intellectual self-cultivation through engagement with high culture—and it could be confessional or transconfessional as the occasion demanded. Relics and imagery inspired by the Nazarene movement were decidedly Catholic.[6] Devotional listening, with its correlate in the "inner religion" of pietism, was Lutheran, which meant that concerts could assume the tone of sermons.[7] All confessions could participate equally in liberal aspirations toward a universal *Deutschtum*—that is, a Germanness that would unify a diverse Habsburg Empire and pacify Germany's Catholic-Protestant divide after the *Kulturkampf* of the 1870s.[8] And behind that confessional facade, Bildung operated as the quiet engine. This concept, too, straddled the sacred and secular: it began as a movement of Rhineland mystics who promoted the divine formation of humans in God's image, and who advanced the conviction that art can freely form the soul, whereas the church, the traditional arbiter of morality, had become too authoritarian.[9] By the late nineteenth century, these philosophies were accompanied by more concrete promise of social mobility through education, with an emphasis on humanistic products of German and Austrian creativity. The *Bildungsbürgertum*—that is, the subset of the middle class who attained their comforts in part through books and not solely through land or trade—earned a reputation for status-conscious displays of their erudition.

While Bildung and art-religion both arose from early Romantic inwardness, these concepts were enacted in such a wide array of behaviors that they resist straightforward definition. The musical ladies who kissed Beethoven's walking stick did so not to express their art-religious transcendence, but to demonstrate their Bildung, the artistic upbringing that made them bourgeois. The bourgeoisie can likewise be difficult to define: a population unified by an antagonistic stance toward the strata above and below, but who denied their own bourgeois identity when the label became pejorative.[10] With the waning powers of the landed aristocracy in the late nineteenth century, the upper bourgeoisie realized with some embarrassment that it was actually the top echelon of society; the German *Besitzbürgertum* (those with property) were keen to be mistaken for the more laudable Bildungsbürgertum (those with higher education), both of whom admired the integrity of the *Mittelstand* (those with a solid but less elite education), and all of these looked down equally on the *Kleinbürgertum* (the petty bourgeoisie, who were the target of anxieties about mass culture and poor taste). These classes all indulged in art-religion in some fashion, but it was largely the Bildungsbürgertum and the Mittelstand who fill the pages of this book. Following in the footsteps of England's literary pilgrims, these educated music lovers visited historic homes with a sense of hushed anticipation, carried pocket-sized biographies that resembled books of hours, serenaded historic sites with choral odes, imagined themselves at the composer's deathbed, and wrote copious odes and tributes that were at times deposited on the floor of house museums.[11] While the high-flown language of their odes might echo the Jena Romantics, this was not the same Kunstreligion studied by historians of philosophy but a predecessor to the parareligion of modern-day celebrity culture.

In isolation, these practices look like acts of affection, but in the aggregate, devotion to composers resulted in a sordid culture of acquisition and display. Devotees collected and traded in intimacy as a form of prestige, a secular mode of salvation in a modern sale of indulgences. They engineered a new economy of piety that made the composer's body transactional. Through pilgrimage, relic collecting, medical writing, and literary fantasies, music lovers sought to possess, scrutinize, and eroticize the body of the departed genius.

Canons on the Ground

At one time, this book was conceived as an explanation for why canons last—that is, an account of the material culture that keeps canons afloat. Yet the stability and authority implied by the word "canon" vanishes the closer one gets to a large population of obscure people who adored composers and their music. As the study of composers' afterlives gets more granular, the boundaries of "the canon" look more elusive, just as "Beethoven" refracted through cultural artifacts is no longer Beethoven.

A rich body of scholarship in musicology has examined canon formation as an emergent process that touched every inch of nineteenth-century musical life.[12] The breadth of those studies suggests a more nuanced definition of "canon" beyond a simple roster of famous people and works: a canon is the centralized production of memory through institutions that enact nineteenth-century concepts like genius, masterpiece, and timelessness. It makes sense that historians have focused on institutions like schools, orchestras, festivals, and especially music societies, as their bureaucracies generated many boxes' worth of archivalia that record the strategic promotion of masterworks. Ironically, when institutions rendered works "timeless" by detaching them from the circumstances of their creation, they affixed them to the new context of nation building: with growing governments came the need for educated personnel, which explains why nation-states funded curricular reforms that widened the reach of Bildung.[13] Today, "the Western canon" is the latest pseudonym for the last vestiges of Bildung, which in music means a disproportionate emphasis on Teutons along with lasting habits of etiquette that preserve the sanctity of listening. In Eurocentric concert culture, "the canon" has become a token of staid authority, inaction, or symbolic violence. The word draws attention to the residues of art-religion's dogmas after they solidified into marble, stone, bronze, and paper—that is, into busts, inscriptions on auditorium facades, and *Denkmäler*, a German expression for monuments both literal and figurative.[14]

For all its value in activism, the concept of "the canon" has become too monolithic to account for the vernacular and eccentric ways that composers are remembered. The term implies an autonomous system of worth that reinforces itself in a linear, predictable process. Take for instance the schema developed by Marijan Dović and Jón Karl Helgason, in which "cultural saints" correlate with beatification and canonization in the Catholic Church. That process begins with the circulation of relics (*translatio*), which go from discovery (*inventio*) to their first display in shrines (*elevatio*) to their final resting place in a sanctified institution (*depositio*); "postulators," or men and women of letters, advocate for canonization; and over time, the artist's vita is mystified when reduced to slogans for ready consumption.¹⁵ This mass production of heritage allows states to mobilize their subjects in an emotive and accessible form of nationalism.¹⁶ As useful as this model may be, it implies that artists' afterlives fit neatly into the borders of a Catholic-inflected secular practice, and it effaces the individuals who carried out this process. On the ground, art-religion and canon formation were unruly, unpredictable, and considerably more idiosyncratic than the monuments and processions that are the familiar stuff of civic history. Art-religious rituals merged Christianity, Hellenism, and the veneration of ancestors. And art's devotees were Protestant, Catholic, and Jewish, which meant that a tapestry of religious convictions shaped whether art-religion aspired to be confessional or universal.

One reason canon formation is so elusive is that its rituals were torn between private and collective memory. The collective side of canonicity is better known and follows Pierre Nora's monumental editorial project of 1984, *Realms of Memory*, which defined French heritage across three volumes of essays about historic sites, people, and objects that have amassed layers of cultural value. In his introduction, Nora critiques museums, archives, anniversaries, monuments, and festivals as a secular mimicry of living religion, a French parallel to the social obligations of Bildung. He calls these "the rituals of a society without rituals," which appear "beleaguered and cold," a deliberate rather than instinctive memory that is "experienced as a duty."¹⁷ (Admittedly, Nora's warnings about archival obsession, about the panic to preserve amid a fear of obliteration, stand in some tension with these three hefty tomes proffered as a duty to public history.) For Jacques Derrida, the archival impulse has deeper roots than public ritual. This is not solely a duty to preserve but an "archive fever," a psychoanalytic drive to exteriorize memory that insulates against the repetitive self-destruction of Sigmund Freud's death drive. Psychoanalysis itself is an archive of the psyche (or a "prosthesis of the inside"), while libraries and historic houses are the external spaces that offer a domicile for archive fever.¹⁸ Both these positions on preservation can explain the devotion to composers, which found expression in public ritual and private adoration, solemn pledge and erotic fixation. To counter the death drive, music lovers had to resurrect the dead composer while masking that death, which Nora calls "illusions of eternity" and Derrida calls "an erotic simulacrum" or "masks of seduction: lovely impressions"—and which I might call the composer embalmed.¹⁹

With their focus on libraries, archives, and other official sites of memory, these provocative critiques do not fully account for the offbeat materials that constructed the canon in more imaginative ways. Already in Goethe's lifetime, high culture became a consumer product: the publisher Friedrich Justin Bertuch marketed Goethe through miniature busts and decorated fans in his Weimar lifestyle periodical, and his efforts were a model for Salzburg's marketing of Mozart, who was stamped on goods as early as 1842.[20] In Vienna, the self-proclaimed city of music, listeners were animated by a potpourri of creative nonfiction and biographical operettas, a predecessor to the biopic genre, which made an industry out of composers' quirks, romantic entanglements, and dramatic deathbeds.[21] One could study Mozart's *Bildungsroman*, his moral upbringing, by collecting a series of trading cards from the Liebigs broth company.[22] One could learn of Beethoven's stubbornness from a story about his visit to a zoo, illustrated with the iconic scowl staring down a lion. This odd fragment of biofiction was offered with a variety of Beethoven-themed poems and essays in a golden-bound souvenir booklet given to the ladies at Vienna's Concordia Ball on Valentine's Day, 1927, the one-hundredth anniversary of Beethoven's death.[23] In the popular novella *Schwammerl*, later staged as an operetta, Schubert can be found grinning with a fork and knife, ready to tuck into his favorite coffee cake.[24] Admirers of Chopin could imagine themselves at the sickbed or deathbed, and postcard collectors can find countless scenes of his frail form at the piano, blanket clad and pale.[25] In fiction, Beethoven could be seen rising from the grave to visit a mediocre orchestra concert; in his birth house, autographs and quills invoked the composer's ghostly hand; at home, music lovers practiced the piano under the blind gaze of his plaster mask and pressed dried leaves from his grave into their keepsake albums. Relics like these—or what Deborah Lutz calls "things that prove embodiment, that have the texture of a life lived"—humanized composers in extroverted ways that differ from the studious and introspective attributes of Bildung.[26] These commodities unsettled early twentieth-century critics, who decried what they called *Halbbildung*, the watered-down manufacture of high culture. To describe all this solely as "canon formation" implies a linear process for a material culture that was diffuse, inventive, and tactile.

Granted, these artifacts make sense only when read against textual sources from the dustbin of musicology: anecdotes and biographical fiction, messages in visitors' books, amateur lyric, and the annals of music societies.[27] Rather than focus solely on familiar names of those who enshrined the past, like Johannes Brahms, Clara Schumann, and other members of what Laurie McManus calls the "priesthood of art," this book unearths lesser-known figures.[28] Woven throughout these chapters is the musicologist Ludwig Nohl, whose verbose tracts on the New German School spread art-religious ideas to a lay readership while ruffling the feathers of more diligent critics. Ludwig August Frankl finds his way repeatedly onto the page: physician, long-serving secretary for Vienna's Jewish congregation, poet, playwright, and passionate advocate for composer monuments. Likewise is the case of Ernst Rudorff, who led dual lives as a composer in the Brahms circle and a steward of Germany's landscape preservation movement. These are individuals about whom we know something, at least, but the cast of characters can be considerably more obscure. Out of the archives, in dense *Kurrent* handwriting, we encounter Wolf Stern, who sent poetic fan mail to Liszt in 1873; Betti Schleiffer, who composed a poem in 1883 that recounted Mozart's deathbed drama in Salzburg dialect; or Hermine Bovet, a local piano teacher who gifted a fanciful account of Beethoven's nativity to his birth-house museum. These tokens of affection made their way to historic houses, which became a medium to communicate with ghosts. Without their archives, we would have far fewer records of art-religion in practice.

When I began this project, I did not set out to uncover a new dimension of Jewish history, or of women's history. Over time, I discovered that both groups have a special place in art-religion. It is well known that many German and Austrian Jews were deeply invested in Bildung, as well as the upward mobility afforded by political liberalism, as a route to assimilation into cosmopolitan society. I have argued elsewhere that this process of assimilation was complicated by the Christian and specifically Catholic forms of art-religion that made musical spaces less secular than they purported to be.[29] Historians have not yet fully addressed how historic preservation movements in Germany and Austria were grounded in antisemitic concerns, particularly with regard to the Jews of liberal Vienna, who were prominent benefactors for monuments and museums. At times the prejudice was blatant, as with efforts to wrest historic sites from Jewish owners in the name of piety. In 1903, prompted by the demolition of Beethoven's death house, Viennese residents launched a petition for the city to seize Beethoven's Mödling apartment from a Jewish landowner so it would not fall victim to "Jewish speculation."[30] More often, the tensions were unspoken yet palpable. The result was a rift in what Bildung entailed: this was not simply a monolithic veneration of composers, but a specifically cosmopolitan and intellectual appreciation for music that found itself pitted against a more populist, fraternal, and collective idea of heritage cultivated in rural locations. Musicologists are of course familiar with these polemics through the nineteenth-century reception of Anton Bruckner, the Upper Austrian whose devotees felt harangued by the Viennese intelligentsia. That sense of pride and defensiveness, of puffed chests and withering glances, underpinned heritage preservation in the region. "Piety" could be more confessional—more Christian, rural, and populist—than might be presumed by its secular context.

Furthermore, many of the devotees who appear on these pages were women. Men were instrumental in heritage projects in all the usual ways: founding museums, serving on boards, and linking the ethos of pilgrimage with the sound of men's choral fraternity. Women participated through the purse and the pen. They gave generously to music societies not only as philanthropists, but as poets who offered inventive lyric that expressed their devotion. Some of the poetry in this book bears a resemblance to the conventional forms of the keepsake album, or *Stammbuch*: short, tender utterances that convey friendship and conviviality.[31] But a significant number of women wrote fantastical or erotic poems, some of which span several pages. It would be wrong to say that women simply "participated" in art-religion. Through poetry, they transformed it into something whimsical, potent, and intensely visual.

One can only guess how composer devotion might have enlivened the experience of musical sound. For the most part, the subjects of this book have taken their listening and playing habits with them into obscurity. It appears that the autobiographical listening examined by Mark Evan Bonds spilled beyond the borders of music criticism, given that music played a vivid role in biofictional novellas and operettas that outfitted canonical repertoire with stories and texts in the composer's first-person voice.[32] Pilgrims invited us to eavesdrop on their inner ear when they left incipits of their favorite works in museum visitors' books. Perhaps these beloved works, performed at the summer festivals that drew them to historic sites, were experienced with a sort of dual autobiography, as personal memories of travels layered atop well-known narratives of the composer's persona. But this is speculative. A focus on the material culture of devotion has its limits: we may not hear composers' works anew, as histories of formal music criticism invite us to do. But we can better understand why those works alone, divorced from the composer's story and presence, were rarely enough.

In short, the closer we look at canon formation, the more we are compelled to redefine what a "canon" means in the first place. The participatory nature of art-religion made canons into something more diffuse than a mouthpiece of institutional authority, a closed loop of aesthetic value, or a sounding monument. On the ground, in the nineteenth-century moment, these were imaginative and intimate relationships with composers long gone.

Spirits and Talismans

If literary and visual culture already made composers feel human, then why did devotees seek out tactile encounters with bodies? This book is not solely about art-religion, but also about relic culture, which I define broadly as the impulse to preserve, study, and display the composer's earthly traces. Recent approaches in musicology and material studies do not quite explain the practices charted in this book. Relics lie outside the domain of musicology's focus on reception, which demythologizes the many politicized convolutions of a persona in an act of hermeneutics.[33] Nor are relics addressed by musicology's kaleidoscopic material and corporeal turns, which have modeled ever richer ways to read around the musical work as text by focusing on paratexts, embodied experience, and instrumental technologies.[34] In the humanities, a broader material turn has uprooted human agency to expose a subaltern world of things. There are things aplenty in this book, but I do not believe they are well served by Bill Brown's oft-invoked "thing theory," which parses things from objects according to use or neglect, or by object biographies that treat select relics as intergenerational celebrities.[35]

Recently, music studies has looked beyond the coolly methodical deflation of human agency in Bruno Latour's actor-network theory (ANT) to the activist animism of the so-called new materialists, like Jane Bennett, who emphasize the "thing-power" and "vitality" of nonhuman matter.[36] New materialism would demand that I center relics and dead composers as protagonists, and some have indeed argued that dead composers enact agency on the living through parareligious rituals and artifacts with powers beyond inert matter.[37] Yet I hesitate to endow relics and composers with posthumous agency. While new materialism intervenes in the crises of the present, its liberatory anthropomorphism can be dissonant with material culture of the past.[38] When commodities were envoiced in the eighteenth century, for instance, they circulated in what Nicholas Mathew calls "a new human economy" in which "people became beautiful objects and objects acquired beautiful voices, in a never-ending exchange," and these lively commodities masked the asymmetrical politics that render envoicing itself an act of mastery and domination.[39] Rosalind C. Morris argued forcefully that ANT and new materialism conceal their family resemblance to the ever-resurgent Western fascination with the "fetish" as Reason's other.[40] In his condescending yet influential work of early anthropology, Charles de Brosses reported in 1760 how West African spiritual leaders consecrated objects so that communities could "worship them in an exact and respectful manner, address their wishes to them, offer them sacrifices, carry them in procession if it is possible, or wear them on their persons as great marks of veneration."[41] Across three centuries, de Brosses's "fetishism" took on a convoluted afterlife, from Marx's project to demystify labor that was obscured by capitalist materialism, to Freud's psychoanalysis of desire displaced onto disembodied objects. On the fringes of that intellectual afterlife, a widespread devotional culture enacted the very fetishism that Western intellectuals critiqued. What might de Brosses have to say about performers who realize canonical works with deferential *Werktreue*, collectors who wear relics close to their heart, pilgrims who lay wreaths at a monument's feet, and donors who describe their philanthropy as sacrificial offerings at the altar of art? To frame composer relics as vibrant matter would rehearse the fetishism of the nineteenth century.

The protagonists in this book are not relics, then, but a network of obscure people who participated in fetishistic thinking—that is, people who understood artifacts, houses, and musical works as talismanic containers for the composer's essence. As talismans, composers' remnants came to resemble the anthropological artifacts that fed Europe's "exhibitionary complex."[42] On the one hand, composers' house museums were cozy environments that make both history and music feel human. These rural houses differed from the urban anthropology museums and exhibitions that displayed exotic treasures to stage Western imperial progress. While museums today are more thoughtfully curated than their cluttered nineteenth-century counterparts, they still evoke the studied miscellany of the curiosity cabinet. At Schubert's birth house in Vienna, the iconic spectacles lure visitors with the (failed) promise of locking eyes. At the Bach Museum in Leipzig, visitors enter an atmospheric "treasure room," named as such by exhibit curators, which playfully critiques relic culture: here curios from Bach's coffin are lit by high beams against a dark wall emblazoned with the famed exclamation, "yes, wonderful things!" (Howard Carter's reaction to Tutankhamun's tomb). Even the most updated collections use objects to construct a tangible Bildungsroman, a biographical life cycle that progresses toward artistic maturity. Meanwhile, out of sight in the archival catacombs, a host of weirder items are retired from display, catalogued informally as "artifacts of memory" (*Erinnerungsgegenstände*). The provenance letters for these heirlooms can expose a fine line between collecting and hoarding. Through *Musealisierung*, the impulse to possess and display, Europe's voyeuristic interest in the fetish folded back onto its own heritage.

The desire to reanimate composers shows how a mainstream Catholic vocabulary (*relics, pilgrimage, saints*) placed acceptable labels on experiences that resemble those at the margins of organized religion. The first occult

INTRODUCTION

movements of late nineteenth-century Germany inherited from Anglophone Theosophy a mistrust in modernity, in which forms of progress that brought material wealth were thought to impoverish the spirit. The events charted in this book coincided with several challenges to scientific materialism: the séance and stage medium as an international sensation in the 1850s, the earliest studies of occult phenomena and parapsychology in the 1870s and 1880s, and a renewed interest in the soul and spirit among turn-of-the-century psychoanalysts.[43] Interlocking with these projects were the many so-called life reform movements that sought a utopian return to nature through vegetarianism, nudism, light therapy, and other challenges to conventional medicine.[44] While a thorough account of the connections between composer cults and occult movements would require a book of its own, traces of an overlapping history surface in these chapters. At times, the links are explicit, as in the case of a stage medium who dictated Beethoven's Tenth Symphony, and a design for a utopian Beethoven temple by the life reformer (and later eugenicist) Fidus. More often, the connections are shadowy and suggestive. In Bonn in 1903, Joseph Joachim's quartet performed the sobbing Cavatina of the String Quartet, op. 130, on the composer's own historic instruments inside his birth house; the instruments were said to be in poor condition, and their raspy tone added a grain to this voice beyond the grave, which left listeners in tears.[45] Beethoven haunted a rather different performance in 1927: Bonn's music director recalled that, as thousands of people gathered for an outdoor orchestra concert in the market square, a thunderstorm descended then suddenly dispersed "as if touched by a magic hand" at the moment the Fifth Symphony began, and "a palpable, otherworldly shiver went through the crowd of devotees, and even the poorest among us felt that God was very near."[46] Concert halls and historic houses could be spaces for spiritual communion.

Relic culture was not just about collecting, then, but about a Western attraction to alternative spirituality and exotic talismans that found voice in a language of Catholic devotion. This explains the prevalence of the word "piety," or *Pietät*, in institutional mission statements and handwritten odes. Just as commodity fetishism, for Marx, was the ghostly force that concealed the reality of labor, Austro-German "piety" toward art obscured the capitalist structures, and the attendant impulse to *own*, that kept heritage projects afloat. The hallowed connotations of piety concealed the transactional nature it inherited from the sale of indulgences. The formula is as follows: art offers salvation, salvation can be bought, to buy salvation is philanthropy, philanthropy is pious, and piety masks a desire not only to love composers intimately, but to possess them.

Getting Possessive

Piety signals not only the holy, but the holier-than-thou. As art-religion became public and performative, distinct from the studious inwardness (or *Innerlichkeit*) of Bildung, a climate of competition separated true from false devotees. Piety appears in this book in three main contexts: the obligation to gather for rituals that lauded the genius; frictions between institutions, cities, and nations that tried to outdo one another; and frictions between devotees who competed for intimate connection with composers amid the noisy collectivity of heritage.

The public face of piety was "monument fever," or *Denkmalwut*, as critics called it. On a remarkably regular basis across Germany and Austria, a sea of hats filled city squares for the anniversaries of artists' birthdays or death days, and for ribbon cuttings for new statues—or in the words of Ryan Minor, "a monument was not simply a thing, but a thing to do."[47] These occasions followed a program that became a kind of secular liturgy: a procession to the site, the laying of wreaths, the recitation of poetic odes, and the performance of a *Festchor* or *Festcantate*. (To further blur the edges of the canon, it should be noted that warhorses of the repertory were often framed by these now-forgotten cantatas. In some cases, our only record of that ephemeral genre survives in piano four-hand arrangements that allowed one to relive the event at home.)[48] The sacred dimension of these rituals could at times be quite literal. In Vienna, monuments to composers were erected near churches, and a lengthy (albeit failed) project to build a shared monument for a pantheon of composers in the Karlskirche shows that secular figures in a sacred space were not thought blasphemous.[49] Choral rituals are explored by Ryan Minor and by Alexander Rehding: Minor argues that festival music was a discourse through which citizens participated in the nation, while Rehding focuses on a contradiction of German timelessness, which looked ahead toward utopia and backward at a commemorated past.[50] This book shows how aspirations to sublimity and nationhood were tempered by an interest in the local and humbly material origins of the artwork. The same crowd that admired a statue, strewing it with wreaths and rubbing its feet shiny, might also admire the plain birth rooms and writing desks where genius was said to originate. Praising composers for their humility, even if that praise was expressed in bugling double dots, suggests the Christian value of poverty, the renunciation of material comforts. What unified the monumental and the intimate was a theatrical performance of prostration and humility, as rituals lauded both the composer and the assembled crowd with a self-congratulatory tone.

As composers were feted in public ways, music lovers liked to imagine that the composer belonged to them alone. The eagerness to own relics, and the competitive stance in devotional writings, prefigure what scholars of celebrity call parasocial relationships, or the fantasy that the absent celebrity is a friend or lover. While the parasocial was first developed as a theory of modern media relationships,[51] historians of celebrity have traced a comparable desire for intimacy to earlier periods when luminaries were celebrated so publicly that their admirers got possessive. For Joseph Roach, a fascination with the celebrity body began with the effervescent *it* factor of seventeenth-century actors, whose presence was amplified by uncanny afterlives in hair, flesh, clothes, and bones. These "staged synecdoches" took on a life of their own, inviting the imagination to reassemble them.[52] Later, among Romantic poets such as Lord Byron, two print media encouraged intimacy with the celebrity: biographies designed to make poets come alive, and poems consciously crafted to create a personal bond with readers. Tom Mole calls this a "hermeneutic of intimacy" by which heartfelt encounters, experienced through mass media, were necessary to "[palliate] the feeling of alienation between cultural producers and consumers."[53] In a period when books could "assume the properties of flesh and blood," as Samuel Taylor Coleridge put it, readers took part in a new subculture of obsessive bibliophilia.[54] Romantic poets, while aware of their own celebrity, worried that biographies could invade privacy and undermine the artist's Romantic autonomy. But by the mid-nineteenth century, the lives of poets conditioned readers' love of poetry rather than the other way around. At best, the earliest biographies combined "reverence and iconoclasm, the elevated and the down-to-earth"; at worst, friends of recently deceased artists traded in anecdotes as social currency and turned vitae into tabloids.[55] Affection was more tangible after poets died, when their houses and graves became sites of pilgrimage that drew attention to the body's absence. Gravesites drew literary pilgrims into states of mournful reflection that Paul Westover calls "necromanticism," while cozy artist houses, which acted as mausoleums and reliquaries, sought what Nicola Watson calls an "illusion of intimacy" where one expects to "find the writer at home."[56]

The wealth of Anglophone scholarship on celebrity culture implies that parasocial inclinations began in England, later migrating to German-speaking regions. Given that studies of English literary culture have outpaced their Continental equivalent, it is difficult to trace whether devotional behaviors had a more homespun German history in the Austro-German anecdote industry, a local penchant for portrait collecting, and the commodified cult of Goethe.[57] Even so, the nineteenth century was a period of Anglo-German symbiosis in musical life; English tourists brought their parasocial longing with them on journeys to the Swiss Alps, the Rhineland, and the sparkling cities of Salzburg and Vienna.[58] Their travels help to explain a German attitude of embarrassment at English excess that made "piety" an antidote to distasteful or false devotion. As German and Austrian city officials observed the lucrative potential of tourism, they turned a piecemeal cottage industry into an international market for historic sites. By the late nineteenth century, with a growing self-perception of Germans as the people of music, German-speaking regions came to boast more composer museums than anywhere else in Europe.[59]

This museum industry complicated the status of private collectors: some eagerly donated composers' effects as a public duty, while others mistrusted museums as sterile.[60] The contested journeys of Beethoven's belongings exemplify this transitional moment in collecting culture. Beethoven's fame meant that admirers asked for locks already during his life, and just after his death, his effects were strewn about Austria in an auction that was later dismissed as a token of Viennese impiety.[61] The auction register shows that someone, we know not whom, bought a wide assortment of household goods.[62] Anton Schindler, Beethoven's infamously self-serving secretary, decried the sale of relics, but behind the scenes he pocketed numerous mementos. It was a sound investment: in 1846, he sold his Beethoveniana to the king of Prussia in exchange for an annuity for life.[63] Later, when the birth-house museum sought to reassemble Beethoven's relics, they encountered some resistance. A childless owner of a lock of hair, for instance, did not want his relic "in profane hands" and hired an intermediary to find a music-loving family with heirs.[64] During Beethoven's lifetime, the poet Theodor Körner was said to have worn a letter from the composer beside his heart; a century later, collectors still held their relics close to the chest.[65]

Already in the early nineteenth century, the dark side of collecting was attributed to the English, who had gained a reputation for impiety. In 1818, the Waterloo elm—that is, the tree where the Duke of Wellington oversaw the Battle of Waterloo—was uprooted to engineer a chair, a bust, snuffboxes, toothpicks, and other assorted relics. That desecration prompted the painter and diarist Benjamin Robert Haydon to opine, as early as 1830, what he called the "English disease." Haydon found the "selfish, domestic, individual, and confined" hoarding of British relic hounds to be just as alarming as the overblown monumentality of French commemoration. The English, he claimed, "can't let a thing remain for all to enjoy. . . . On every English chimney piece, you will see a bit of the real Pyramids, a bit of Stonehenge! . . . You can't

INTRODUCTION

admit the English into your gardens but they will strip your trees, cut their names on your statues, eat your fruit, & stuff their pockets with bits for their musaeums [sic]."[66] (What looks here like a lucid appraisal of English colonial hoarding is not so clear-cut: elsewhere, Haydon expressed firm confidence in English imperialism.)[67] The "English disease" explains why the 1829 travelogue of Vincent and Mary Novello, published as their "pilgrimage to Mozart," culminates in a quest for relics at the feet of Constanze Mozart. For Mary, Constanze and her sister were already relics, or "the nearest approach to his earthly remains," but Vincent demanded more tangible souvenirs: "She had given away nearly everything to the numerous persons who had applied to her at different times for a memorial of him. Relics she gave me—a small portion of the little Hairbrush with which he arranged his Hair every morning, a part of a Letter addressed to him by his Father."[68] Splintering objects like Mozart's hairbrush was a practical way to meet the demand, and the approach continued for many decades. Following the lead of Mary Hart, who sold off pieces of Shakespeare's chair in the late eighteenth century, Franz Xavier Jelinek, who directed the archives of the Internationale Mozart Stiftung, disseminated shards of Mozart's cradle and forgeries of his autographs.[69] Visitors to Beethoven's birth house dismantled the cradle and pocketed bits of the floorboards, forcing museum staff to cordon off the birth room. Some decades later, after a major renovation, the museum itself sold pieces of the staircase in the gift shop.[70] It is too simple, then, to frame relic culture purely as a tender extension of living friendships. When every music lover wants to possess something real, they destroy the relics they seek.

Naturally, relic culture and pilgrimage extended far beyond Anglophone and Germanophone regions, and the practices charted in this book can be traced across Europe.[71] What was distinctive about Germany and Austria was their conflicted relationship with their own culture of veneration. In Richard Wagner's semiautobiographical novella "A Pilgrimage to Beethoven" (1840), the antagonist is a superficial Englishman concerned with selfish relic collecting, while his antithesis is the pious German pilgrim.[72] Wagner's stance ignores that the English helped to engineer German heritage in the first place. English donors funded museums and festivals, English pilgrims put German museums on the international map, and English publishers translated biographies in droves and added English-language texts to Beethoven's instrumental works so that his music found its way into every piano bench.[73] Germans and Austrians cloaked in piety that which they found distasteful and irresistible. By the twentieth century, those guilty pleasures were subsumed under a provocative label: kitsch.

From Piety to Kitsch

The satire of art-religion in my preface encapsulates this book's arc, which moves from piety to kitsch, from faith in to mistrust of art-religion. I begin with a study of homes and haunts, which were established as museums in a climate of competition that telescoped across levels of society, from individuals who boasted of their pilgrimage to distinguish themselves from idle tourists, to state-funded music festivals in Germany and Austria in a rivalry for Beethoven's legacy. Historic sites took on the function of shrines: in chapter 1, we see Beethoven's birth room arranged in the fashion of a Christian Nativity scene, and in chapter 2, Mozart's *Magic Flute* cottage transplanted to a mountaintop as a destination for traveling men's choirs. Both Bonn and Salzburg called themselves sites of pilgrimage that sheltered art from the alleged impiety of Vienna, with a distaste for big-city commemoration that counters the usual narrative of Vienna as the "city of music."[74]

These first chapters demonstrate why devotees and institutions were so invested in art-religion. First, it offered a language to assert one's status as a disciple of high culture while denigrating dilettantes. That aim was articulated through pilgrimage, relic collection, and philanthropic relationships with institutions that competed to possess the composer. Secondly, art-religion allowed music lovers to imagine an affectionate relationship with composers, borrowing from practices of memory for loved ones. Institutions, likewise, performed their affection through ritual processions, speeches, odes, and cantatas, all the while trading in relics in a mirror image of anthropological acquisition and display.

The book's next chapters examine how relics of composers' bodies circulated in medical and visual culture. The desire to draw composers close found its way into an intellectual tradition of pathography (pathological biography) and pseudosciences that sought to measure genius. In chapter 3, I show how both piety and curiosity animated music-loving doctors and anatomists who valued relics as specimens. Their influence is felt today in the continued scientific scrutiny of composers' bodies as if to embalm them through sheer descriptive detail. Chapter 4 offers a close reading of a single object: Beethoven's plaster mask, a mass-produced relic that adorned parlor walls in a cult of the face derived from Christianity. In literature and visual culture, this object reflected competing visions not only of Beethoven's persona but of the very face of death, from the celestial transfiguration of the nineteenth-century "beautiful death" to the grisly contortions of modern warfare.

INTRODUCTION

The final chapter of this book illuminates a turning point in the history of art-religion. Rather than conclude with a kaleidoscope of parallels between secular religions past and present, which would imply (falsely) that celebrity culture is neatly transhistorical, chapter 5 turns to moments of rupture when the first skeptics of composer cults sounded an alarm. The first decades of the twentieth century saw a mistrust of the once laudable Bildungsbürgertum. Civil servants by day, musicians by night, genuflecting before Beethoven without understanding him—these were, in the eyes of some critics, the uncreative people to whom art-religion was an empty exercise in social climbing.[75] If Bildung could no longer be entrusted to the Bildungsbürgertum, then it was increasingly unclear who benefited from the cultivation and erudition of Austro-German art music. Did Bildung belong to cosmopolitans and assimilated Jewry who were so often derided as false disciples? Should high culture be reclaimed by the radical left for whom these composers, especially Beethoven, offered a model for brotherhood and revolution? Or by modernists like Arnold Schoenberg who began as outsiders peering into the halls of the academy, translating ideals of a pure, unmediated aesthetic experience into the language of the avant-garde? By traditionalists who likewise sought an unmediated experience, but for whom only classics of yore deserved that honor? Or was Bildung a target for populists, such as the charismatic mayor Karl Lueger, who harnessed the appetites of a petit bourgeoisie galvanized by their exclusion from high culture?[76]

In this unstable moment when art-religion lost its aura, we find the first writings on kitsch and mass culture. This early discourse on kitsch grappled not only with the tasteless trivia of the petit bourgeoisie, but with a mistrust of the Bildungsbürgertum as manic masses who flocked to the temple of art. I argue in my last chapter, then, that with the commodification of Bildung came the *Verkitschung*, or cheapening, of art-religion. Unlike simple definitions of kitsch as a cantankerous reaction to mass culture, I argue for a more nuanced understanding that revolves around the concept of anachronism: a temporal misfit that draws attention to its own disenchantment. Here, the anachronism in question was nineteenth-century art-religion, whose former sublimity had collapsed into a cheap imitation of religion, a kitsch replica.

Guilty Pleasures

Well before this turning point, it was possible to chuckle at the excesses of Austro-German piety. The nineteenth-century ancestor to kitsch was a warm humor, lighthearted and facetious, which reveals much about our current relationship with material culture. In 1820, when Washington Irving recounted his visits to British writers' houses to an American readership, he beheld Shakespeare's chair in the birth house in Stratford-upon-Avon, where visitors were invited to sit and absorb an echo of genius. In a tongue-in-cheek tone, he wrote:

> I am always of easy faith in such matters, and am ever willing to be deceived where the deceit is pleasant and costs nothing. I am therefore a ready believer in relics, legends, and local anecdotes of goblins and great men, and would advise all travelers who travel for their gratification to be the same.[77]

Irving's jab at the tourism industry, with all its familiar deceptions, conveys more gentle amusement than shame. Compare this with a sparkling anecdote from late nineteenth-century Germany that epitomizes the guilty pleasure:

> Returning from Schumann's grave, I soon stood before a nondescript house in the Rheingasse with the following inscription: "Beethoven's Birth House." With a beating heart I entered the damp hall, climbed a life-threateningly narrow and dark wooden staircase and was led by the owner or renter of the house into a bare, shabby room, whose badly damaged walls and small crown-glass window betrayed considerable age. "In this parlor Beethoven was born," said my guide, as resolutely as if he had been there. I am quite devout in all piously hallowed sites, self-consciously devout or even superstitious, when I want to be, and I bear no mistrust for the walking sticks and tobacco boxes that are shown to me as favorite articles of deceased celebrities. Thus, with bare head and touched heart, I beheld the hallowed and rather filthy room in which Beethoven let out his first cry. . . . Endangering my life I groped again through the pitch black hen-house stairs into the open air and, shortly thereafter, I was not a little surprised to read an inscription on a house in the Bonngasse: "Here Ludwig van Beethoven was born." Appalling. In my excitement I had forgotten the quarrel two years ago, in which two different houses in Bonn claimed the honor of being Beethoven's birth house. . . . Thus I had climbed the shaky, winding staircase in the Rheingasse, inwardly sunk onto my knees before the birth site of Beethoven, only to discover five minutes later that I, with all my holy experiences, had been duped! With hindsight it seems downright funny. But at the time, the bucket of cold water that drowned my flushed excitement was rather embarrassing.[78]

This episode was recounted by none other than Eduard Hanslick in 1885. Readers familiar with his stance on aesthetics may find this surprising. It was Hanslick, after all, who distanced musical works from cults of personality in his best-known treatise, *On the Musically Beautiful* (1854), and like others in the Brahms circle, he derided the cult that had formed around Wagner.[79] When Hanslick shifted his attention to journalistic writings about concert life, he traveled to music festivals in cities that boasted historic gravesites and dwellings.[80] Hanslick's tone is challenging to interpret. On the one hand, there is playful self-deprecation when he laughs at his own misplaced piety and exposes his pilgrimage as a guilty pleasure. All the same, his insistence that he is "self-consciously devout or even superstitious" seems defensive, as if teasing snobbish readers who find piety beneath them.

Are these accounts by Irving and Hanslick the first glimpse of hyperbolic piety on the edge, or lighthearted satires of local color? It is worth stepping back for a moment to interrogate where we as musicologists fit in this spectrum, with our self-evident love of certain musics and our more private affection for composers' eccentricities, handwriting, glasses on a pedestal, and other traces of humanity. Perhaps the discipline of musicology, wedded as it is to a careful balance between analytical close readings of music and historical frameworks, finds its footing in this two-pronged approach because it deflects any suspicion of cult fervor. This is why museums remain sites of leisure for music lovers, separated from the archival back rooms where researchers toil. I speculate that German materiality today carries a much older stigma of the guilty pleasure, and that stigma is felt keenly by Anglophone musicologists caught in the orbit of German studies, who labor under some pressure to justify that focus by doubling down on the intersection of notes and contexts. When musicologists do turn to material objects, it is with a sense of what Benjamin Walton calls "quirk shame," the guilty pleasure that runs as an undercurrent through the ensemble of reflections on "quirk historicism" published in *Representations* in 2015.[81] The authors of that special issue articulate a pronounced anxiety that objects and ephemera serve largely as a shortcut to technicolor storytelling.

Quirk shame has a historical precedent. Hanslick's memoir, with its critique of art-religion as a confession for the gullible, shows how relics could be both coveted and cheap, sacred and degraded. There is always a kernel of kitsch in the guilty pleasure. Prominent symbols of Western art music—the golden concert halls fringed with busts, monumental editions and festivals, apotheoses and geniuses and claims of transcendence—began as a noble counterpart to the commodification of tourism and relics. But today those very symbols appear kitschy and over-the-top in their own way; as Charles LaPorte put it, "maybe Victorian pieties jar with us because we have different pieties."[82] It is only when we encounter the nineteenth century fully, with all its quirks, that we can begin to interrogate the guilty pleasures at the heart of Western music culture.

1

Beethoven's Nativity

In 2018, a traveler named David wrote a pithy note in the visitors' book for the Haydn House in Vienna: "It is not a visit. It is a pilgrimage."[1] Over a century earlier, an Englishman named Frank Ernest Williams wrote the words "on pilgrimage" at Beethoven's birth house in Bonn.[2] This persistent language of genuflection comes as little surprise to those who have attended candlelight vigils at Graceland or poetry readings at Jim Morrison's grave in Père Lachaise.[3] Nor is it surprising that secular pilgrimage became widespread in the mid- and especially late nineteenth century, that historic houses were founded with increasing frequency at that time, and that these sites satisfied an ongoing hunger for tangible contact with the intangibility of "genius." There is more to the story, though, than a loose conflation of tourism with pilgrimage, because piety was not only heartfelt but also strategic. Visitors came away moved, to be sure, but they also found currency in art-religion as a means to feel like artistic insiders. The same was true at municipal and state levels. The rhetoric of pilgrimage allowed German and Austrian music societies to attract philanthropists, and small cities, in particular, competed with urban centers by presenting themselves as sanctuaries of heritage preservation. Those frictions are especially revealing for a discipline like musicology, which has tended to overemphasize European capitals, whose relationship with memory was considerably less urgent than that of small cities and towns.

This chapter examines the expressions and strategies of devotion that coalesced around two of Beethoven's houses, which met a divergent fate. Beethoven's birth house in Bonn was made a museum in 1889 and presented itself as shrine of Nativity whose pilgrims wrote provocatively about their encounters with Beethoven's spirit, at times evoking the language of a visitation or séance. His death house in Vienna was demolished in 1903 in the name of architectural upgrades, and a funereal ceremony became a flashpoint for debates about historic preservation, memory, and material presence. In both cases, these houses were understood as surrogates for the composer's body, but in Bonn that body was embalmed, while in Vienna it was buried and mourned. Nearly every composer of the Western canon has at least one museum to their name, and each of these museums has a compelling history of its own. Beethoven's houses stand out for two reasons. First, his birth in Bonn and his career in Vienna encouraged a tug-of-war between German and Austrian heritage projects, amplifying tensions in a period that saw not only the Franco-Prussian War, but also Austria's resentful economic dependence on Germany as a primary trade partner.[4] Those tensions were veiled by the sacred language of *Ehrenpflicht*, the obligation to honor. Secondly, historic houses offer a colorful microhistory of art-religion in practice. Museums were repositories for devotion on the ground. Without their visitors' books, archives, festivals, and the buzz of conversation they invited in the local press, historians would have little information about how pilgrims thought and behaved.

The chapter begins in Bonn with a closer look at piety and Ehrenpflicht, which emerged in part from the landscape preservation movement that developed in the Rhineland shortly before Beethoven's museum was founded. Documents from the Beethoven-Haus archive show how the museum's earliest curators presented it as a site of Nativity, and visitors responded in turn. Their writings, which they sent to the museum as gifts or inscribed in the visitors' books, offer a glimpse of how pilgrims felt and behaved in hallowed rooms. Meanwhile, in Vienna, the demolition of Beethoven's death house stirred up a controversy that exposed a city torn between piety and pageantry, with an ambivalent relationship to its own past. A comparison of these two houses shows not only the understandable differences between small and large cities, but also a critical tension between competing ideas of history, in which the musical past is either living or dead, continuous with the present or frozen in time.

Preservation, Piety, Ehrenpflicht

In Germany and Austria, historic preservation efforts resounded with two words: piety, or *Pietät*; and honor duty, or *Ehrenpflicht*. The word *Ehre* (honor) originated with Christian ritual, worship, and gifts to the divine, and by the late nineteenth century, it took on a moral dimension, referring to duty, ethical pride, and dignity.[5] This makes the compound word *Ehrenpflicht* tricky to define, as *Ehre* itself could be a synonym for *Pflicht* (duty). The combination of the two denotes both an obligation *by* honor (that is, being honor bound to do right), and an obligation *to* honor. The latter meaning becomes apparent when we read the word in context: it surfaces most often in reference to symbolic acts of veneration such as festivals, tributes, and of course historic preservation. This explains why the word first appeared with some frequency in the 1850s, when heritage projects gained steam, and its usage increased again shortly after the First World War, when grief at Europe's destruction redoubled the urgency of preservation.[6] The word surfaces regularly in internal memos of the Beethoven-Haus founders, as well as local newspapers that reported on the house's "rescue," such as a local notice that explained in 1889 how "it would have been an *Ehrenpflicht*, above all, to divest this site from its profane use."[7] Another newspaper invoked the more powerful and imperative debt of *Ehrenschuld*, a guilt-ridden obligation to honor, by announcing with some relief that "a heavy *Ehrenschuld* so long weighing on the shoulders of our city has finally been lifted."[8]

It is notable that the entire city felt that burden. Concepts like Ehrenpflicht were linked with competition between cities, particularly smaller ones whose economies stood to benefit the most from tourism. This explains why Salzburg became known as the *Mozart-Stadt*, the city of Mozart, Bonn as the *Beethoven-Stadt*, Weimar as the *Goethe-Stadt*, and so on. If pilgrimage was defined by a special journey to the backcountry, then rural birth towns and courts were preferable to the large cities where some artists established their careers. Already in Beethoven's lifetime, small cities had begun to sell prints of famous birth houses in local shops, and that is how the dying Beethoven came to cherish a print of Haydn's birth house in Rohrau, which was a parting gift from Anton Diabelli. According to Gerhard von Breuning, who relayed the memories of his father, Stephan, Beethoven had marveled at how such a great man could hail from a quaint cottage; the sentiment is reminiscent of the comments made by pilgrims to Beethoven's own birthplace seventy years later.[9]

The word "piety" gave these sites a certain cachet, while "impiety" was a barb to accuse cities or nations of failing to honor. After the demolition of Beethoven's death house was announced, an anonymous journalist for the liberal *Neues Wiener Tagblatt* bemoaned the building's fate in powerful terms: "When one hears . . . that Beethoven's death room cannot be preserved, and instead that it must yield to the requirements of modern times, one gets the impression that conservation is virtually impossible and that we are witnessing again the fate of great men who live in large cities: the octopus of the big city envelops and crushes sites of memory, whereas in small cities piety has free rein."[10] This passage is more evocative than most, but the sentiment was common. Viennese city officials developed a reputation for neglecting local artists during their lifetimes and continuing to spurn them posthumously. Cities across Europe were swept up in the so-called monument fever that I discussed in the introduction, erecting one statue after another while eyeing the progress of their neighbors.

Competition stiffened in the year 1870, when Beethoven's one-hundredth birth year coincided with the Franco-Prussian War. The international celebrations were preserved in an 1872 album edited by the Bohemian Jewish writer Hermann Josef Landau, whose aims for the collection expose Austro-German tensions.[11] Landau's stated aim is to show the nobility of these commemorations, which exhibited the "German mind, German spirit, German thought and German power, without appearing laughable in the eyes of the civilized world."[12] A second aim was to affirm Austria's commitment to Beethoven's legacy: Landau goes on to explain that Beethoven "occupies a position not only in German music, but also in German-Austrian," and although the composer was "the most German national musician," his true homeland (*Heimat*) was Vienna.[13]

An assortment of festival prologues, allegorical plays, and concert programs in this volume praise Beethoven as the balm to heal North and South, uniting Prussians and Bavarians as brothers. In a theatrical allegory in Wiesbaden, the flowing figure of Germania ritually demilitarizes two soldiers, one Prussian and the other Bavarian, at the foot of a statue of Beethoven, who "heals with conciliation that which weapons injuriously cleave."[14] The event ends with a performance of the "Ode to Joy" with singers bathed in artificial sunbeams. It made sense to hail Beethoven as a pan-German messiah, or what Ruth Solie has called a "secular humanist": in this period, his Ninth Symphony was thought to teach ethical lessons that would bridge the Catholic and Protestant divide.[15]

But in some cities, this optimism was undercut by a tone of defensiveness. Bonn's poetic prologue for the inauguration of its concert hall, the Beethovenhalle, overflows with Rhineland pride. In contrast, Vienna's prologue for its parallel festivities braced itself for German accusations of decadence, claiming that its celebrations were the finest in Europe because Vienna is "not so frivolous as those chattering from certain sides maintain."[16] The prologue then launches into an emphatic refrain: it would be frivolous if Vienna cherished Beethoven only in 1870 rather than as a part of the city's everyday fabric, or if Vienna lost its musical finesse amid the din of war "at a time when a strong fist is preferred over a keen mind," and so on.[17] There is a marked insecurity in these boasts that reflects the disappointment of the festival itself: Vienna's celebration took place with nary a celebrity as a result of an oversight by the planning committee, who naively invited members of both sides of the New German School controversy (Johannes Brahms, Joseph Joachim, and Clara Schumann on the one hand; Franz Liszt and Richard Wagner on the other), which led all of them to decline.[18]

Meanwhile, the events of 1870 drew attention to Vienna's lack of a Beethoven monument. Despite its reputation as *Musikstadt Wien*, the city of music, Vienna was slower than others to erect statues to composers.[19] The delay may be explained in part by a hostile attitude toward the liberal Jewish bourgeoisie, who were seen to seize control of Bildung when they financed monument projects and Vienna's Ringstraße renovation. The Jewish poet Ludwig August Frankl—a music lover who served on various monument committees—met with disdain from critics like Ferdinand Kürnberger, who saw those efforts as self-aggrandizing, and whose dissent made these projects into a polemical minefield until the non-Jewish Nikolaus Dumba took the reins.[20] After Frankl's involvement in the monument on Beethoven's grave in Währing, followed by a comparable tomb for Gluck, he became vice president of a committee that ultimately erected a Beethoven monument in Heiligenstadt in 1863. To rally the philanthropists, he disseminated a poem that characterizes the tensions of Ehrenpflicht:

> In his birth city, where the Rhine flows,
> The master's bronzen visage has long loomed;
> The German *Volk*, beholden to genius,
> Erected a pillar of glory for him. . . .
>
> In luxurious Vienna, so musically rich
> No monument yet shimmers for him aloft. . . .
>
> That lively city full of sound and dance—
> Has it merely gravestones for the Genius?[21]

> (*In der Geburtsstadt, wo der Rheinstrom wallt, / Schon lange ragt des Meisters Erzgestalt; / Das deutsche Volk, dem Genius verpflichtet, / Hat ihm des Ruhmes Säule aufgerichtet. . . . // In üpp'gen Wien, dem musikalisch reichen / Kein Denkmal schimmert noch für ihn empor. . . . // Die heit're Stadt voll Klang und Tanz, hat sie / Grabsteine nur für das Genie?*)

Vienna did eventually erect its response to Bonn's monument in 1880, which a local newspaper notice called "an *Ehrenschuld* of our city, this cradle of German music."[22] Eduard Hanslick, who served on the monument committee together with prominent members of the Society for Friends of Music, explained in the *Neue Freie Presse* that Beethoven belongs more firmly to Vienna than to Bonn: "Great men are indeed born in small places, but only large cities can raise and perfect them." Furthermore, he criticized Bonn's monument design: foreigners, he said, can be heard whispering "that's meant to be *Beethoven*?" when they approach the monument in Bonn, but Vienna's would pose no such difficulty.[23] The resulting likeness sat with a look of defiance on a tiered pedestal, and even the stone was subject to competition: it was rumored that the committee had petitioned to quarry stone from Beethoven's native Rhineland, but resistance from the community in Bonn redirected them toward a native Austrian source.[24] From this point onward, under Dumba's leadership, Vienna etched its identity as the city of music into the streets, parks, and facades.[25]

As veneration escalated across borders, commemoration in one area could highlight neglect in another. The city of Bonn established a reputation for Ehrenpflicht starting with its monument unveiling in 1845, and thereafter the city hosted semiregular Beethoven festivals and a large birthday celebration. But the more visitors were drawn there, the more complaints emerged about the derelict state of the birth house. Some expressed dismay that the building had since become a tavern and that the alleged birth room served as a changing room for a vocal ensemble. Passing pilgrims were not sure which house to visit, as two different buildings bore the label of the birth house. When Eduard Hanslick reviewed the music festival in 1885, he found himself duped into squandering his piety on the wrong site. His amusing anecdote, quoted at length in the introduction, captures the disillusionment of forgery. Hanslick had entered the filthy room "with bare head and touched heart" and "inwardly sunk on my knees before the birth site of Beethoven" only to discover his mistake minutes later, which led him to remark: "Truly, the municipal authorities of Bonn ought to confiscate one of the two birth house signs."[26] The authorities, in this case the mayor, noted in an internal memo that "so crazy a fellow [as Beethoven] continues in the afterlife to badly damage our city's reputation."[27]

This sense of embarrassment prompted the rescue of the Beethoven-Haus. The project was spearheaded by none other than Joseph Joachim, the famed violinist and composer who was active in the Brahms circle. It was a natural next step in Joachim's ongoing role as what Laurie McManus calls a "priest of music" in Johannes Brahms's circle, whose members cultivated a language of art-religion to position themselves as protectors of musical "purity," moral uplift, and authentic deference to the past, or Werktreue.[28] Joachim was further hailed as "a paragon of authenticity" in the 1850s, according to Karen Leistra-Jones, because his spontaneous and improvisatory performance style captured the composer's living presence. While members of the New German School were accused of calling themselves Beethoven's heirs for profit,

Joachim was said to channel composers directly and to efface his own ego.²⁹ As a philanthropist, Joachim continued that mission of faithful Werktreue, preserving the composer's traces with head bowed.

By 1889, Joachim had assembled a team of local philanthropists known as the Beethoven-House Society, or Verein Beethoven-Haus (hereafter called the Verein), which sought to purchase the property, relocate its inhabitants, and restore it to some semblance of its original state. Predictably, the language of priesthood informed the Verein's fund-raising drive. One of several mission statements dictated by the secretary Wilhelm Kuppe reads: "The preservation and consecration [*Weihung*] of the birth houses of great geniuses and art heroes as monuments of piety and active veneration is a duty [*Ehrenschuld*] of all nations.... Beethoven's birth house has been purchased and shall serve not only as a Beethoven museum after its stylistically accurate restoration, but also as a center of collection and support for the care of musical art."³⁰ To extend its priesthood, the Verein granted honorary memberships to those deemed worthy: Brahms, Clara Schumann, Reger, Verdi, and even Bismarck himself, whose support for the cause was said by Germany's chancellor, Bernhard von Bülow, to serve as a model for "every German, even if he has never touched a piano key."³¹ Memberships of this kind could also be granted in exchange for donations of Beethoveniana, making all those who contributed feel like initiates into a pious rite, or as one donor of medallions put it, he would "preserve and cherish the site" as an "honorable duty [*ehrenvolle Pflicht*] that I shall undertake with the most heartfelt inner joy."³²

For all his aversion to the New German School, some aspects of Joachim's art-religion followed in the footsteps of Wagner, whose headquarters at Bayreuth had become a model for pilgrimage. Prior to Bonn's first chamber festival, Wilhelm Kuppe expressed his hope that the same visitors who "make pilgrimage to Bayreuth in great flocks" would also flock to Bonn as a center for chamber music performance.³³ In a local newspaper, one announcement turned this wish into an imperative of Ehrenpflicht: "if one can pay 200 marks for a seat in Bayreuth, then it is one's duty also not to forget the great composer [Beethoven]."³⁴ The Beethoven Verein, like many music societies in this period, borrowed one of Wagner's primary strategies for self-promotion, which Nicholas Vazsonyi has called "marketing martyrdom": a proclaimed disinterest in money that veils commercial intent.³⁵ Vazsonyi explains this apparent contradiction as an instance of Pierre Bourdieu's cycle of economic transformation, when symbolic capital (art) transforms into material capital (money) and vice versa.³⁶ Donors feel affinity with composers and regard themselves as disciples whose money is an offering at the altar; their donations maintain heritage sites and fund summer festivals; those sites attract pilgrims who bring another revenue stream; the music performed at those festivals strengthens donors' affinity with the composer—and on it goes, as one form of capital transforms into the other and back again.

Bayreuth was not the only model for Beethoven heritage in Bonn. There was an adjacent movement at work that remains neglected by musicologists: the landscape preservation initiatives of the Rhineland, or what was called *Naturdenkmalschutz*, the protection of natural monuments. This movement was quite a bit more adjacent than one might think, given that its torchbearer was the Dresden music professor and composer Ernst Rudorff, who was a member of Joachim's circle and a former student and friend of Clara Schumann. In his correspondence, Rudorff kept his circles somewhat separate; he rarely discussed his landscape projects in his letters to Joachim or Clara, or his musical affairs with his colleagues in the landscape movement. All the same, it is fair to speculate that Rudorff's interests hailed from a common source. In his capacity as a musician, he founded a Bach Society in Cologne in 1867, then proceeded to serve on the editorial board for the *Monuments of German Musical Art* (*Denkmäler deutscher Tonkunst*), a multivolume score edition that sought to identify and preserve capstones of German music. Meanwhile, he founded the German Association for Homeland Protection (Deutscher Bund Heimatschutz), a bureau that grew out of his writings in which he lamented the environmental and ethical deterioration of the Rhineland.

With its medieval castles and verdant hills, the Rhine valley had long been a paradigm of German Romanticism, or what was called *Rheinromantik*. Rudorff claimed that these landscapes served as the fundamental tone (*Grundton*) that underlay German music, the truest expression of the German spirit. (By no coincidence, among his many nature-themed compositions is a choral work titled "Ave Maria on the Rhine.")[37] Rudorff was alarmed when the region was scarred by the mining industry, and he led an environmentalist intervention against this "disfigurement" of the landscape's "physiognomy." But the real object of his disdain, as Thomas Lekan has shown, was the triviality of tourism that cheapened the "moral" benefits of nature.[38] Among these ills Rudorff cites funicular railroads, the desecration of ruined castles as playgrounds, and the garish illumination of waterfalls with fireworks. At the core of Rudorff's antitourism was, as Lekan puts it, "a nineteenth-century Romantic travel ideal that often clashed with the democratization of leisure: an ideal that favored the independent, male, upper-bourgeois lover of nature who undertakes travel for the sake of Bildung rather than entertainment."[39] Rudorff insisted that in order to truly flourish, a society must develop "a genuine, living piety for nature."[40]

This is where Rudorff's impact on Joachim's project, just one decade later, is most evident. When the Verein dictated an announcement to the *Bonner Zeitung*, it couched the project in opposition to trivial tourism. Joachim's "smoothly executed act of piety" served to rescue the house from profane misuse:

> No longer shall the glory connected to this hallowed site be exploited by the market cries of the industry that once occupied it; no longer shall the visitor to Bonn, who enters that antiquated house with an exalted bearing, be repulsed by raucous mongering or the offensive sounds of questionable music and dance performances. . . . Just as the Goethe house in Frankfurt, the Mozarteum in Salzburg, the Shakespeare house in Stratford, so too shall our Beethoven house in Bonn become a pilgrimage site for educated humanity.[41]

In this passage we find a clear echo of Rudorff: the antipathy toward cheap entertainment, the anticapitalist aversion to the "market cries" of Bonn's Marktplatz down the street, and the articulation of a more refined class of visitor, an "educated humanity" that would exercise the same connoisseurship and Bildung as Rudorff's ideal traveler. When the museum opened its doors in 1889, its aim was not only to preserve, but to cultivate "living piety," to borrow Rudorff's phrase. A report disseminated to members in 1904 admitted that "we don't want the hallowed place of Beethoven's birth to be seen only as a second monument for Bonn's greatest son, as a dead museum—the great life that was first brought forth in this modest house shall continue to have an ever blossoming, newly living impact."[42]

Joachim demonstrated this approach during the chamber festival of 1903 when he hosted a ritualistic musical performance inside the house for a small audience of sponsors. Together with his string quartet, he performed the Cavatina of op. 130 on Beethoven's own quartet instruments, which he reported to have a shallow tone suitable only for the most intimate venues, and this concert became a touchstone in the museum's history, an "act of consecration" that made an "unforgettably deep impression on all participants."[43] Later, after Joachim's death, a fund-raising flyer for his bust implies a spiritualist séance, as if he had assimilated Beethoven's swan song into his own: as the "deeply melancholy sounds of the Cavatina, op. 130 were coiled around us with a kind of transfiguration that could only be conjured by Joachim and his ensemble, many listeners must have asked themselves if [Joachim] himself . . . were not relaying the genius of Beethoven to his community for the last time."[44] A striking caricature by Franz Stassen from around this time makes it clear that Joachim was not only a faithful "priest" of art-religion, but a kind of spirit medium (fig. 1.1).[45]

Abb. 65. Franz Staffen: Jofef Joachim.

FIGURE 1.1. Caricature of Joseph Joachim by Franz Stassen, ca. 1900, in Karl Storck, *Musik und Musiker in Karikatur und Satire: Eine Kulturgeschichte der Musik aus dem Zerrspiegel von Dr. Karl Storck* (Oldenburg, 1910; Laaber: Laaber Verlag, 1998), 69.

That same year, Anton Rubinstein chose Bonn's newly erected Beethoven-Halle for his farewell concert, and he framed his performance in language borrowed from Joachim: "They have erected a temple to the godhead here in Bonn, and I hurried forth to present my sacred offering."[46] Performance was the antidote, it seems, to a "dead museum," and that is why the earliest call for donors in 1889 insisted that the Verein sought to "animate these rooms with Beethoven's genius" by filling them up with "letters and relics that speak of him with mute eloquence . . . along with everything that mediates sensory and emotional contact with him."[47] Joachim not only consecrated the space but revived the composer's spirit with a mode of restoration that felt warm and alive.

The result was a museum torn between two missions. An episode in its annals makes the problem clear. Since the museum opened, Beethoven's string quartet instruments had been on semipermanent loan from Berlin, but in 1923, a Prussian bureaucrat sent a curt missive demanding their return. The issue, he claimed, was that they sat mute behind glass, decaying and unused.[48] Granted, the instruments' condition had never been pristine: on the rare and ritualistic occasion that they were played, their "thin tone" (as Joachim put it) made them more of a curiosity than viable concert instruments, which explains why distinguished visitors like Prince Joachim Albrecht of Prussia were invited to play them solely behind the scenes.[49] Eventually, the hubbub went up the chain to the former Prussian minister of culture, who had heard Joachim's Cavatina performance in 1903 and ordered that the instruments remain in Bonn; his successor Gustav von Goßler was quoted as saying, "Beethoven's string instruments belong in the birth house of the master, where his piano also stands. Do not give them back to Berlin!"[50] A final clause in the Verein's response is among the museum's most powerful expressions of priesthood: these instruments were not for concert use, but "valuable relics," and "as relics they are of particular, even immeasurable worth. The members of the board see themselves as priests of a holy sanctum, which they serve with joyful devotion."[51] With this, the bureaucrat backed down, lauding the museum as a "national sanctuary."[52] At its core, the kerfuffle was not only about instruments, which make complicated relics indeed. Nor was it solely the growing pains of a young institution as it matured into a cultural hub. This dispute revealed an internal contradiction in the museum's primary charge: the duty to preserve does not always suit the desire to reanimate.

Performing Devotions

Those who have visited this museum know the birth room well: the narrow stairs, the velvet ropes, the quiet hush at the threshold of the attic, the lone bust on a pedestal. Some may be disappointed to learn that it is unclear whether Beethoven was born in that room. The attic came to be associated with the composer's birth in part because its modest dimensions were reminiscent of Christ's manger, and curators encouraged the public to see it as a shrine.

For many, this room was the museum's biggest attraction. The Verein's 1889 call for donors advertised that "much remains in its original condition, particularly the birth room with its deeply moving simplicity."[53] In its announcement in 1890, the *Musical Times* likewise speculated that "many will surely shed tears on beholding the wretched garret—a small lean-to attic in the roof—in which Beethoven is said to have been born."[54] And visitors did appear to come away moved. Writing in the museum visitors' books, many cited an emotional connection with the space as a "holy site," marveling at its modesty.[55] A couple from Cologne contributed a brief verse: "Earthly greatness may arise from any little hovel. We learn this anew from the place where we now stand."[56] Another wrote that "Eternal Beethoven on high—born in a *lowly* room—struggled to achieve the *highest* heights of human achievement through genius and—industry!"[57] Most poignant of all: "No church has made such a deeply moving impression on me as this birth room."[58]

It was rare for nineteenth-century museums to leave a room so bare. Most rooms in the Beethoven-Haus preferred the density of salon hanging, a floor-to-ceiling style of display made famous by the Hermitage Museum in St. Petersburg; the result was a visual superfluity, as Beethoven's face stared down from every angle. The birth chamber, in contrast, was more relic than room. This was a space into which visitors could peer but not tread, empty but for a bust and a few wreaths. In gift shop postcards, the room was photographically manipulated to show celestial sunbeams streaming down (fig. 1.2). It is no wonder that someone wrote in the visitors' book: "Just as the sun, from millennium to millennium, sends its rays into the small darkness of your birth site, so your sounds ring forth exuberantly to the whole world, you godlike King of eternal generations."[59]

The simplicity of the space invited creative reimaginings of Beethoven's birth story in the press, and in 1898, the museum acquired an object that made its Christology explicit.[60] Inspired by a visit to the house in 1895, the Viennese painter Friedrich Geselschap planned a large-format canvas that would depict Beethoven's birth like a Christian Nativity scene, and while he tragically committed suicide before the painting was complete, his watercolor sketch was bequeathed to the museum in his will (fig. 1.3). In Geselschap's image, the infant Beethoven is lowered gently into his cradle by his mother Maria Magdalena, who is clad in Marian blue. Choirboys serenade the scene, attended by Ludwig's father as St. Joseph. By far the most striking, though, are the two allegorical figures who loom over the cradle: a heavenly muse crowns baby Beethoven with a laurel wreath, while an ascetic figure of fate waits in the shadows with her menacing crown of thorns, the embodiment of Beethoven's artistic glory at the cost of Christological suffering. Given that Geselschap was a Viennese Catholic, he might well have borrowed from the icon of Our Lady of Perpetual Help, which shows the Madonna and Child flanked by angels who bear instruments of the Crucifixion. A more direct inspiration could have been the 1845 commemorative poster by Johann Peter Lyser, which showed the composer sandwiched between two muses, one crowning him with laurels and the other offering him thorns.[61] These dual crowns, known as the *Lorbeerkranz* and *Dornenkranz*, had become a poetic trope in popular Beethoven literature and poetry of the period.[62]

FIGURE 1.2. (*Top*) Second floor of the Beethoven-Haus museum in 1910, reproduction of a photograph by A. G. Siegburg. (*Bottom*) Postcard of the birth room, ca. 1900. © Beethoven-Haus, Bonn, B 1424a and B 653b.

Curators positioned the painting at the threshold of the birth chamber, where it shaped the impressions of visitors. Here they followed the wishes of the executor of Geselschap's estate, who wrote that the sketch should find "a place of honor in the birth room of that great man, and thus an honor for both Beethoven and Geselschap that will surely arouse hallowed excitement in every observer of this sanctified site."[63] Poems in the visitors' books—a trove of handwritten odes that I explore in more depth shortly—show the influence of this allegory. Take, for instance, an inscription from 1911:

> Triumph through art!
> The thorns must spur
> To jubilantly attain the rose.
> No doubt you have suffered much, master—
> And through that crown, you achieved immortality![64]

A more colorful example hails from the little-known Margarete Stadler, who published a "musical fairytale" in December of 1912 that reenacts Geselschap's "fairies." The evil muse taunts Beethoven's mother with a cackling premonition that the baby shall "wear a crown of thorns" and die a lonely death; meanwhile, the good fairy shows his mother a vision of the monument in Bonn and later comforts the adult Beethoven in his hour of need. The story concludes with the adult Beethoven's turn to austerity and martyrdom as he relinquishes earthly love, composes his Ninth Symphony by divine dictation, renounces "earthly music" on his deathbed, and walks off into the lush valley of the afterlife with a violin under his arm.[65] The museum's displays encouraged this association of birth and death, promise and suffering, pity and solace: alongside Geselschap's painting at the threshold of the birth room,

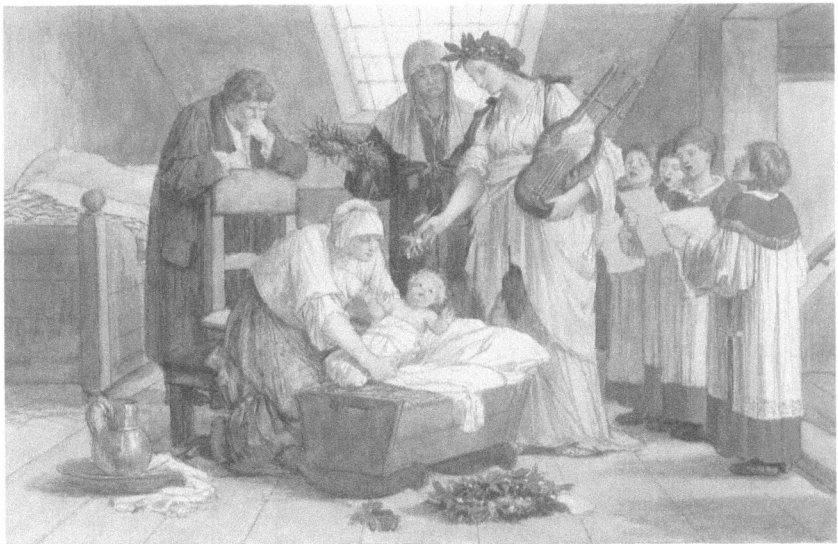

FIGURE 1.3. Friedrich Geselschap, *Beethovens Geburt* (*Beethoven's Birth*). Watercolor study. Dresden, 1895–98. © Beethoven-Haus, Bonn, B 109/a.

curators installed Beethoven's death mask, his last testament in jagged script, and a framed invitation to his funeral.

By far the most interesting response to Geselschap was a lengthy poem gifted to the Beethoven-Haus in 1903, written by a sixty-one-year-old piano teacher named Hermine Bovet who hailed from a nearby town.[66] Bovet's verse traces the fate of dual crowns through Beethoven's lifetime and imagines the good muse summoning a cast of sprites, nymphs, and fairies to help the composer carry out his sacred work, first as a child prodigy and later as a heroic composer-commander. Just when Beethoven gains recognition, the thorns "buried themselves deep in his brow, marred with the creases of affliction." Bovet's poem reflects the strange amalgam that characterized Beethoven's literary and visual reception around 1900: a Protestant work ethic, a Christological fate, Hellenic muses, and the colorful creatures of art nouveau designs. At the poem's close, the flight of fancy lands back on earth when Beethoven's ghost gazes down on his birth chamber.

For all its ecstatic overtones, this poem had a more practical aim. Bovet sent her ode to the Verein three weeks before the chamber festival, and her postscript hints that she hopes the Verein might publish it for that occasion. When she signs her name with "music writer" rather than piano teacher—and, indeed, whatever writings she produced have since fallen into obscurity—she demonstrates how devotees could engage in subtle forms of self-promotion. In the visitors' books, too, those who wrote their names rarely included professions *unless* they were music writers, scholars, chapel masters, or composers. Through poetic offerings, these devotees demonstrated that they were musical insiders.

When I conducted research for this chapter, even the archivists at the Beethoven-Haus expressed some excitement as they wheeled in the earliest visitors' books from storage. These books had not been examined for some time, and it is understandable that their contents might fly under the radar: in late nineteenth-century museums, it was common practice to simply record one's name, place of origin, date of visit, and profession if it was musically apposite, but no more.[67] At first, these pages look to be long lists of names (fig. 1.4). A closer study reveals a wealth of aphorisms, poems, and even musical incipits that surface every three to five pages, which shows that some visitors treated this as Beethoven's Stammbuch, a friendship album for a ghost. The messages are tremendously varied. Some betray a tone of self-consciousness as if expecting their contributions to be read, or what Jürgen Habermas called "privateness oriented towards an audience."[68] Others are so cryptic and intimate that they give the impression of a spontaneous outpouring. As a result, these books are multifaceted historical records that serve alternately as sources (self-consciously crafted for posterity) and as traces (not

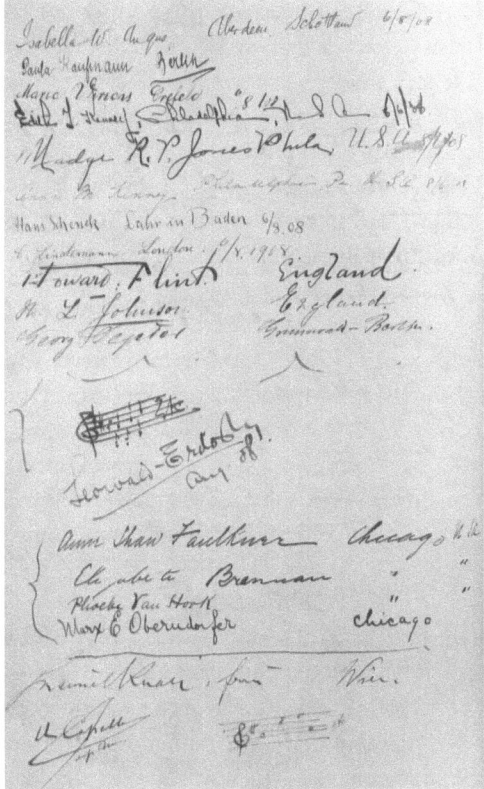

FIGURE 1.4. Page from the Beethoven-Haus visitors' books. August 6, 1908. © Beethoven-Haus, Bonn.

intended to be read). While it may be impossible to determine the sincerity or performativity of these utterances, they offer an unusually concrete picture of the "popular imagination," a common expression in cultural studies that can be difficult to pin down.

It is not possible in the span of a few pages to capture the delightful multiplicity of messages in the visitors' books. For all their miscellany, there are a few commonalities. First, and unsurprisingly, they show how these historic rooms amplified an experience of Beethoven's tangible presence. Secondly, they bring to life some of the claims made by K. M. Knittel in her study of a large body of "pilgrimage to Beethoven" anecdotes.[69] These stories share a common set of narrative tropes: a pilgrim overcomes obstacles on a quest to visit the near-inaccessible Beethoven, which culminates in heartwarming moment with the composer, who stamps the pilgrim's musical offering with approval. Knittel argues that it is the *author* rather than Beethoven who becomes the hero. When Beethoven is elusive to all but the privileged few, these imaginary visits enact a fantasy of intimate friendship with the deceased celebrity. Visitors' books became a space where music lovers could not only perform their devotions, but also outperform each other.

Occasionally, the tone of competition was blatant. Take for example a petulant entry from 1918: "Dear Master: today I came up here to visit you—but I'm hopping mad that I had to first flip through this pompous book.—What would you say about this voluminous rabble of names and titles? . . . For the time being, a heartfelt farewell. I must now tour around for a bit."[70] This hostility toward other guests, derided as the "rabble," did not go unnoticed. In the margins, a fellow visitor protested: "Dear friend, whoever you are, on this holy site you ought to show some humility!"[71] Such explicit antagonisms were rare, and the visitors' books could just as easily create a community of devotees. In 1910, a poem written by an English visitor was almost immediately translated into German in a gesture of goodwill.[72] Travelers who followed in their friends' footsteps could flip backward to find each other's names, and some engaged in musical dialogues with each other, parroting each other's excerpts or adding on the next successive movement in a work.[73] When visitors showed off their status as insiders, they did so in more subtle ways: they named themselves pilgrims, recalled pious feelings in the house, offered prayers for the museum and its curators, wrote emphatic odes to Beethoven's greatness, and demonstrated their expertise with musical quotations.[74] The most common format for these entries was a single rhymed verse, which was the standard language of friendship albums, and examples range from the sublime to the ridiculous. Most took the task seriously: "Beethoven-Haus, the site of German greatness / You cradle of that man who created a world / Stand unharmed in the most distant future / Just as he does, he who expects the kiss of divinity."[75] Others were irreverent: "Beethoven, you are / my dearest composer / Thus I rhyme here this little verse / As is the custom when venerating / But now I must depart again / For I still need to behold your house."[76] The most interesting messages of all betray a self-consciousness about album conventions, as if visitors yearned to distinguish themselves from the platitudes of other guests. A key example is this note from 1922: "How can one do anything but quiver with devotion through—ugh. Beethoven and *I* compose? Remove your hat, you pass through holy rooms."[77] Commands to treat the house as a church ("remove your hat" or "I bow with prayer and awe") were quite common, and they show how piety trickled down from cities to institutions to devotees.[78]

Perhaps the most touching type of entry is the outpouring of affection from those who saw themselves in Beethoven. Two notes from deaf admirers show how Beethoven's biography, not only his music, could produce a

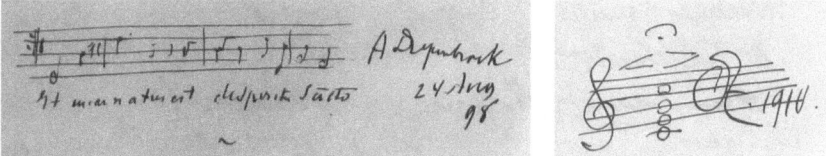

FIGURE 1.5. Atmospheric musical entries in the Beethoven-Haus visitors' books. (*Left*) Passage from the credo of Beethoven's *Missa Solemnis*: "Et incarnatus est de spiritu sancto" (He was incarnate by the Holy Ghost), August 24, 1898. (*Right*) A hairpin, fermata, and C-major chord create an intriguing imaginary soundscape, August 12, 1910. © Beethoven-Haus, Bonn.

following of its own. In almost inscrutable handwriting that seems intended for Beethoven's eyes only, someone wrote: "I was told that I resemble you, master of tones. My face is *not* like yours, but rather my *hearing*. I always wanted even once to hear your Ninth. Like you, I am deaf. The difference: You could create it even without hearing, and that is a great deal more."[79] Less than a year later, another visitor chimed in: "As someone who is hard of hearing, I stand with the deepest emotion on the holy floor of the birth house of my greatest comrade in suffering!"[80]

It is clear on every page that this house brought Beethoven to life. Visitors commented frequently on the "ghostly, bewitching" presence of Beethoven in those rooms.[81] One wrote that "the soul of Beethoven fills the entire house"; another confessed that "here I feel the greatness of your spirit" and that a "wave of emotion fills my soul."[82] One instance is quite tactile: "In this house time seems to have no power. . . . One feels the eternal in these rooms so close and naturally as the air that one breathes."[83] There is a fine line in these passages between piety and the paranormal.

The silence of the space could invite another form of interaction: it encouraged some to imagine their favorite music by Beethoven and to emulate the composer's inner ear, his ability to conjure sounds in the mind. A Bonn bookseller named Hans Maria Saget gifted a poem in which dreamlike music rushes through the historic rooms like clear waters from a spring. His reverie ends with the decaying sound of an ellipsis: "Beethoven plays . . ."[84] Fragments of Beethoven's music waft across the pages of the visitors' books. Some were symbolic, like the frequent appearance of the Fifth Symphony motif or the "Ode to Joy" melody that acted as metonyms for Beethoven's name. Others might have had personal significance, such as excerpts from accessible piano and chamber music that may recall fond memories of playing.[85] Some passages affirm the sacredness of the house; in 1898, for example, someone selected a hushed, otherworldly passage from the credo of the *Missa Solemnis*. Yet others offered more cryptic excerpts that give the impression of ghostly strains (fig. 1.5).

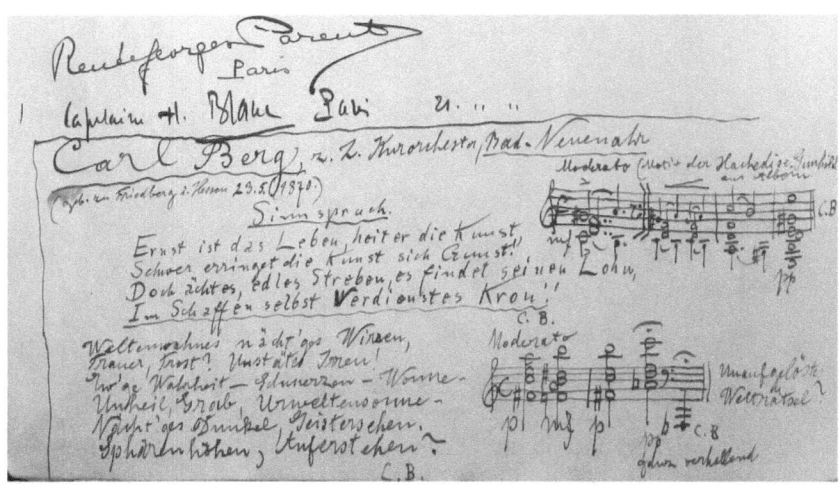

FIGURE 1.6. Carl Berg's longest entry in the Beethoven-Haus visitors' books, September 21, 1920. Berg contributed two simpler entries, both musical in nature, in 1900 and 1918. © Beethoven-Haus, Bonn.

While countless visitors thought of their trip as a pilgrimage in some form, there are a few individuals who performed this identity in more intense ways. To conclude this section, I spotlight two such pilgrims who exemplify Knittel's arguments: the composer Carl Berg and the writer Margarete Koelman (pen name Irene Wild). Both pilgrims wrote unusually detailed entries in the visitors' books on multiple occasions, returning to offer up their work to Beethoven's spirit.

The composer Carl Berg, who directed the orchestra at the popular Rhineland spa destination of Bad Neuenahr, has fallen into obscurity.[86] Judging from the entries he left in the Beethoven-Haus visitors' books in 1900, 1918, and again in 1920, he authored a tone poem that was suffused with Wagnerian leitmotifs. In painstaking detail, Berg copied a selection of his leitmotifs and poetic descriptors into the visitors' book, including a recurring "Alboin motif" that indicates that the work was based on the medieval King Alboin of the Lombards (inspired, perhaps, by the same libretto that Hugo Wolf took up and quickly abandoned in 1876).[87] Berg's final entry included two richly harmonized motifs (fig. 1.6). In the first, which represents Alboin's shield, a luminous half-diminished seventh chord oscillates between chromatic neighbor tones before it settles into a dominant seventh—a clear nod to Wagner's harmonic language. The second entry invokes the language of Friedrich Nietzsche with its provocative label, "unsolved world riddle? Quite shadowy." The harmonies here are considerably more adventurous than those in the Alboin motif, and their mixture of chromatic motion and planing makes this world riddle even more elusive and experimental than Richard Strauss's polytonal equivalent, his clash of B major and C in the final moments of *Also Sprach Zarathustra*. (For the curious listener: Berg's riddle begins with an F-sharp half-diminished seventh chord, which resolves to a B7-flat 5 chord that begs to resolve further to E major. Instead, he evades this resolution with a strange stepwise amalgam of elements from these first two chords, and the progression evaporates into thin air when it floats up a third to an inversion of a D half-diminished seventh chord. The middle member of this chord, a lone A-flat, rings out in the bass, leaving us with a powerful question mark, indeed.) Here we may seem to stray rather far from Beethoven. But Berg, I would argue, acts as a perfect pilgrim. Knittel has shown how pilgrimages to Beethoven culminated in a compositional offering to the great master, and Berg offers his own works to Beethoven's memory. What's more, he draws attention to himself as the hero in this journey. Not only did he introduce his birthplace and birthdate alongside his name, but he compulsively labeled every motif and stanza with his own initials, C.B., to remind us that these are the fruits of his labor.

Several devotional poems by Margarete Koelman paint an even clearer portrait of a pilgrim to Beethoven. Her poems are suffused with affection for the faraway celebrity, an ancestor of what media theorists call parasociality.[88] Koelman, whose maiden name (Friedländer) suggests that she might have been Jewish, married a Prussian bureaucrat whom she then followed to Hannover and Breslau. After his death in 1904, she moved to Berlin and became an active translator and writer who published poetry under the pseudonym Irene Wild.[89] She was a contributing member of the Beethoven Verein who visited Bonn regularly to enjoy the chamber music festivals, some of which she reviewed as a music critic. Over the course of a decade, she left four unpublished odes in the visitors' books that encapsulate the pilgrim in all its forms: she praises the holy site while invoking Ehrenpflicht to insist that it be venerated properly; she reimagines Beethoven's life, death, and afterlife in vibrant verse; she communes with the composer's face and gaze; she touts her own appreciation of his works as true devotion while her concert neighbors are dilettantes; and above all, she yearns for Beethoven's presence with erotic charge.

Koelman first appears in the visitors' books in 1903 with an ode to the museum. As was common in these books, she urges the reader to "remove your everyday shoes" because "this is truly holy land."[90] Beneath this short verse, which was quite characteristic of an album, she then copied a longer rumination on Beethoven's face that shows the depths of her devotion. As the lyrical subject places a flower offering beneath Beethoven's bust, she seeks solace in a "serious and sad" visage "furrowed dark from the labors of mankind." Like a votive, her flower grants her protection: "Lovingly then I step back / so that your closeness may watch over my sleep, / I lower my head to your noble visage: / My great dead one, do not leave me!"[91] We know that Koelman enacted this ritual herself on at least one occasion. Several years later, in 1912, she laid a poem on the birth-room floor on the occasion of Beethoven's ninetieth death day; in it, she reimagines the thunderous deathbed scene in which Beethoven is struck by sudden inspiration for his Tenth Symphony, his "death song," which only nature could overhear.[92]

One year after Koelman copied her first poems into the books, her husband passed away, and her affection for Beethoven became more intense. Whether Koelman displaced more personal forms of longing onto Beethoven is speculative; suffice it to say, her poems from these years are charged with an erotic desire to merge into Beethoven's being. In 1906, she left a Christological poem in the visitors' book that culminates in a prayer: "Let his melodies / resonate in you. / When holy days move you— / look upon his face. / When dark hours / wring the heart, / draw power and solace, / the most exalted life, / from his spirit's lofty flight!"[93] While this passage invites others to join her in devotion, the poems from 1911 and 1912 take a personal turn. In one, we see an instance of competition, as Koelman derides the inadequate appreciation of her peers:

Recently I held your Ninth Symphony,
You great lonely one, in my hands
And quietly turned the pages
Upon which my master's hand once rested.
I pressed my lips to the notes
Which house my holiest reflections,
A treasure that centuries could not replace.

Indeed I was not alone with this treasure
—the others, though, took little care,
Buried in erudition
That which my lips bade silently in prayer.[94]

(*Jüngst hielt ich deine Neunte Sinfonie, / Du großer Einsamer, in Händen / Und wandte leise Blatt um Blatt, / Auf dem einst meines Meisters Hand geruht. / Ich drückte meine Lippen auf die Zeichen, / In denen sich das Heiligste mir birgt, / Ein Schatz, den nicht Jahrtausende ersetzten. / Wohl war ich nicht allein mit diesem Kleinod—die Andern aber hatten wenig Acht, / Vergraben in Gelehrsamkeit, / Was meine Lippen stumm, anbetend taten.*)

Like Knittel's pilgrims, Koelman places herself at the center of this ode. She

FIGURE 1.7. Albin Egger-Lienz, *Ninth Symphony*. Postcard. Vienna, Brüder Kohn, ca. 1900. Historical Picture Postcards, Universität Osnabrück, Collection of Prof. Dr. Sabine Giesbrecht. www.bildpostkarten.uos.de.

distinguishes her pure love of Beethoven—sealed with a kiss, like the osculatory devotion of Catholic relics—from that of listeners who pick apart his music, presumably a reference to analytical concert guides, lectures, and other teaching tools of Bildung.[95] To read this poem out of context, one would assume that Koelman adopts a Catholic frame of devotion to denigrate Bildung, the burying of art in erudition associated with cosmopolitan Jewry. Yet as her maiden name is a common Jewish surname, we might speculate that her poetry was a passionate expression of her own assimilation. One year before she wrote this poem, Koelman published a short story that presents a grotesque caricature of two Chinese men, one assimilated and one a new immigrant. Her devotional writings to Beethoven reflect an established Jewish population that supported Catholic-inflected forms of art-religion to distinguish themselves from eastern Europe's new arrivals.[96]

The longest and richest of Koelman's poems is striking not only for its Christological metaphors, but for its parasocial longing to connect with a version of Beethoven that is both alluring and aloof. Contemporary representations of the composer often showed him lost in musical thought, an object of admiration that could not gaze back; this is why so many postcards and ex libris plates (that is, illustrated frontispieces for home libraries) showed his stony-faced mask, as I discuss at length in chapter 4 (fig. 1.7).

Images like these haunt Koelman's fantasia, which captures the struggle to reach an indifferent Beethoven, who is preoccupied by otherworldly matters. The poem is so evocative that it is worth reproducing in its entirety:

Beethoven
(Vision upon hearing one of his last works.)

It was in those days,
When you still lived, Beethoven,
When your great soul
Bled to death with loneliness
And when no tone was audible to your ear,
Except the sounds from within.
      You sat at the piano
And played your last fantasies,
Great, wild and strong:
The yearning of a god,
Who searches through the eons for a soul,
Who does not fear his powerful breath,
Who frees him from his loneliness
Who just loves him and is his.

I lived in those days and I heard you.
Your immortal work looked at me
And drew me utterly to you.
I had never seen you face-to-face,
But I felt what your face was like.
You sat alone
And played, utterly turned inside yourself,
And listened to your inner ear,
To your own deeply moving songs,
Meanwhile the crowd gathered
Festively around you
And, some amazed, some unsettled
They heard your work.
—But you saw no one.

I crept into that world
And stood spellbound at your threshold.
I closed my eyes
And let the surge of your tones
Wash over me like an eternal ocean
Of waves that will never dissipate,
Eternal, inexhaustible, always creating themselves anew

From depths that no mortal eye could see.
    There I stood,
The world collapsed beneath me,
But I did not stir. It decays away
And the ocean of ever-eternal earthly sounds
Carries me out of the muddled dream of being.

It grows still. The master had finished,
The crowd rejoices with applause. I startle.
    How the sound clangs
In a final chord that rattles the gods!
      Now the herd quiets down. The master
      sits alone,
He no longer hears the voice of daily life,
    His eye
Loses itself far off in the unknown.
The air grows still around him. I stand yet lost
At his threshold,
And with closed eyes, softly probing
I feel my way to him.
And again I drop to my knees before him
And I call out: Precious Master,
Let me kiss the dust on your shoes!
But he does not hear me. Now he raises
His eyes
And behold: his lost gaze lights up,
A divine sunbeam flashes,
For an immortal found a human soul
That does not fear
His powerful breath,
That lovingly frees him from his loneliness
And lives in him and is his.[97]

(Beethoven / (Vision beim Anhören eines seiner letzten Werke.) / Es war in jenen Tagen, / Da du noch lebtest, Beethoven, / Als deine große Seele / Einsam verblutete / Und deinem Ohr kein Ton vernahmbar war, / Als nur der Klang aus deinem Innern. / Du saßest am Klavier / Und spieltest deine letzten Fantasieen, / Groß, wild und stark: / Die Sehnsucht eines Gottes, / Der durch Äonen eine Seele sucht, / Die sein gewaltger Odem nicht erschreckt, / Die ihn aus seiner Einsamkeit erlöst / Und ihn nur liebt und sein ist. // Ich lebte damals und ich hörte dich. / Mich hatte dein unsterblich Werk geguckt / Und ganz zu dir gezogen. / Dich selber hatt' ich nie von Angesicht gesehn, / Doch füllt ich, wie dein Antlitz war. / Du saßest einsam / Und spieltest, ganz in dich gekehrt, / Und lauschtest mit dem innern Ohr / Den eignen, tief ergreifenden Gesängen, / Indes die Menge festlich um dich her / Versammelt war / Und halb verwundert, halb erschüttert

Dein Werk vernahm. / —Du aber sahest keinen. // Ich schlich mich still herein / Und blieb gebannt an deiner Schwelle stehen. / Die Augen schloß ich / Und ließ die Wogen deiner Töne / Über mich hingehn wie ein ewig Meer / Von Wellen, die sich nie verströmen können, / Unendlich, unerschöpflich, immer neu sich schaffend / Aus Tiefen, die kein sterblich Auge sah. / Da stand ich, / Die Welt ging unter mir zugrunde, / Doch rührte ich mich nicht. Weg sie vergehn / Und mir das Meer urew'ger Weltenklänge / Mich tragen aus dem wirren Traum des Seins. // Still wird es gesetzt. Der Meister hat geendet, / Die Menge jubelt Beifall. Ich erschrecke. / Wie gellt der Ton / In eines Göttersturmes Schlußakkord! / Nun weicht die Schar. Der Meister / sitzt allein, / Er hört die Stimmen ihres Alltags nicht, / Sein Auge / Verliert sich weit in unbekannte Fernen. / Still ist's um ihn. Ich stehe noch verloren / An seiner Schwelle, / Und mit geschlossnen Augen, leise tastend / Such ich mir meinen Weg zu ihm. / Und in die Kniee gleit ich vor ihm wieder / Und rufe: Teuerer Meister, / Laß mich den Staub von deinen Schuhen küssen! / Er aber hört mich nicht. Doch hebt er jetzt / Das Auge / Und siehe: licht wird sein verlorner Blick, / Ein Götterstrahl blitzt auf, / Da ein Unsterblicher die Menschenseele fand, / Die sein gewalt'ger / Odem nicht erschreckt, / Die liebend ihn aus seiner Einsamkeit erlöst / Und in ihm lebt und sein ist.)

This poem is ecstatically devotional, with a tangible and embodied sense of eros. With sensory language and a nonlinear sense of temporality, Koelman longs for Beethoven's music, body, and presence. Beethoven's music—which emanates from his own living body and ripples through the cosmos beyond his death—leaves her transfigured as the earth crumbles beneath her feet, and the strength of her reverie sets her apart from the clapping "herd" who jolts her back to the mundane. Most potent of all is the last stanza, when the sullen mask springs to life and seems almost to see her, as if she, the long-posthumous devotee, were Beethoven's lost soulmate. Koelman fantasizes that her relationship with Beethoven is mutual, a love across the ages. She appears here as a bride of Christ.

While the concept of parasociality invites valuable comparison with celebrity studies, the nineteenth century had its own vocabulary for love across great distance. When I paged through these effusions in the Beethoven-Haus library, I stumbled across a line of morse code inscribed in 1925 by J. Baruda of the Royal Army Service Corps.[98] The word I decoded made my hair stand on end: *mizpah*. This Hebrew word from the story of Jacob in Genesis (31:49) means "watchtower," which implies that God watches over those who are separated from each other. In the nineteenth century, *mizpah* was adopted into the English language to denote poignant closeness or affection across great distance; it signified the two-part rings or necklaces that lovers wore as mementos when separated, titled countless poems about the absent lover, and found its way onto public monuments and tombstones (and, curiously, the grand archway that greeted passengers at Denver's historic railway station).⁹⁹ There is every chance that Baruda simply missed his family back home in England, but in the context of this repository of tenderness for Beethoven, its pages dotted with devotional expressions, Baruda's choice of *mizpah* implies an aching bond with the composer. However exaggerated we might find the erotic poems by Margarete Koelman, or the eager self-promotion of Carl Berg, their entries show how Beethoven could stir feelings of mizpah.

The affection of these devotees adds a richly human dimension to an otherwise faceless history of heritage preservation. What museums offered, for a fee, was a sensory relic experience. In those rooms, visitors did not need to trade in locks of hair or cradle shards for an intimate encounter with the deceased. Lately, historic house museums have earned a reputation as frozen monuments or beautiful corpses; some describe them as mausoleums, petrified and lifeless, while curators search for innovative ways to revive them.¹⁰⁰ But a century earlier, visitors to such houses did not find them dead at all. When the luminary's effects could channel the spirit and "speak of him with mute eloquence," any mode of display, however static, was already interactive. To absorb those traces of Beethoven, pilgrims had to leave behind a trace of themselves.

"All of Vienna Is a Beethoven House!"

It is one thing to feel Beethoven's presence in a house, and quite another to treat a house as a stand-in for human remains. At the turn of the twentieth century, Beethoven's death house in Vienna, called the Schwarzspanierhaus, acted as a surrogate body. What that meant for historic preservation was not straightforward because some felt that bodies should be buried and mourned, while others wanted them embalmed and displayed. This is why the Schwarzspanierhaus was given a funereal send-off in 1903, and debates about its demolition show conflicting philosophies of preservation in a lively city on the move.

It makes sense that Vienna and Bonn would differ in their approach to memory. Bonn had developed a tourist industry that tapped into the poetic fantasy of the untouched Rhine valley as a *Heimat*, or homeland, where time stood still.¹⁰¹ In the 1890s, to stage Germany as continuous with the Holy Roman Empire, Kaiser Wilhelm II restored medieval castles and erected memorials throughout the Rhine region.¹⁰² Meanwhile, when Vienna's population surged in the wake of industrial revolution, city administrators and landlords were incentivized by generous tax exemptions to raze and rebuild. These changes were framed not as a desecration of the past, but as the natural life cycle of a growing city.¹⁰³

My focus on Vienna's demolitions diverges from narratives of self-fashioning as the so-called city of music. Late in the century, Vienna inscribed its musical heritage into facades and street names; its exhumation and reburial of Beethoven and Schubert in 1888 prompted a composers' grove of *Ehrengräber*, or honorable graves, that constructed its "collective cultural memory," as Reuben Phillips calls it.[104] But Vienna's identity was not nearly as homogenous or teleological as it may appear. The demolition of Beethoven's death house, and the outcry that ensued, shows how the city was not a collage of static "sites of memory" (*lieux de mémoire*), an oft-cited expression by Pierre Nora that tends to be divorced from its initial bite. For Nora, heritage can be cannibalistic. A mindless habit of archiving leads society to embalm itself, as "sites of memory" displace the oral histories that live in "environments of memory," or *milieux de mémoire*.[105] In Vienna, those who demanded sites of memory had to negotiate a lived environment of rapid modernization and growth.

When the owners of the Schwarzspanierhaus—the centuries-old Stift Heiligenkreuz, led by the abbot Gregor Pöck—determined that demolition was financially sensible, the initial wave of protests was louder abroad than in Vienna, where the decision met with quiet acceptance. To pacify Beethoven lovers, the Stift held a farewell ritual, or what one paper called a "death ceremony for the old, condemned house."[106] On November 15, 1903, a week prior to demolition, a few hundred devotees gathered in the pouring rain to hear a series of speeches, a choral ode, and a touching performance of the last quartet, op. 135, which onlookers heard faintly through the windows of the death rooms themselves. A closer view was reserved for the city's elites, by invitation only (fig. 1.8). The Viennese were known for their morbid sensibilities; in his granular memoir, Stefan Zweig recalled that "even funerals found enthusiastic audiences and it was the ambition of every true Viennese to have a 'lovely corpse,' [in dialect, *a schöne Leich*] welcomed with a majestic procession and many followers; even his death converted the genuine Viennese into a spectacle for others."[107] With this farewell ceremony, the Stift Heiligenkreuz made Beethoven's death house into a "lovely corpse" and recapitulated the two reburial events of 1863 and 1888, offering the composer a fourth funeral.

The highlight for many was an opportunity to tour the doomed apartment one last time, as if viewing the body at the wake, and to chip away at the wallpaper, rugs, door frames, and floors of the death room to take home a sacred shard.[108] (This explains why photographs taken a few days before demolition show pitted walls and splintered floors.) Even the Stift Heiligenkreuz participated in this relic collecting when they removed pieces of the room, such as door frames and parquetry, with the intention to embed these into a

FIGURE 1.8. Front page of the *Illustrirtes Wiener Extrablatt*, November 17, 1903. ANNO/Österreichische Nationalbibliothek.

new Beethoven archive; this would serve as a compromise for protesters who wanted to transplant the entire apartment into the Wien Museum, like the poet Franz Grillparzer's rooms a few years prior. The pieces of Beethoven's room, or "sarcophagal fragments" as one critic put it, were transported with solemnity to the Vienna City Museum.[109]

Anthropomorphizing the building as human remains was a clever strategy. The Stift Heiligenkreuz deflected attention from their decision by framing demolition as inevitable and by encouraging the public to mourn. Even the choice of the op. 135 was calculated: the insistent "it must be!" of the finale envoiced the house with an iron resolve.[110] A poetic prologue read at the event went so far as to suggest that Beethoven's spirit was trapped in those walls: "The cloud seeks a way out of the light, / Not to rest, but to build a new path! ... Here eternity sinks down upon this house, / The walls yield, a spirit finds its way homeward."[111] In his official speech, Abbot Pöck tried to recast the word "piety" as conservative and backward looking: "The inevitable necessity that the old disappears to make room for the new has resulted in the sacrifice of this revered house, so rich with historical reminiscences. . . . As painful as it may be, piety may not and cannot obstruct the development of things when it merely protects the pious remembrance of great men."[112]

Not everyone was persuaded. For German correspondents like Anton August Raaf, this performative event followed a lineage of tasteless spectacle that claimed famous sons in death without nurturing them in life. Raaf leveled this accusation at Vienna's two earlier reburials in ever more ostentatious tombs, "and now, adding insult to injury, his death chamber is turned to ruins!"[113] For one reporter, Beethoven's bust in the corner took on an "almost living expression," and it "seemed almost to ask, bewildered: what did I do to deserve the honor of this mass visit, when I was left to die alone?"[114] Rebuttals appeared in the Viennese papers, too, which debated whether Beethoven's spirit still resides in those walls, or whether it had since been flushed out by living tenants; one countered that there remains in those rooms "not a single atom of Beethoven himself, no invisible breath from his lips."[115] Others regretted the loss of layered histories in the house, such as the coffee shop that had long been a hub for the nearby medical school of the University of Vienna; the café was frequented by anatomists Josef Hyrtl and Carl Langer von Edenberg, whose analyses of composers' skulls are discussed in chapter 3.[116]

Debates about the demolition informed the philosophies of historic preservation that shaped state policy for decades to come, starting with the essay "The Modern Cult of Monuments," by Alois Riegl, which appeared in the same year that Beethoven's house fell. Riegl's ideas emerged from his prominent role as chair of the state preservation bureau, which passed a series of new laws in 1905. His approach was cautious: he advocated against sentimental motivations and the tendency to anthropomorphize architecture, such as the irrational fear of wounding a building's physiognomy. Riegl's successor, Max Dvořák, was alarmed by the "lack of piety" that led historic Vienna to fall victim to "the duties and obligations of modern city building at the expense of the past."[117] His more stringent policies, spurred into urgent action by the devastation of the First World War, sought to preserve Vienna in amber. It remains so today.

I conclude this chapter with poignant reactions to this demolition that show how the Viennese wrestled with Ehrenpflicht. The best-known response hails from Stefan Zweig's memoir *The World of Yesterday* (1942). As he reconstructs fin de siècle Vienna, Zweig describes how true lovers of theater and music were pitted against the forces of Viennese impiety, which compelled devotees to preserve shards of beloved halls "in costly caskets, as fragments of the Holy Cross are kept in churches." When Beethoven's death house was torn down, and when the beloved Bösendorfer Saal was demolished a decade later, it felt like "a bit of our soul that was being torn out of our body."[118] This explains why music lovers collected stones from the rubble. An unknown mourner for the Bösendorfer hall pillaged a stone from the wreckage and labeled it with an epitaph, and three stones from Beethoven's death house received similar treatment. A chunk of plaster from the death room was mounted with a photograph and plaque; a stone from the courtyard was painted in copper and decorated with the incipit of op. 135; and most unusual of all, an admirer harvested stones from the debris and commissioned a three-dimensional carving of Beethoven's life mask surrounded by a halo of light rays, which he called "a relic for my family." In an affixed memoir, he recalled his pensive trip to the ruins, standing in the rubble as the heavens shone into the roofless death room.[119] In the absence of historic preservation, it was left to Beethoven's devotees to preserve his remains in idiosyncratic ways.

Ultimately, these relics offered a kind of solace and acceptance, and this was the sentiment that characterized a striking poem published shortly after the demolition by the little-known writer Hermann Penn. He concludes with these lines:

> In all the battles of my life,
> Of my furious wrestling, struggling, striving,
> Your song resounds, it brings peace—
> Adelaïde—Adelaïde!——
> And if this house must disappear,
> It is easy to find solace:
> *All of Vienna is a Beethoven house!*

There your spirit comes and goes,
There your song rings in all circles,
There is no festival without your melodies,
Amid all our thoughts, our doings,
You live on, Beethoven, among us![120]

(*In all' den Kämpfen meines Lebens, / Des heißen Ringens, Mühens, Strebens, / Dein Lied ertönt, da kommt der Friede— / Adelaïde—Adelaïde!——— / Und muß auch dieses Haus verschwinden, / Ein guter Trost ist leicht zu finden: / Ganz Wien ist ein Beethovenhaus! / Da geht Dein Genius ein und aus, / Da klingt Dein Sang in allen Kreisen, / Kein Fest gibts ohne Deine Weisen, / Inmitten unsres Denkens, Tuns, / Lebst Du, Beethoven, unter uns!*)

Penn's encomium suggests that in the "city of music," Beethoven lives on in sound, not in matter. He invokes an environment of memory, where culture is continually reborn, rather than the hushed piety of a shrine, where time stands still. Even so, Penn's quotation of "Adelaïde," Beethoven's best-loved art song in that period, suggests a tinge of *Sehnsucht*, as if he were searching for Beethoven's ghost in the rubble. This poem maps onto the structure of the original text by Friedrich von Matthisson: the pining cry across distance ("Adelaïde—Adelaïde!") is met with consolation, as the speaker admits that even if his love is unrequited in life, the blue flower of longing will grow from his grave and shine forth his lover's name. By mirroring this poetic arc from pining to solace, Penn implies that piety does indeed reign in Vienna, and each performance of the composer's music operates as a metaphorical flower on his grave.[121] Even in Vienna, where the treatment of material traces was so different from in Bonn, music lovers could still experience a sense of mizpah, yearning for Beethoven's presence across time and space.

2

Mozart on the Mountaintop

In the previous chapter, Vienna faced some derision for its neglect of sacred sons. The earliest seeds of that stance were planted not in Bonn, but in Salzburg. When Mozart's monument was unveiled there in 1842, complete with festivities that overflowed the city streets, Viennese officials felt pressure to keep up. Over the course of four decades, the city made three attempts to erect a Mozart monument, none of which gathered steam. But the project reached a turning point in 1883, when some called for a posthumous Richard Wagner monument. In response, others demanded that it was Vienna's duty to "erect a statue *first* to great masters of the fatherland, who created a wealth of immortal works in Vienna."[1] Even when those demands were heard, and a Mozart monument was finally underway, administrative errors and the unexpected death of the architect delayed its unveiling in the Albrechtsplatz until 1896. The result was an air of defensive embarrassment. Nikolaus Dumba, a leading arts patron and industrialist who spearheaded many of the city's commemorative projects, gave a speech as committee chair that made a surprising argument: a longer wait, he claimed, is an even greater honor, as only frivolous people rush to commemorate those who have scarcely left this world.[2] Implicit in that statement is a comparison with Salzburg, which had labored to fashion itself the Mozart-Stadt, the city of Mozart, since shortly after the composer's death. The birth city had hosted one event after another in Mozart's honor: the monument in 1842, the birthday in 1856, and a series of eight music festivals from 1877 to 1910 that laid the foundation for the annual Salzburg Festival we know today.

During the debate about Mozart's Viennese monument in 1883, the poet Ludwig August Frankl, whom we met in the previous chapter, inscribed several poems into the Mozart Album, a collection of celebrity contributions amassed

FIGURE 2.1. Postcard of the *Magic Flute* cottage on the Kapuzinerberg in Salzburg, ca. 1882–92. Photograph by Würthle und Spinnhirn, Salzburg. © International Mozarteum Foundation, Salzburg, Archive.

by the International Mozart Foundation.[3] In an excerpt from "All Souls' Day," originally written in 1834, Frankl calls urgently for a monument—and make haste, he says, before Mozart's spirit appears as an angry stone guest, like the Commendatore in *Don Giovanni*.[4] (The same metaphor later haunted Viktor Tilgner, the designer of Mozart's monument, who embossed a tableau of that very churchyard scene on the pedestal.) Frankl's second rallying cry, penned in 1883, seems to fear the retribution of an angry spirit. He begins: "When the body was still warm you cried out together, 'A monument to him!'" But allow me, he laments, to paint a picture of Mozart's funeral: "Who gives the dead their escort? No one! No wreath, no lamplight! . . . Alone, the carriage rattles along." And in closing: "Do you know where he was hastily buried? Have you atoned for the deeds of your forebears? In vain! We wait and we waited—how can you rest so thanklessly? . . . You people of Vienna, think twice!"[5]

We have already seen how Frankl had an ax to grind, and texts like these are among the most vivid expressions of Ehrenpflicht, the duty to honor, which I explored in the previous chapter. But in the case of Frankl's poems in the Mozart Album, the texts are just one part of the story. Equally important is the *site* where this album was housed, which shaped the experience of those who leafed through its pages. The collection was displayed in the so-called *Zauberflötenhäuschen*, or the *Magic Flute* cottage, pictured in figure 2.1 at its unveiling in 1877. It is an odd sight to behold: a rustic cabin outfitted with a monument that dwarfs its modest dimensions.[6]

Allegedly, this was the site where Mozart finished his *Magic Flute*. As the story goes, in 1791, librettist Emanuel Schikaneder had locked a procrastinating Mozart inside this garden house to enforce his deadline. Almost a century later, in 1873, the house was found rotting in the Viennese Freihaus, a residential and commercial development owned by the Starhemberg family that eventually fell into disrepair. This prompted the musicologist Carl von Sterneck, president of the International Mozart Foundation, to purchase the structure and hoist it to the peak of the Kapuzinerberg, which overlooked Salzburg. When the house was dismantled for transport, its underside had decayed so badly that it needed substantial restoration.[7] It was relocated to a picturesque spot fringed with woods, a belated imitation of the shack in which Goethe inscribed his famous "Wand'rers Nachtlied II," which had attracted literary pilgrims as early as 1839. Soon the cottage became a small museum that simulated Mozart's work space, with a replica of his rough-hewn desk and chair. On that desk lay the Mozart Album, which contained Frankl's poems, among much else. The unveiling ceremony in 1877 was a grand event, with choral ovations, speeches, and wreaths laid by prominent donors.[8] What made Mozart's cottage distinctive among historic sites was its cultivation of ritual, with regular ovations that combined elements of Catholic pilgrimage with the Enlightened brotherhood of the opera's freemasons. Year after year, for decades, men's choirs journeyed up the mountains to pay their respects. Their hymns made this cottage a resonant and distinctly masculine site of Ehrenpflicht.

This chapter traces how Mozart's *Magic Flute* cottage was transformed from shack to shrine, which made it a fanciful space to work through questions about Mozart's creative process and his value to Austro-German heritage. Like the cabins of Henry David Thoreau and Gustav Mahler, this house became a shrine to the muse, and it fostered a new category of nativity sites that celebrated the "birth" of the work.[9] Reenacting creative labor was more urgent in Mozart's case, as his biographers were torn between two models of inspiration: Protestant hard work versus Austrian ease, or in this case, messy sketches versus note-perfect dictation from the divine.[10] The picturesque garden house, which invited comparisons of Mozart with Beethoven the nature lover, affirmed a myth of hermetic authorship that originated with a well-known forgery. In 1815, Friedrich Rochlitz mistakenly passed off a spurious letter as genuine in his *Allgemeine musikalische Zeitung*, in which Mozart confesses his need for solitude to find the inspiration that "fires my soul."[11] Six decades later, this transplanted cottage portrayed Mozart as more serious, solitary, and Beethovenian than was borne out by biographical evidence, and the ritual sanctity of its choral odes mirrored patterns in his music's critical reception—most notably *The Magic Flute*, whose fairytale whimsy was derogated in favor of themes of Enlightened brotherhood. Fifty years before the founding of the Salzburg Festival proper, the city understood itself as a pastoral sanctuary that could shelter Mozart and his music from the metropolis, and return the cosmopolitan artist to a more natural state of rustic Enlightenment.

Making the *Mozart-Stadt*

Salzburg was a pathbreaker in musical pilgrimage. This was the first city in Europe to erect a composer monument, to restore a composer's birth house as a museum, and to sell products branded with a composer's name.[12] It was exceptionally easy to craft the Mozart-Stadt because, like the Rhineland, this baroque jewel had been a tourist destination for the early German Romantics. With the support of the archbishopric, the city boasted a lively culture of concerts hosted by music societies such as the Dom-Musik-Verein, the ancestor of today's Mozarteum. But with the exception of the Novellos, few early visitors to Salzburg traveled in Mozart's name.[13] He was the prodigal son who had fled to Vienna; the city's preferred forebears were more faithful residents like Heinrich Biber and Michael Haydn, Franz Joseph's brother. Even Franz Schubert made no mention of Mozart in his 1825 account of his travels to Salzburg, but he paused to shed a tear before Michael Haydn's tomb.[14]

Salzburg's identity as the Mozart-Stadt can be attributed to Constanze Mozart, whose curation of her husband's legacy surpassed the norms of widowhood. In 1841, she raised funds to create the Mozarteum, advocated for a monument to Mozart, and assisted her new spouse, Georg Nikolaus von Nissen, with his biography of her late husband.[15] The festive unveiling of the monument in 1842 set the bar for Beethoven's unveiling in Bonn three years later, and it was the city's first encounter with the commercial prospects of Mozartiana. The journalist Ludwig Mielichhofer, who authored the event's festschrift, remarked in the *Salzburger Zeitung* that "everything was christened with the name Mozart," including "Mozart festival programs, Mozart biographies, Mozart portraits, Mozart festival concerts, etc. In the stores one saw Mozart busts, Mozart models, Mozart pipes, Mozart figurines.... In the inns one could even find Mozart rooms, Mozart-bread, Mozart-wine."[16] (This tradition is still palpable in Salzburg: the beloved *Mozart Kugeln*, the marzipan bonbons on the shelves of duty-free stores in European airports, were invented by the Fürst confectionery for the music festival of 1884.) Just prior to the monument unveiling in 1842, Mozart's birth house was labeled with gleaming letters, a first step toward the building's eventual conversion to a museum in 1880. The city came alive with visitors again in 1856 for the festival of Mozart's one-hundredth birthday, which boasted four days of choruses, speeches, and commemorative performances of the composer's music.[17] It is remarkable how Constanze's efforts to preserve her late husband's memory so thoroughly transformed the city's identity.

After Constanze's death in 1842, Mozart commemoration was kept alive through a series of institutions: first the Dom-Musik-Verein and Mozarteum, which were later consolidated into the Internationale Mozart Stiftung (International Mozart Foundation), later renamed the Internationale Stiftung Mozarteum and hereafter abbreviated as IMS. Much like Bonn's Beethoven Verein, the IMS put Salzburg on the European map. Its members constructed a concert hall and archive, made the birth house into a museum, reclaimed Mozart memorabilia that had been strewn across Europe, spearheaded the first complete edition of Mozart's works, and organized several music festivals.[18] It was the first of these events in 1877 that prompted the rescue and unveiling of the *Magic Flute* cottage.

It is no coincidence that the festival was founded one year after Wagner's premiere at Bayreuth. The parallels between the two were unspoken at first, but they became more explicit with each successive event until an organizer in 1913 called Salzburg the "Austrian Bayreuth."[19] When the IMS was tasked with choosing a music director and conductor for the Salzburg series, they brought in Hans Richter, who had just directed Wagner's *Ring* cycle premiere in 1876. As if to create his own Bayreuth, Richter later supported an initiative to erect a Mozart-Festspielhaus in time for the one-hundredth anniversary of *Don Giovanni* in 1891. The committee conscripted the Austrian architect Karl Demel to design a fifteen-hundred-seat hall atop the Mönchsberg that would exclusively showcase Mozart's operas; his sketch looks like a curious hybrid between Bayreuth and Vienna's Staatsoper (fig. 2.2). Like Bayreuth, the hall would be hoisted above the fray of the "raucous goings-on of the everyday" in Salzburg's city center, just as Wagner maintained that his *Ring* cycle should be performed "in some beautiful solitude, far away from the fumes and industrial stench of our urban civilization."[20] While the project was abandoned in 1891 for lack of funds, the question was revisited in 1920 with a more inventive design that would blend seamlessly with nature. Located at Hellbrunn—the seventeenth-century grounds with their famous water-powered automata, stony grottos, and comical trick fountains that splash visitors in the face—the proposed hall would feature organic arches and turrets that emerged from the hillside like stalactites. After this admittedly eccentric proposal fell through, the committee located its concert space inside the former equestrian hall, as

FIGURE 2.2. (*Left*) Wagner's festival house at Bayreuth. In Gustav Adolf Klöden, *Unser deutsches Land und Volk: Bilder aus der schwäbisch-bayerischen Hochfläche den Neckar- und Maingegenden*, ed. Fedor von Köppen (1879), Bavarian State Library, Germ.g. 231 ns-2, p. 215. (*Right*) Design for a Mozart-Festspielhaus on the Mönchsberg by F. Fellner and H. Hellmer, 1890. © Archive of the Salzburg Festival.

if to establish a bourgeois court atop an aristocratic blueprint. These failed designs show how the festivals claimed respite from urban impiety, and reconciled Mozart's spirit with the city he had never much enjoyed in life.[21]

An influx of tourists to the region meant that the 1877 event revolved as much around Salzburg's scenery as its music, with a program that resembled a Baedeker travel guide. Attendees were invited on tours of Salzburg and Hellbrunn, boat rides along the River Salzach, mountain excursions, and an echoing choral concert at an Alpine gorge. The same was true of the second festival in 1879, which became the pinnacle of Thomas Cook's tour through the region. A London advertisement gave equal weight to the festival's scenic attractions, concerts, and the "procession of artistes in the morning to the Mozart house."[22] It helped that Salzburg already had a history of pilgrimage to its medieval churches and relics, including the head of St. Erentrudis and a crucifix with a live, growing beard. The path up the mountain to the cottage was dotted with stations of the Cross, and the cloister across the way was said to enhance the spiritual atmosphere of the cottage.[23]

Pilgrimage was not a concept imposed from the top down. Visitors saw the site as sacred, and we find their homilies in letters and poems that festivalgoers sent in thanks to the IMS. The most extensive example is a thirteen-page ode to the Mozart-Stadt by Louisa Lergetporer, a visitor from Vienna.[24] In rhymed stanzas, she chronicled Mozart's Nativity-like birth in Salzburg, his upbringing and career, his cottage, his dramatic death, the 1877 festival, and the cottage ovation. In a miraculous turn, the clouds part, and a sunbeam alights on Mozart's bust when attendees shout, "hoch!" (a ritual exclamation of exultant praise). Some years later, during an 1885 ceremony that recapitulated the 1877 unveiling, a journalist likewise recalled that a sunbeam landed on the bust during the rendition of "O Isis und Osiris" (perhaps he had in mind the melodic leap at the line, "dark night is banished by the sunlight"). The conflation of Christian and Egyptian religions in this account is typical of art-religion: "Had the holy spirit itself taken the form of a dove to perch on the roof of the hut, it would have made no greater impression than this serene greeting of the ancient Egyptian god of light Osiris, who seemed to wait backstage behind the clouds for his sacred password."[25]

Lergetporer could have chosen to end her poem with miracles, but instead she emphasized the city's self-image. This was no ode to Mozart, but to Salzburg's Ehrenpflicht, starting with its first awakening in 1842 and culminating in the pilgrimage to the cottage. Her final stanza closes with a fanciful speech that lauds Carl von Sterneck: "hail him who so nobly conceived this, the Baron Sterneck, may thou receive our honor!"[26] This Viennese visitor, whose own city was Mozart's chosen home, saw Salzburg as the true Mozart-Stadt.

From Shack to Shrine

How, exactly, did this cottage become a place of miracles? The transformation of a modest garden shed into a shrine required some fund-raising; a closer look at the mechanics of philanthropy shows how art-religion operated in practice. Take for instance a pamphlet compiled in 1880 by the museum archivist Johann Horner, which reported "international pilgrimage" to the site, which he hailed as a "temple of art."[27] On its decorative cover, the pamphlet declared that it was a commemorative festschrift for the anniversary of Mozart's 124th birthday, an ostensibly arbitrary anniversary that gave Sterneck an occasion to solicit donations. As in the previous chapter, philanthropists were eager to feel like disciples whose transactions were sacred offerings.[28]

In Vienna, this house was no shrine. Aside from its modest plaque, the cottage was largely ignored in the courtyard of the Freihaus. In a memoir about her childhood as a Freihaus resident, the wife of Alfred Heidl, general secretary of the IMS, recounted with some disdain how this noble city-within-a-city had since become an impoverished neighborhood for eastern European immigrants. She recalled childhood games in Mozart's cottage, which morphed at dusk into the witch's hut from Hansel and Gretel, or Bluebeard's castle, or an enchanted ruin haunted with ghosts. Once she learned that Mozart had composed there, her games became more pious, and she stood on the composer's stool to conduct his arias as they tinkled from a music box. But the site was profaned: the cottage was rumored to house the occasional lovers' tryst, and it was rented for use as a craft studio, flower shop, and animal pen. One fateful day, she was dismayed to discover the house teeming with live rabbits: "Although I couldn't fully appreciate Mozart's greatness back then, it was unfathomable even to my child's consciousness how one could perpetrate such an impiety, to degrade with a rabbit hutch this site where the genius had once sat."[29]

For Sterneck, the hut was no hutch but a "relic" and a "treasure." When it was announced in 1873 that the Freihaus would be razed, its owner, Camillo von Starhemberg, planned to relocate the cottage to his noble estate in the countryside. But Sterneck intervened, insisting in a letter that the "birthplace of the *Magic Flute*" belongs to the public. Sterneck was well aware of how sacred tourism might bolster the local economy. When he pitched the plan to Salzburg city officials in 1874, he compared the cottage to Goethe's newly restored house in Frankfurt, which managed to charge an entrance fee with only a simple plaque on the wall: "It is an indisputable fact that this is a lucrative operation." It helped that these spaces did not need costly decoration, since modesty was their chief attraction: "the apartments and work spaces of great German figures were never valued for their grand architecture, their splendor, or their style, for indeed these spaces are often tiny garrets." Profit was compatible with art-religion because international pilgrimage to this cottage would "bring further sacredness to Salzburg's *Mozartkultus*," or Mozart-devotion.[30]

Once its ties to historical place were severed—recorded only in illustrated magazines that sought to "preserve [the cottage's] likeness for posterity"[31]— the site became a cross between relic, reliquary, shrine, monument, and substitute for Mozart's remains. Outside, the cottage was crowded by a monument so tall that some said it had an "unpainterly" effect.[32] Inside, the museum display was cluttered to the eye. An inventory of the museum's contents from 1879 describes a room packed with framed portraits and a large commissioned painting, artifacts that narrated the pious activities of the IMS, the much-coveted Mozart Album of celebrity autographs, a visitors' book, golden decorations and wreaths, and souvenir postcards for sale; by the time it was photographed decades later in 1932, the displays had been tamed somewhat (fig. 2.3).[33] In a corner, the curators reconstructed Mozart's work space as if he had just stepped out, arranging an inkpot and quill on his desk and chair; the original furniture, while on display for a time, was later replaced with replicas so these "priceless relics" could return to the "family museum" of the Starhemberg estate.[34] The layered materiality of this cottage, which behaved as both relic and reliquary, is made plain by an intriguing artifact. In 1908, a music lover in nearby Hallein fashioned a miniature cottage that functioned not as an architectural model, but as a decorative box (fig. 2.4).

FIGURE 2.3. Photograph of the interior of the *Magic Flute* cottage from 1932. ANNO/Österreichische Nationalbibliothek, Bildarchiv, L 32072b-C.

FIGURE 2.4. Handmade wooden box in the shape of the *Magic Flute* cottage, from Hallein, 1908. © Salzburg Museum, K 3606-49.

Like a matryoshka doll, the reliquary needed its own reliquary. Already in 1874, Sterneck noted that the cottage was in a fragile state, and by 1877 the IMS secretary Johann Horner (Sterneck's righthand man) led a fund-raising campaign to turn it into a temple. He solicited a watercolor sketch from the famed Wiener Werkstätte architect Josef Hoffmann, who had designed costumes and sets for a performance of *The Magic Flute* at the new Hofoper in Vienna in 1869, followed by Wagner's *Ring* cycle premiere in 1876.[35] Hoffmann's vision was an atmospheric nod to his own Egyptian stage sets. He proposed to hide the cottage within a boxy, coral-colored structure rimmed with the same Horus motif he used for the opera, and upon entering the windowless structure, visitors would behold the cottage illuminated by an eerie glow through a glass ceiling (fig. 2.5).[36] With the hut transformed into Sarastro's temple, visitors might enact the plot of the opera as they beheld the site of its creation. Hoffmann's colorful proposal was soon abandoned in favor of a more Catholic design: a reliquary-like glass structure, or what Horner called a pavilion, detailed with golden emblems that showed scenes from the opera. In yet another proposed design, the lower half of the cottage is visible beneath an ornately decorated wooden frame (fig. 2.6).

The story of Horner's fund-raising for this project is a long one, with twists and turns that offer much to historians of philanthropy. In this period, the landscape of arts philanthropy was divided between the courtly patronage model and a new focus on collective support from less affluent subscribers. Horner's first instinct was to contact a few solitary millionaires, namely the Bonanza Kings (or what he termed the "Millionaire Quartet"), a group of four Irishmen who became sudden celebrities in 1873 when they struck it rich in the Nevada silver mines. Horner invited these men to the Salzburg Festival of 1879, and behind the scenes he expressed the hope that they might finance the entire project: "for them, the money is obviously a mere chimera . . . and perhaps one or another of them is an art patron or a yet-undiscovered admirer of Mozart."[37] When this failed, Horner's next impulse was to plead with the many celebrity musicians who had contributed leaves to the Mozart Album, hoping that those who have "piety for the spirit of Mozart" might participate in a competitive lottery to win the honor of contributing an emblem to the reliquary.[38] And when that, too, met with a dead end, he turned instead to the middle-class audiences who populated Salzburg's music festivals. His new drive carried a tone of urgency: donating would be a "commandment of piety" because "this precious relic, which has become a true temple of Mozart devotion for thousands upon thousands of art-loving pilgrims, will soon crumble to the ground if it is not outfitted with a protective shell with worthy decor."[39] When that, too, did not garner enough support, Horner used every

FIGURE 2.5. Josef Hoffmann's watercolor design for an Egyptian temple enclosure to protect the *Magic Flute* cottage from decay, 1882. © International Mozarteum Foundation, Salzburg, Archive.

FIGURE 2.6. Protective cover design for the *Magic Flute* cottage by W. Schutz, 1877. © International Mozarteum Foundation, Salzburg, Archive.

trick in the book. He conscripted Sabilla Novello, the daughter of Vincent and Mary, who had made their Mozart pilgrimage fifty years earlier, to serve as the project's ambassador in England. He wrote personal letters to those who had attended the unveiling; some, like the music critic Albert Weltner, replied with effusive letters about their sacred experiences. He networked with local artists, some of whom donated the proceeds from their concerts in the hopes of a spot on the festival program.[40] And he paid special attention to wives and widows, who were among the most enthusiastic philanthropists of the century.[41] Despite leaving no stone unturned, after four years he had raised only half what he needed. Infrastructure is rarely a popular cause, even for devotees to art.

As it turned out, the whole undertaking was for naught: in 1883, a logistical report found that sealing up the house in glass was a bad idea. Not only would it clash in a "tasteless" way with its forested surroundings, but the greenhouse effect would hasten damage from the sun.[42] The authors of the report noted that a historic site simply can't be treated as an object, like a lock of hair or a quill, where "believers are moved to devotion through their holy proximity alone." Rather, visitors want to "behold the original appearance of the relic as free as possible from foreign ingredients," which instills "pious contemplation."[43]

The authors of this report were not the only ones to point out the problem. At the height of fund-raising in 1880, an anonymous satirist imagined a newspaper notice from one hundred years in the future:

> It is well known that, around 100 years ago, a cover was installed over the immensely valuable wooden cottage in which the immortal Mozart supposedly spent his entire life (—what coarse manners must have prevailed in the 18th century!—); but since this cover, as well as five others installed gradually on top of each other, has once again fallen into ruin, the International Mozart Foundation has decided to allocate 1/10th of its income (17,000 fl.) to encase the Mozart house along with its various covers in the middle of an airtight and watertight cement mound, which will then provide the extraordinarily practical opportunity, in another hundred years, to host an exhumation festival, whose rare and incomparable celebration will see the compilation of a festschrift with fitting biographies.[44]

What makes this satire so incisive is that it mocks, almost systematically, every one of commemoration's ills. There is the strange conflation of bodily remains and buildings, the festive exhumations of Beethoven and Schubert, the sense of grandeur even for mundane spaces (that is, the house where Mozart "supposedly spent his entire life"), the overwrought festivals with their overwritten festschrifts, and the careful engineering of miscellaneous reasons to celebrate. In 1880, imagining life in 1980, this author predicted Pierre Nora's critique of preservation as both arbitrary and obsessive.[45]

In light of this satire, it seems fitting that Mozart's cottage did not remain on the mountain for long. The site experienced a last hurrah at a festive reopening in 1925, but pilgrimage ceased for obvious reasons in 1939, and by 1950, it had found a new resting place in the backyard of the Mozarteum, where chamber ensembles performed occasional tributes at its doors. It lay there covered in moss until 2022, when it was again dismantled and rebuilt in Salzburg's Freilichtmuseum, a kind of open-air zoo for historic houses; it has since been situated in what looks to be a permanent resting place in the courtyard of Mozart's residence at the Makartplatz. (The cottage was once again found to be rotting, and some sections were replaced; as with the ship of Theseus, it is unclear how much of the original remains.) The strange story of this site exposes a problem for music institutions like the IMS, which are obligated to steward whatever historical remnants we value at the time. When institutions guess posterity wrong, their efforts are the brunt of satire, and they invite us to ponder how our own descendants will react to the stuff we dutifully preserve.

Choral Fraternities

While many house museums could claim the status of a shrine, the *Magic Flute* cottage was among the only such sites to host a ritual. From the unveiling in 1877 until the Third Reich, men's choral societies journeyed almost annually to the peak of the Kapuzinerberg to laud the cottage, typically singing the priests' chorus from act 2 of *The Magic Flute*, "O Isis und Osiris, welche Wonne" (fig. 2.7). In total, from 1877 to 1939, the site saw seventy-two choral visits from fifty-eight different societies. Eight of these hailed from the Habsburg Empire, fifteen from the south of Germany, seventeen from other regions in Germany, and eleven from significantly farther way—San Francisco, New York, Stockholm, Sofia, and several locations in Switzerland. Fourteen of those ensembles made repeat visits.[46] This was a masonic pilgrimage of traveling men's choirs, despite that freemasonry on the continent was supposed to be adogmatic. What Sarastro's priests and Catholic pilgrims shared was the sound of solemnity, a ceremonial aesthetic associated with men's choral singing in an era of festivity. Those sounds elevated the cottage far more than its monument or museum, and they enabled people to live inside the opera in an unusual way, acting out the role of priests of Mozart's memory.

The history of men's choirs can shed light on why this ritual arose. Men's choirs had their deepest roots in Germany, where they were associated with political ceremony and nation building. In Austria, this political function met with distaste—Metternich, for one, banned men's choirs until 1848, calling them a "German poison"—but the practice took firmer root in singing and

FIGURE 2.7. Example of a men's choral ovation at a ceremony for the reopening of the *Magic Flute* cottage in 1925. *Das interessante Blatt*, June 25, 1925, 5. ANNO/Österreichische Nationalbibliothek.

folk festivals that were more local than nationalistic.[47] At choral festivals, men's choirs were more likely to sing songs about nature, seasons, and drinking than patriotic odes, which could be politically divisive.[48] These festivals explain why a turn-of-the-century catalog of men's choral repertoire contains numerous songs about travel, including a variety of "singer marches" and texts about foreign lands, *Wanderlust*, mountaineering, and greetings to singers arriving "from the North" or "from the South."[49] In 1845, when Wagner chose the reedy resonance of a men's choir for the pilgrims of *Tannhäuser*, he signaled not only his concurrent role as director of the men's choral society of Dresden, but also the widespread association of men's choirs with a traveling band of brothers.[50]

There are historical reasons, too, why Mozart's "O Isis und Osiris, welche Wonne" became a staple of men's choral repertory. Men's choirs were a stone's throw from the rituals of the freemasons: in Zurich, the first men's choral society was founded by freemason Hans Georg Nägeli, who authored treatises that promoted communal singing as Bildung, modeled on the educational philosophies of fellow freemason Johann Heinrich Pestalozzi.[51] When the nobility suppressed the lodges out of alarm at their growing influence, artists' clubs and men's choral societies became an alternative headquarters for gentlemen of letters. "O Isis und Osiris," the most explicitly masonic moment in *The Magic Flute*, was the sound not only of pilgrimage and brotherhood, but of Enlightenment and Bildung. Its luminous hush seems almost to stop time, creating a space of ritual, and its text setting alternates between ceremonial grandeur and introspection, an anthem for societies founded on self-cultivation and interiority, or *Innerlichkeit*. The chorus opens with a nod to church homophony, as a harmonic pun on a plagal cadence seems to recite "amen" for Isis and Osiris in turn. At the exclamation "what delight!" the chorus sighs in parallel unity. The brothers find themselves lost in a cloud of uncertainty at "the dark night," but the darkness is purged by a fanfare on the words "banished by the sunlight." The chorus unifies in feeling once again as it sings "soon, soon, soon" with mounting impatience, the basses beating like a heart. Mozart's freemasons, like the choral societies that grew out of that tradition, resound as a collective brotherhood bound by interiority, sharing heartbeats and sighs.

Muted and sensitive choral writing better suited the *Magic Flute* cottage than did the jubilant conventions of most festival odes, which lauded the attendees as loudly as the luminary.[52] At the unveiling ceremony in 1877, a commissioned *Festchor* managed to combine both modes of commemoration, starting with intimate ritual and building toward the monumental assertion that "we are here!" This single-movement work for mixed choir and winds, called "The Artist's Genius" ("Des Künstlers Genius"), was composed by Max von Weinzierl, the director of the Wiener Männergesangverein and prominent composer of occasional works.[53] His ode moves from hushed interiority to bugling jubilation to complement the arc of the text, which was authored by the Viennese humorist and playwright Dr. Märzroth, the nom de plume of Moritz Barach. Each stanza is structured around a hypophora—a common rhetorical strategy in festival odes in which a text immediately answers its own question—which invites a lively dialogue between voices. Early in the piece, the spirit of Mozart as Genius incarnate floats through Austria's Alpine landscapes, and in the last two stanzas, he alights on the doorstep of the cottage:

> Who is it that enters this hut
> With a gently loping pilgrim's tread?
> Who transforms even the poorest room
> Into a king's throne with golden trim?
>
> That is the hand of the Genius,
> Before which prince and servant bow,
> For, smiling, He embraces us all!
> Thus let us, O Genius, receive your greeting![54]
>
> (*Wer ist's, der in die Hütte tritt / Mit leichtbeschwingtem Pilgerschritt? / Wer wandelt selbst den ärmsten Raum, / Zum Königsthron mit güld'nem Saum? // Das ist die Hand des Genius, / Dem Fürst und Knecht sich beugen muss, / Weil Alle lächelnd Er umschliesst! / D'rum sei uns, O Genius, gegrüsst!*)

To capture the natural surroundings, Weinzierl modeled his opening gestures on late nineteenth-century landscape symphonies. The work begins with a hymn-like woodwind chorale, with reeds acting as shepherds' pipes in a faraway pianissimo, and at times he dips into a more expressive harmonic universe of chromatic passing tones and half-diminished suspensions in lieu of dominant sevenths. As the text becomes increasingly assertive—no longer posing its rhetorical "who?" but answering definitively, "that is the hand of the Genius!"—Weinzierl responds with the characteristic bugles of double-dotted homophony and perfect authentic cadences. The final refrain imagines the spirit of Mozart communing with the assembled masses, lauding those gathered at his feet. The choir repeats, "thus let us, O Genius, receive your greeting!," which can be more musically (if less literally) translated as, "thus let us, O Genius, be blessed!" Along a rousing ascent to a pedal-point dominant at the word "Genius," the word "us" intones predominant harmonies with insistent eagerness: "Thus let us, let *us*, O Genius, be blessed" (mus. ex. 1). That same year, Moritz Barach inscribed an ambiguous note in the Mozart Album that betrays the fine line between lauding Mozart and patting oneself on the back. It is hard to know whether he praised Mozart's music or

MUSICAL EXAMPLE 1. "D'rum sei uns, O Genius, gegrüsst!" (Thus let us, O Genius, receive your greeting!). Max von Weinzierl, text by Dr. Märzroth (Moritz Barach), "Des Künstlers Genius," 1877. Landesbibliothek Coburg, Ms. Mus 944.

MUSICAL EXAMPLE 1. (*continued*)

MUSICAL EXAMPLE 1. (continued)

MUSICAL EXAMPLE 1. (*continued*)

his own when he wrote, on the occasion of his own work's premiere, "only that which moves the hearts of the people has the right sound!"[55]

The rituals at this cottage betray a shift in the conception of Mozart's craft. Starting in the early nineteenth century, anecdotes and biographies recast the composer as a nature lover who pined for Salzburg while cooped up in his Viennese apartment.[56] Decades later, the cottage became a space of re-enactment, where an imaginary version of Mozart found his inspiration in the woods. At first, the house was slated for display in one of Salzburg's palace gardens, but it was decided instead to model the site on Goethe's writing shack on the Kickelhahn, into which he had carved his famous "Wand'rers Nachtlied II." Goethe's cottage had been a literary attraction since 1839, when the nearby town of Ilmenau began to draw crowds as a spa destination; after the structure was destroyed by fire in 1870, it was painstakingly rebuilt, which meant re-carving the poem in a forgery of Goethe's hand. The replica was constructed in 1874, the same year that Sterneck rescued the *Magic Flute* cottage; when the site was announced to the papers, IMS secretary Johann Evangelist Engl emphasized its wooded setting, noting that "this little sanctuary" would be sheltered "in the middle of God's exquisite nature, to whom Mozart was always an intimate friend."[57] The festschrift for the 1877 festival affirmed this point: "he, who venerated God's glorious nature above all else, created from this immediacy, from this gaze upon eternal beauty, the richest and most beautiful pearls of his immortal sound-muse."[58] A similar sentiment was affixed to the back of a souvenir photograph: "Here, with the blossom-sown fields and flowering trees before his eyes, the master could listen to the inspiration of his abundant imagination."[59] All these passages could just as well be about Beethoven, and they ignore that Mozart felt most at home in Vienna, Prague, and Paris.[60]

Visitors responded to the foundation's language, leaving traces of their impressions. A poetic entry in the visitors' book in 1881 meditates on Mozart's music as the sound of nature's freedom: "nature unified the free and the carefree . . . / thus emerged Beauty, and the rose that was once / her symbol: to be found / on the Kapuzinerberg."[61] In Lergetporer's ode, Mozart pines for the mountains of his hometown while in Vienna, and it is not until Schikaneder finds him a garden cottage that he can compose in peace:

Here he was happy, undisturbed, alone;
How many nights he scarcely closed his eyes,
At the piano, alas! he forgot his own existence!
And what he created from his imagination
Arranged itself in a wonderful melody.

O, how wide and light that narrow room
Seemed to him in that moment.
The muse filled him so with a joyous dream,
Like a laurel wreath she braided 'round his brow.
And what he presaged in those hours
Could be grasped only in higher worlds of the future.

Alas, all earthly woes, they must vanish
Where it rings so exquisitely through the pages.
The tones arrange themselves into melodies
That wonderfully penetrate the soul.
Here 'twas, where he wrote his *Magic Flute*,
That exalted triumph of its creator.[62]

(*Hier war er glücklich, ungestört, allein; / Wie viele Nächte schloß er kaum die Lider, / Am Flügel ach! vergaß er selbst das Sein! / Und was er schuf, aus seiner Fantasie, / Reiht sich zu wundervoller Melodie. // O wie erscheint es ihm im engen Raume, / In solchen Augenblick so weit und licht, / Die Muse füllt er ja im sel'gen Traume / Wie Lorbeer sie um seiner Stirne flicht. / Und was in solcher Stunde er geahnt, / Uns nur an höh're, künft'ge Welten mahnt. // Ach alles Erdenweh, es muß entfliehen, / Wo es so herrlich durch die Seiten klingt / Die Töne reihen sich zu Melodien, / Daß wunderbar es in die Seele dringt. / Hier war's, wo er die "Zauberflöte" schuf, / Die hoch verherrlicht ihres Schöpfers Ruf.*)

In the months that followed the unveiling ceremony of 1877, the IMS commissioned a moody portrait of Mozart's divine dictation from the Viennese portraitist Anton Romako. In dark sepia tones that evoke the palette of Eugène Delacroix, *Mozart at the Spinet* shows the composer lost in compositional thought as music-making cherubs tumble from the lid of his instrument and a robed muse dictates from a score (fig. 2.8).[63] Here Romako, like Lergetporer, reconciles conflicting models of creativity as both divine inspiration and a labor of the hands.

This brooding image shows how Mozart presented a problem in the late nineteenth century. The issue was not so much his galant aesthetic, nor his persona as the eternal child, both of which could be Romanticized in terms of purity; rather, Mozart's cosmopolitan career, Italianate sensibilities, and scatological humor made it hard to associate him with German seriousness, or *Ernste*, even as he was upheld as a pillar of German art. The fact that he was Austrian was of no consequence. In an allegorical lithograph from 1868 by Wilhelm Lindenschmit the Younger, a lively array of "heroes of German music" sweeps in rough chronological order from a central pedestal, with the early modern predecessors gazing down from a mezzanine behind them (fig. 2.9). Mozart initiates the arc of history, comparing scores with Haydn on the pedestal. Behind them, Gluck gazes ahead in white-clad purity, Bach

FIGURE 2.8. Anton Romako, *Mozart am Spinett* (*Mozart at the Spinet*). Oil on board. Vienna, 1877. This painting was displayed inside the *Magic Flute* cottage. © International Mozarteum Foundation, Salzburg, Archive.

improvises at the organ, Beethoven scowls, Handel puffs out his chest, Schubert sits at Beethoven's feet while gazing with a dreamy expression, and history continues to curve along the righthand side with Robert and Clara Schumann, Felix Mendelssohn, and an array of prominent female singers and pianists, culminating with Wagner, Liszt, and their crew. It is striking how many *Austrians* appear here. David Lee Brodbeck has shown how Austria

FIGURE 2.9. Wilhelm Ritter von Lindenschmit the Younger, *Ruhmeshalle der deutschen Musik (1740–1867)* (Hall of Heroes of German Music). Munich, Friedrich Bruckmann's Verlag, 1868. Sächsische Landesbibliothek—Staats- und Universitätsbibliothek, Dresden, Mscr. Dresd.n., Inv. 3, II, Bl. 21, Nachlass Schnorr. © SLUB Dresden / Deutsche Fotothek.

sought to unify its Slavic, Hungarian, and Italian identities through universal *Deutschtum*, and claimed German-born composers as their own sons.[64] It made matters more complicated that some Salzburgers identified as Bavarians, which explains why Hugo von Hofmannsthal later pronounced Salzburg the epitome of achievement on "south German soil."[65] Lindenschmit's hall of heroes is perhaps the clearest illustration not only of monumental editorial projects like the *Monuments of German Musical Art* (*Denkmäler deutscher Tonkunst*), but of Franz Brendel's pan-Germanism and historicism, the teleological progression of genius.[66] There is a plainer insight here: these monumental heroes are engrossed in conversation, assembled in a social club like freemasons or men's choirs. Here, pan-Germanism looks quite a bit like choral fraternity.

For Mozart to be assimilated into this pan-German lineage, he had to be understood as a composer of Beethovenian seriousness. This explains the intensity of the Romako painting, the poetic odes to inwardness, and the quiet cottage in the woods. Likewise, *The Magic Flute* was recast as a work of sublime mysteries and solemn lessons; its whimsy was blamed on the Viennese superficiality of Schikaneder, its impresario and librettist. On the Berlin stage in 1845, an operetta called *Mozart und Schikaneder*—a contrafactum of Mozart's own opera-within-an-opera, *Der Schauspieldirektor*—recounted a fanciful origin story for *The Magic Flute*, in which Mozart appears as an ambitious artist foiled by Schikaneder's greed.[67] A similar stance returned when the New German School claimed Mozart as an ancestor. In his 1878 essay "The Public in Time and Space," Richard Wagner held *The Magic Flute* to be an immortal work whose frivolity was a mere coincidence of circumstance.[68] Wagner's claims were amplified by his advocate Ludwig Nohl, who argued in 1861 that *The Magic Flute* reveals the very origins of art-religion, a "purer way" to depict the eternal that was free from the church walls that confined Bach and Palestrina. A few years after the ovation of 1877, Nohl composed a new essay that repositioned *The Magic Flute* as a predecessor to Wagner's *Ring*, which he saw as the two fullest blossoms of German opera.[69]

The cottage on the mountain was no ordinary monument. Its rituals show how ceremonial odes participated in the social lives of men of letters, and evoked feelings of friendship, travel, and brotherhood. The rich homophony of male camaraderie not only was a sonic representation of German seriousness but constructed those very spaces of culture: the IMS was supported by donations from men's choirs, which were known to give charity concerts to finance the festivals in which they later participated. The result was a divide in the spheres of commemoration. In public, heritage sounded like choral fraternity, as a semicircle of male pilgrims trumpet themselves: "let us receive your greeting!" All the while, female poets composed lively recollections of pilgrimage that were filed out of sight. While the repertoire of men's choirs was not always patently political, the brotherhood of Bildung became the engine of pan-Germanism, a hall of German heroes that lauded Mozart as forebear. And when Mozart was reframed as a solitary composer in the woods, his pan-German persona anticipated Salzburg's emerging identity as a city at the crossroads of Europe.

Utopian Salzburg

After the collapse of the Austro-Hungarian Empire in 1918, a cohort of writers and theatrical personalities reconstructed Austrian identity by reviving Catholic and baroque traditions.[70] In 1920, the writer Hugo von Hofmannsthal, along with actor Max Reinhardt and playwright Hermann Bahr, established the Salzburg Festival to "transform the sleepy traditions of Salzburg—Mozart and the church—into the unified *mythos* of Salzburg," as Michael P. Steinberg argues.⁷¹ When Hofmannsthal and his colleagues are the nexus of this history, the Salzburg Festival looks purely ideological, an act of political theater. But when we shift the nexus to a mossy garden shed, we find more continuity than rupture. In light of the eight music festivals that preceded the Great War (which were not "festivals of some sort," as Steinberg calls them in passing, but interesting in their own right), and the droves of pilgrims already drawn to this city, it seems an oversight to suggest that the Salzburg Festival rose from the ashes of World War I primarily as a reaction against Carl Schorske's Vienna, a fin de siècle bundle of Freudian contradictions between modernity and conservatism.⁷² In the long view, the Salzburg Festival was just another sanctuary for art, the latest iteration of nineteenth-century Ehrenpflicht modeled after Wagner's Bayreuth.

Decades before the Salzburg Festival, and the clichés associated with Rogers and Hammerstein's *The Sound of Music*, Salzburg was said to be alive with song. After attending a three-day open-air folk festival in 1862, which featured the newly established Salzburger Sängerbund, Eduard Hanslick described Salzburg as a place where "art and nature come together in harmony."⁷³ And in 1879, at the threshold of the *Magic Flute* cottage, a member of Vienna's men's choir recited a poetic prologue in which men's voices echo across the hills, and the IMS secretary praised "you Salzburger" for "opening before you the sublime book of nature like a prayer book."⁷⁴ Later, Hofmannsthal echoed these sentiments; he was so moved by the festival of 1891 that he wrote: "the whole city [of Salzburg] is filled with softly vibrating, unceasing music . . . that we no longer experience in the metropolis."⁷⁵ When Hofmannsthal's fellow founders pitched their new festival as an artistic utopia that might heal Europe's wounds, they echoed the provincial "idyll" of late eighteenth-century literature—that is, the small town as a sanctuary of bourgeois arts and culture.⁷⁶ Gerhart Hauptmann wrote in 1918 that a festival in Salzburg would be "the most natural and happiest thought that I can imagine. . . . Who, before his death, would not wish to make a pilgrimage there to celebrate peace with his fellow man, and escape this dark and ravaged world?"⁷⁷

Salzburg could rebuild a broken Europe because it synthesized opposites. In his first call for support in 1918, Hofmannsthal described the city as Europe's "heart of hearts," a midpoint between Slavic and Nordic, Germanic and Italian, mountain and plain, age-old and modern, aristocratic and pastoral.⁷⁸ In the same year as Hofmannsthal's essay, the conductor and Mozarteum director Bernhard Paumgartner applied these characteristics to Mozart's music, which he held to synthesize southern and northern, folk-like inwardness and artistic virtuosity, baroque fantasy and clearheaded objective irony.⁷⁹ Shortly before the first Salzburg Festival, the Viennese writer Hermann Bahr published a witty story that imagines Salzburg as a utopian artists' colony.[80] Local journalists championed their city in these terms, as in this newspaper feature from 1902: "The name Salzburg is itself a platform where the German forests, the boldness and harshness of the Alpine lands, the bright sun and beautiful lines of Italy, all permeate and meld together. Mozart, the *genius loci*, combined German depth and intimacy with Roman grace into a higher unity. This spirit lives visibly in the city, which bears the imprint of all architectural style periods, from Gothic severity to the playful charm of the rococo."[81] Mozart's *Magic Flute*, in particular, came to reflect this crossroads. Erwin Kerber, an early director of the Salzburg Festival, described the work's blend of Viennese, Alpine, and Italian variants of the commedia dell'arte as the emblem of Salzburg.[82] In the early twentieth century, Mozart's music was not just a refuge from modernism, but a pan-European medium to overcome the trauma of war.[83]

A long view of the Salzburg Festival explains why we find nineteenth-century concepts embedded in W. H. Auden's wry tribute to Mozart in 1956, the bicentennial year. In collaboration with his partner Chester Kallman, Auden translated and adapted Mozart's *Magic Flute* for a televised performance hosted by the National Broadcasting Company. The production begins with a long "metalogue" read aloud by Sarastro that jests about Mozart's escalating mythos. A transnational Sarastro observes how international festivals make Mozart's modest divertimenti into the difficult music of the ivory tower; once a Viennese Italian who was "gay, rococo, sweet, but not sublime," Mozart is now a German "*Geist* whose music was composed from *Angst*."[84] To justify his rude interruption of the opera, Sarastro explains that he is no advertiser, no music director naming its sponsors: "I come to praise but not to *sell* Mozart." The history of pilgrimage to Salzburg shows this to be a false binary. Praising and selling have long been one and the same.

Epilogue: Alpine Shacks at the Crossroads

If this story seems recondite—a hut on a hill, a procession of pilgrims, a festival founded at Europe's conflicted crossroads—it should be noted that Salzburg was not an isolated case. The following will sound familiar. From 1908 to 1910, Gustav Mahler summered in the South Tyrolian town of Toblach, or Dobbiaco, which lies in Italy a stone's throw from the Austrian border. On a wooded slope that overlooks a breathtaking pastoral scene, in a tiny shack, Mahler composed his Symphony No. 9 and the (unfinished) Symphony No. 10, along with sketches for *Das Lied von der Erde*. The hut was scarcely big enough for a desk, a wood stove, and a piano, making it by far the most rustic of his three composing cottages.⁸⁵ For decades after his death, it lay forgotten in disrepair.

In the 1950s, the newly formed Gustav Mahler Society in Vienna proposed a Salzburg-inspired festival in Toblach. The rhetoric of those founders was likewise laced with competitive Ehrenpflicht, deriding the Italian government for failing to preserve Mahler's legacy. They noted with some annoyance that the nearby Franciscan school had organized a camping trip there and used the cottage to prepare the boys' meals. In 1957, the international branch of the society backed the aspirations of Vienna's local chapter to host the first commemorative concert in Toblach, along with the festive unveiling of a memorial plaque at the hut. Shortly afterward, a museum was founded in the rooms of the Trenkerhof hotel where Mahler had resided. By 1980, the administrative leadership of the resort founded a committee to "do something with Mahler and thereby make Toblach important," and a year later, they established their own Alpine rival to the Salzburg Festival: the Gustav Mahler Weeks at Toblach.⁸⁶ Tensions came to a head when the German-speaking festival committee struggled to convince Italy's leading radio company (RAI, or Radio Audizioni Italiane) to broadcast its events.⁸⁷ The success of the Mahler Weeks has made Toblach a hub for the arts, and visitors today are eager to admire the birthplace of some of Mahler's most adventurous compositions.

But the story does not end there. Mahler's status in Toblach remains complicated, not only because Toblach sits in a border region where Austrian Ehrenpflicht jars against Italian interests, but because the town's inhabitants have had to wrestle with some residual indifference to Mahler, whose name affixes a Viennese insignia to a region with a proud identity of its own. It is a compelling place: a close-knit community with a population just over three thousand, where the Dolomites jut like shark's teeth from an emerald lake. The mountains seize not only the eye but the local economy, which is rooted in winter sports and hotel dynasties of intergenerational hospitality, such as the inns of the Trenker family and the Santer family. It was the Trenkers who owned the apartments and cottage in Mahler's time, and in 1981, with the advent of the Mahler Weeks, the Santer family purchased the land surrounding the cottage and restored it to a condition that allowed visitation. But to pay their respects to Mahler, pilgrims first had to venture through an unrelated attraction: a wild-animal park populated by owls, wild boars, and bison, in which the cottage was nestled. Even as the site was registered for historic preservation in 1998, Mahler lovers voiced their dismay at the indignity of this animal kingdom, echoing the 1909 memoir in which Mozart's cottage teemed with rabbits.⁸⁸ Tired of the upkeep and uproar, the Trenker family distanced themselves from Mahler's legacy, and in 2022, the Santer family divested from the animal park, conducted a painstaking restoration, and designed a digital tour of Mahler's surroundings as a resonating landscape.[89]

I asked a local why Mahler heritage stood on such shaky footing here until the 1980s, and the answer was candid: antisemitism and anticommunism clouded Mahler's reputation and delayed a sense of Ehrenpflicht.[90] It is telling evidence of how the politics of a composer's lifetime can inform the historic preservation that follows. Just as Bonn and Vienna contested over Beethoven's legacy, and Mozart was transplanted to Salzburg, Mahler sought out the sublimity of the natural world in locations across Austria that found him too eccentric, and too Jewish, for comfort. Perhaps it is worth stepping back from the tangled politics of these crossroads to imagine Mahler and Mozart in simpler terms: as writers in need of quiet, seeking out four wooden walls, a roof, a desk, and a chair.

3

From Relic to Specimen

This chapter begins with two hearts: those of Beethoven and Chopin, both poised between relic and specimen. Of the two, Chopin's heart is better known because its contested status caught the attention of the media. After the composer's death in 1849, his heart was preserved in cognac and returned to his native Poland, where it was embedded in a marble tomb at the Holy Cross Church in Warsaw. In 2008, doctors petitioned to scrape the heart for a sample to confirm their theory that Chopin died of cystic fibrosis, rather than tuberculosis, the fashionable disease of nineteenth-century Paris.[1] The request had some precedent, as scientists had already analyzed a skull that allegedly belonged to Mozart and had tested Beethoven's hair for lead levels (in a dramatic twist, some of those locks were not Beethoven's to begin with, as I explain later).[2] The Polish government saw medical testing as a desecration of remains, and they denied the biopsy. Finally, in 2014, they allowed a team to photograph the heart, and the debut of this swollen, greenish organ was met with a flurry of retrospective diagnoses.[3] Some may be tempted to frame the government's initial refusal as a clash of science and religion, but that binary is too simple. This chapter argues that, despite their secular milieu, doctors who collect and measure the remains of composers perpetuate nineteenth-century art-religion. Scrutinizing every crevice and fold is an act of veneration that blurs the line between relics and specimens, while medical writing can serve to resurrect the dead.

Medical secrets are tantalizing and risky. When James Q. Davies uncovered evidence of Chopin's urinary problems in a thesis by the composer's medical-student flatmate, he was torn between rich insights into music-medical discourse, such as intersecting theories of autonomic nerves and pianistic touch, and a temptation to repathologize a composer whose body may be best left alone.[4] The risks of retrospective pathology are clearer in the case of Beethoven's heart. In 2014, several physicians consulted with a musicologist to coauthor an article in a medical journal that interprets Beethoven's works as a "musical electrocardiogram."[5] While the lecture-recital that spurred this article was no doubt illuminating as public musicology, its claims look tenuous in the cold light of print. The authors admit that there is no evidence that Beethoven suffered from cardiac arrhythmia, nor from heart disease of any kind, but this signals an opportunity, they argue, for retrospective diagnosis through music analysis, which makes syncopations and gasping rests into symptoms. They go on to suggest that works written in times of stress, a known trigger for arrhythmia, are more likely to be symptomatic, and they draw parallels between the contours of modern electrocardiograms and select passages from Beethoven's works.

A glance at the article's bibliography shows it to be in good company: this participates in a thriving practice called "pathography," in which doctors posit retrospective diagnoses and analyze artworks through the lens of maladies. The genre is by no means homogeneous, and many pathographies arise on the practical grounds that celebrities, whose fame generates a hefty corpus of life writing, offer a trove of information about the history of medicine.[6] Even so, the genre has changed remarkably little across the generations. Claims about Beethoven's arrhythmia in 2014 echoed arguments made a century earlier by a German physician named J. Niemack, who suggested that Beethoven's deafness amplified his perception of his own heartbeat, and that halting breaths in his melodies indicate his arrhythmia. Already in 1922, Niemack was rebuked for these unprovable arguments by a more cautious pathographer.[7] A comparable study was published as recently as 2020 alongside essays by some of the most prominent voices in Beethoven studies, and to this day, pathography remains the most speculative content to appear regularly in peer-reviewed medical journals.[8] If we agree with Mina Yang that pathography is a tabloid culture that appeals to the public's appetite for forensics, then the genre risks an association with true crime, conspiracy theories, and gossip.[9] In the 1930s, the psychiatrist Mathilde Ludendorff posited that Mozart was poisoned by a secret pact between Jews, freemasons, and Jesuits; her theories were promoted by music's most prominent pathographer, Dieter Kerner, whom Otto Erich Deutsch deemed a "fanatic."[10] Even when they are carefully researched, pathographies can cling to concepts that have been the bane of musicology: the biographical fallacy that listens for physical symptoms in artworks, the universalism of the Western canon, and a lingering interest in human difference grounded in European exceptionalism. As such, this practice and its persistence should be better understood.

For doctors still invested in the enterprise of pathography, the word "pseudoscience" is likely to sting. Yet I would argue that even the most scrupulous pathographies have inherited a fraught lineage. Today, practitioners of medicine operate in different circles from those of osteoarcheologists, but in the nineteenth century, a substantial community of doctors embraced the phrenological methods of Franz Joseph Gall, who located character and aptitudes in the contours of the skull. The term "pseudoscience" was first formally defined by Gall's skeptics, notably François Magendie, for whom phrenology's offenses were twofold: it concealed latent theological convictions that stymied the powers of observation, and it sought to amuse the general public, in contrast with rigorous scientific observation that could meet with ennui.[11] Pathography, phrenology, and the related practice of physiognomy (the study of faces) made the nineteenth century an era of "geniology," to borrow a useful neologism by Darrin McMahon. In this period, "genius" was newly understood not as an external force, or *daimonos*, but an ability inherent in the body, which led anatomists' inquiries to creep deeper into the brain, from the skull's surface, to its capacity, to the brain's folds, and ultimately to the psyche.[12] Implicitly, craniometry compared society's echelons out of a wonder at creative difference and the divine spark, which gave rise to what Joseph Straus calls the "saintly sage" topos of disability.[13] Geniology encouraged another kind of wonder: the collection of curiosities in the tradition of the Renaissance *Wunderkammer*, which led specimens to function as relics, or objects with talismanic properties. This was another form of hidden magic that raised critical suspicions of pseudoscience. Underlying these methods is a desire to possess the dead, and in so doing, to make the divine tangible. To understand genius, doctors first had to hoard it.[14]

Why does the interest in composers' bodies run so deep? To explain this striking continuity with the past, this chapter turns to the enmeshed social worlds of music and medicine, in which doctors collected relic-specimens that they regarded with both curiosity and posthumous affection. The related practice of pathography enacted a newly aesthetic approach to medicine that found a holistic resonance between diagnosis, relics, and compositional style. As composers were exhumed and examined in the late nineteenth century, it became a special act of piety to lure the composer's character out of the bones, making the most of this disruption of eternal sleep. But by this point in the nineteenth century, amid an escalating skepticism of phrenology as pseudoscience, those at the helm compensated for the field's ambivalence by detailing every inch of these bodies to aid future generations in cracking the code. Anthropometry was, in short, an extension of the duty to preserve. The detailed accounts of illness that still appear in pathographies have retained that nineteenth-century sense of Ehrenpflicht, as if to embalm the composer's body in textual formaldehyde.

The Music-Loving Doctor

In a 1962 essay on chamber music, Theodor W. Adorno recognized a demographic hiding in plain sight: "That physicians tend to love chamber music and have a talent for it may well be explained as a protest against a profession that makes unusual demands on the intellectual who takes it up. It calls for sacrifices of a kind otherwise required of manual laborers only, for touching nauseating things and being on call, not master of one's own time. The musical sublimation of chamber music makes up for this. It would be the mental activity a doctor feels deprived of."[15] Given that doctors overcome the aversion to "touching nauseating things" early in their medical training, some might scoff at Adorno's hemophobia. He was wrong on another count, too: a large subset of doctors sought corporeal explanations for genius, at times collecting composer specimens, which contradicts the premise that music offered a lofty "sublimation" and an escape from the flesh. Where Adorno is right is the observation that doctors have historically been invested in German idealism and Bildung. Doctors constituted a distinct subset of men of the professions who delighted in string quartets in their living rooms, collected musical relics, and published for the lay reader.[16] Michael Hau has shown how prominent doctors of the late nineteenth century combined the analytical gaze of expertise with a holistic, intuitive gaze borrowed from the art connoisseur, appreciating both artworks and bodies as a coherent whole. To defend the principles of Bildung, and to push back against alternative health-reform movements that threatened conventional medicine, these doctors published prolifically to offer "authoritative aesthetic advice to the public based on their expertise as physicians."[17] For Penelope Gouk, music and medicine became "sister disciplines" because they shared an ethical pursuit of the good, of human wellness and excellence, with the assumption that Eurocentric and specifically German art music "is intrinsically 'good,' in both a moral as well as an aesthetic sense."[18]

Standard music histories have neglected this demographic, but the moment one looks for musical doctors, they appear in every corner. Composers maintained close friendships with doctors, such as Johannes Brahms and Eduard Hanslick with pianist-physician Theodor Billroth, Felix Mendelssohn with anatomist and string player Jacob Henle, and the Schumanns with surgeon Ernst August Carus.[19] Anatomists who were eager to study composer specimens were likewise embedded in musical networks: Carl von Rokitansky assisted with the autopsy of Beethoven's hearing organs in his youth, then went on to become president of the Imperial Academy of Sciences and a leading Austrian politician; the brain anatomist Theodor Meynert, who collaborated with skull collectors Josef Hyrtl and Carl Langer von Edenberg, was born into a family of musicians; and Julius Tandler, an anatomist who gave numerous phrenological lectures on composers' brains to the Vienna Medical Association, went on to become a public figure in both domains (as state minister of health, for instance, he planned fiftieth-birthday celebrations for Arnold Schoenberg on a committee chaired by Alma Mahler-Werfel).[20] A full account of musical doctors fills a curiosity cabinet of its own.[21] Their names appear largely in anecdotal surveys of music and medicine, such as an essay for a medical school bulletin that compares the physiognomies of music-loving doctors with those of famous composers.[22]

Musical doctors were active during a transitional period in the history of medicine, which explains why the genre of pathography was so eclectic. On the one hand, the nineteenth century saw new forms of specialization, as professional societies inscribed less porous boundaries around disciplines and founded the first peer-reviewed journals. When insular societies competed with enterprising publishers to disseminate their knowledge, they both shaped and were shaped by a public whose attention to scientific findings defined what credibility and authority should look like.[23] Even doctors who worked as specialists maintained the late eighteenth-century archetype of the well-rounded man of letters, the naturalist who moves freely between fields, collects curiosities, and enlightens the public with charismatic displays of expertise. While music-medical writing has its roots in the early modern period, what differentiated the approach of nineteenth-century doctors was a narrative element grounded in the anecdote, the intimate secrets and telling episodes collected in volumes that were styled after display cabinets and vitrines.[24] Pathographers not only consulted anecdotes as data but at times adopted this storytelling form to describe their encounters with medical curiosities and secrets. As a result, medical writings for the lay reader were remarkably kaleidoscopic, weaving together the latest insights in pathology, criminology, and cranioscopy with anecdotal evidence from patient interactions and idiosyncratic opinions on art criticism.

A paradigmatic example of the genre hails from Billroth, who was famed equally for his innovations in abdominal surgery and his fine musicianship on the piano and violin. Billroth spent many evenings reading through chamber works with Brahms, and his friendship with Hanslick led him to try his hand at concert reviews, which meant that Vienna's concerts were often related to the local papers by "Hofrath Billroth." In the 1890s, he developed a series of essays on music and medicine that were posthumously stitched together by Hanslick, and the intended readership was clear: "I don't endeavor to write anything academic," Billroth explained in a letter to Hanslick, "but rather to correct misconceptions among dilettantes."[25] (His book offers a precedent to the best-selling *Musicophilia* by Oliver Sacks, who was the latest in a long line of doctors who behaved as charismatic public intellectuals.)[26] Billroth's volume, called *Who Is Musical?*, examines topics as varied as music and the nerves, auditory hallucinations, music's effect on animals, the perception of rhythm, synesthesia, aesthetic taste and appreciation, and (Beethoven's) deafness and the inner ear.[27] When Billroth the music critic appears, his arguments diverge from empirical reasoning. In the same essay where he suggests that culture filters music perception and future generations "will enjoy other things of which we today have no idea," he also upholds Germanocentric notions of genius. Some composers who were famous in their day, he explains, were only later understood to lack originality; such is the case for Cherubini, Spontini, and Berlioz, whose "melodic and rhythmic inventions I find dull." Berlioz's tinkering with timbre pales in comparison with the "real musical kernel" that was better developed by Weber, Meyerbeer, and Wagner, who sprouted from the seeds of Haydn, Mozart, and Beethoven.[28]

Billroth was neither the first nor the last doctor to make arguments grounded in Teutonic historicism. In his defense, the medical content of his essays reflected the current state of knowledge, while many of his contemporaries published texts that were less authoritative. Consider a book published in 1910 that distilled two centuries of physiognomy, phrenology, and pathography into a reader's digest of outdated theories, authored by Oswald Feis, a German Jewish gynecologist, violinist, and trustee of Dr. Hoch's Conservatory in Frankfurt.[29] Echoing phrenological theories from a century earlier, Feis recalls that "I myself have observed that the skulls of musicians have a protrusion in the left temple" and "many musicians are known for an unusually wide and large skull: Wagner, Bruckner, Marschner, Beethoven, Schumann, among others."[30] Feis goes on to catalogue the eccentric habits, nervous disorders, and addictive behaviors of composers, admitting that the link between creativity and degeneration may require further study.

While a thorough account of degeneration would surpass the scope of this chapter, it is worth noting the dominant influence of Cesare Lombroso on pathographical texts like those of Feis. Lombroso's career as a criminologist was interleaved with essays on disturbed geniuses such as prophets, graphomaniacs, revolutionaries, and mad artists, and music appears frequently in his work because it exemplified his primary interests: the heredity of talent (Lombroso cites the Bach family, among others, despite that what really ran in families was artisanal careers); hyperaesthetic sensibilities such as Berlioz's agitation; melancholy and hallucination as a fount of ideas for troubled souls like Robert Schumann; and race, of which his primary example was a preponderance of Jewish musicians with nervous disorders.[31] Lombroso's influence on psychopathography—the branch of pathography that put composers on the couch, most famously Richard Wagner with his depraved silk fetish—has been well charted.[32] Lombroso's speculative approach to scientific writing showed aspiring pathographers how to posit diagnoses based on artists' daily habits.

In 1925, the physician Richard Waldvogel published a treatise on Goethe and Beethoven that demonstrates the eclectic argumentation that was typical of the genre.[33] To square degeneration with greatness, Waldvogel suggests that these artists suffered from tuberculosis and syphilis, two diseases thought to heighten the nervous system. (Here it should be noted that pathographers' interest in tuberculosis and sexual deviance was not only common, but amplified by the figure of Frédéric Chopin, whose pathographies are so numerous that they must be addressed in a study of their own.)[34] Admitting that syphilis can make female readers blush, Waldvogel scours the family histories of Goethe and Beethoven for evidence that their illnesses were hereditary, which leads to his grand theory: "genius arises in the hereditary process particularly at that point where a highly gifted family begins to degenerate"— that is, childless geniuses always emerge as the last swan song of a degenerating family line, inheriting the diseases that spur their creativity. (Goethe's illegitimate offspring are neglected in this account, presumably because they might sully the artist's reputation.) For Waldvogel, degenerating artists found their outlet in "egocentric behaviors" like alcoholism, and when artists like the martyr Beethoven gave up those needs in the name of art, that renunciation was said to amplify creative acuity.[35] Waldvogel's treatise exemplifies how pathographies could interweave theories from medicine, anthropology, and criminology, held together by the premise that genius resides in the body and overflows into art when prompted by suffering.[36]

Texts like these show not only the influence of Lombroso, but also the roots of the "saintly sage" archetype, which had become so established by the interwar period that it warranted a social theory of its own. In two books of 1928 and 1932, the psychiatrist Wilhelm Lange-Eichbaum linked pathography with martyrology and proposed that genius substitutes for religion in a secular age.[37] Like saints, artists have a "nimbus," a halo of magic, that can be conceptualized with a matrix arranged by the type of fame (such as prophet, philosopher, innovator, and artist) and the feelings that a genius can instill (such as *majestas, facinans, sanctum,* and *tremendum*). Mozart and Schubert serve as the primary examples of facinans, the impact of enchantment on beholding pure beauty, while Beethoven exemplifies tremendum, the impact of the uncanny, gruesome, and self-annihilating.[38] Pathography's function is to explicate the intersection of "artist" and tremendum, at the axis of which one finds "the work born of suffering, ingratitude, suffering as reward, early death, suicide, mental illness."[39] In Lange-Eichbaum's sprawling theory of genius-saints, we find pathography's first awakening, its self-awareness as a practice grounded in veneration and not solely in medical curiosity.

There is more to the musical doctor than a history of hobbies and leisure. The labyrinthine development of pathography—which I have sketched here only in its barest outlines—reflects a divide in the medical profession between curiosity and care. On the one hand, music-loving doctors took up a belated charge to tend to the dead as imaginary patients. Their writings show not only a perceived role as champions of Bildung, but a private sense of compassion for the artist's suffering. Pathographies were acts of atonement for the shortfalls of composers' own doctors; in the retrospective diagnosis one finds a covert apology, a script for how to save the ailing artist in their time of need. That sense of care is tempered by a more destructive curiosity about how genius operates, which leads pathographers to reanimate the body by reconstructing its every function, while anatomists pull it apart into remnants—organs in jars, exhumed skulls, and hair. When doctors traded in human remains, and elevated relics of genius above specimens of anonymity, they grappled with a uniquely invasive form of piety.

Relics of Genius

In 1798, two centuries before the dispute about Chopin's heart that opened this chapter, a letter appeared in *Der neue teutsche Merkur* that framed the scientific study of skulls in terms of honor and homage. The author, meditating on recent lectures by Franz Joseph Gall, wrote that "when the mortal remains of admired and beloved persons are valuable as relics to their devotees, they can at least find solace in the fact that the crania of those mentioned, as well as those of many other important individuals, can be found in the interesting skull collection of Dr. Gall, where their heads not only are protected from the ravages of decay, but may be instructive and useful after their deaths."[40] It is surprising to see phrenology—the study of localized capacities that reads the face and skull as texts—here offered up as a way to honor the dead. The contents of Gall's cabinet (103 luminaries, 69 criminals, 67 insane, 35 "pathological cases," and 25 "exotics") were foraged in part from prisons, asylums, and graveyards, which earned his followers the nickname of "resurrectionists."[41] His project was so strongly opposed in Vienna that the authorities exiled him to Paris in 1805; there, many of his fellow scientists critiqued him for generalizing a handful of specimens to entire populations and for assuming, without evidence, that skull shape predicts the contours of the brain.[42] (Granted, those same colleagues had even less accurate theories of brain function; today, historians recognize that Gall's core premise of localization, however misguided in practice, was foundational to neuroscience once it was refined by Paul Broca.) In Vienna, the animosity was more visceral. Not only did Gall advocate a materialistic definition of the soul that verged on heresy, but his collection stoked a more basic fear that anyone could end up on the shelf.[43]

Across the globe, skulls have been ethically contested specimens. James Poskett has traced how the global dissemination of phrenology's materials—first skulls and plaster casts, and later books, periodicals, and photographs—led to intercultural encounters between divergent concepts of death, burial, and honor.[44] Skulls were valued as evidence of racial theories, war trophies of empire, or relics of antiquity. To collect these specimens could be an act of piety or desecration; the British Egyptologist Orlando Felix, for instance, distinguished his own respectful forms of tomb raiding from ordinary plunder, presenting himself as a good Christian invested in biblical genealogy, not wholesale.[45] In Europe, skulls could be materials for the scientific study of talent, talismanic objects that harbored divine mysteries, and mementos of personal friendship with the deceased. The boundaries between those functions were porous. The anatomist Samuel Thomas von Soemmerring mounted his recently deceased friends in his personal library on velvet cushions; he was one of many anatomists who collected their own friends and colleagues in a dismembered brotherhood.[46] Anthropometry was, in short, a project of Ehrenpflicht, a means to honor the deceased by shielding them from the "ravages of decay" to make them "instructive and useful," as Gall's advocate claimed.[47] But behind every velvet pillow and glass reliquary lies a tale of plundering, or what some called "relic snatching" (*Reliquienhascherei*), that shows scientific piety to be both fragile and arbitrary.[48]

Artists' remains took on special properties in a period when divine genius was thought to radiate from flesh and bone. That premise was popularized by Johann Caspar Lavater, the late eighteenth-century founder of the doctrine of physiognomy, an ancestor of sorts to phrenology. Given that Lavater was a Zwinglian pastor famed for his eschatological essays, his writings on physiognomy read like sermons, and his stated goal in studying genius was to uncover the Christ-like in man, or what he called the "Man-God," that might help Christians fulfill *imitatio*.[49] Geniuses, he claims, are uniquely able to channel the divine spirit; it is "as if [the artist] were dictated to by a *genius*, an invisible being of a higher kind, as if he himself were this being of a higher kind."[50] For Lavater, the mark of genius was not only the high forehead but "fiery eyes" that shone with the divine, like the eyes of celestial light that were common miracles of the saints.[51] While Enlightenment secularists found his sermonizing unnerving, Lavater's methods reached an uncommonly wide public thanks to his innovative subscription series, which took subventions to analyze readers' own faces.[52] Lavater shaped not only how artists were represented in portraiture, but how their contemporaries responded to their faces in life. Beethoven's acquaintances noted the composer's "fiery eyes," the "high and mighty forehead," and the "high curvature" of his brow that made his head a "sarcophagus of art music."[53] One critic was disappointed to find that Beethoven's features did not align with his music, and in the same period, contemporaries of Schubert found his rotund appearance ill suited to his compositional grace. Eventually, Lavater's ideas found a new life among antisemitic critics who searched composers' faces (and music) for Jewish traits.[54]

The high forehead and fiery eyes were complemented by a bulging left temple, the locus of musical acumen according to Gall and his collaborators. Phrenology was wedded to music from the start: Gall's earliest lectures recounted a young musical savant whose protruding temple revealed the root of his ability. Even before Gall published his findings, his ideas were applied to music in an 1801 article by Georg August von Griesinger, a personal friend of Beethoven and Haydn and biographer of the latter. Griesinger noted a triangular widening of the brow in the faces of composers prominent in their day, such as Beethoven, Haydn, Mozart, and Salieri, and he suggested patterns in the "national skull" that distinguished musical peoples like Germans and Bohemians from "wild peoples" devoid of musical talent.[55] When Gall took up residence in France, his proponents focused their attention on celebrity singers, comparing their heads and bodies in fine detail.[56] Sixty years later, as composers were exhumed, their skulls were still described in terms that echoed Gall's earliest advocates.

Skulls were not the only tools to measure the body: Gall's collection was populated by plaster masks that were later valued as relics. In 1812, a sculptor took a cast of Beethoven's face that has become the most iconic representation of that composer; the vibrant afterlife of that object is the subject of the next chapter. For now, it is worth noting that this mask was created because there were only two degrees of separation between Beethoven and Gall. Musicologists are familiar with Beethoven's friendship with the piano makers Nannette and Andreas Streicher and have charted their prolific correspondence and the weekly salons at the Streicher home that offered an intimate venue for Beethoven's works. Less often discussed is the Streichers' close friendship with Gall, who—in his early days, before he turned his attention to phrenology—served as the family physician, and who benefited from the couple's support throughout his career, even after his exile from Vienna. Historians speculate that the Streichers introduced Gall to Mozart, Beethoven, Schiller, and other major figures in their milieu; in fact, the Streichers' salon is one reason among many that Gall took an interest in measuring aptitudes, and why he later oversaw the production of Schiller's death mask.[57] The Streichers' interest in Gall's theories led them to commission portrait busts of living geniuses that would line the walls of their salon room. For that task, Gall recommended his collaborator Franz Klein, a sculptor who had begun his career attending anatomy lectures and working on animal specimens for the Pathological Anatomical Museum of the University of Vienna.[58] When Gall was building up his first collection in Vienna, he employed Klein to create over one hundred plaster casts. It is no coincidence that Klein was the one to cast Beethoven's face as a model for the bust that would decorate the Streichers' atrium.

This web of connections shows that Klein's cast served a dual purpose: it was both a tool for portraiture and a specimen in the anatomist's cabinet. A century later, the mask fulfilled its intended purpose when anatomists of the Third Reich studied Beethoven's physiognomy. Walther Rauschenberger, among others, published racialist analyses of not only Beethoven's mask, but also the faces of Wagner, Schubert, Weber, and Wolf; in 1943, even after Bruckner had become a figurehead for the National Socialists, Rauschenberger evaluated whether the composer had traces of Black lineage.[59] Plaster masks began their life not as innocuous models for portraiture, but as specimens, and their meanings became ever more convoluted when the pseudoscience of difference took its darkest turn.

This flash-forward in history conveys the longevity of Lavater and Gall's ideas, which were promoted through a tautology: the template for what genius should look like is a canon of geniuses, which means the yardstick measures itself. (When a genius did not fit the bill, Lavater blamed the portraitist for inaccuracy.)[60] Later in the century, composers' exhumations were met with tautological readings of skulls that conveniently located the markers they sought. Those who were invested in these projects were devotees or personal friends of the deceased. Cranioscopy offered them a tactile and intimate reunion.

To understand the affection that animated these relics, I turn to Goethe, whose dynamic relationship with the skull of his friend Schiller set a precedent for the remains of Beethoven, Haydn, Mozart, Schubert, and Bach. Goethe was among the most enthusiastic purveyors of Lavater's physiognomy, and his influence on later generations made him what Richard T. Gray has called the "foundling father" of physiognomy.[61] That expertise led to an unusually close and personal confrontation with the theories he had admired from afar. In 1826, the mayor of Weimar sought to relocate Schiller's body to the ducal crypt to ease overcrowding at the cemetery, and he exhumed several bodies from the area where Schiller was said to be buried. Goethe was called on to identify the correct skull, so naturally he chose the one with the broadest brow. When the literary circles of Weimar heard that King Ludwig of Bavaria wished to see Schiller's skull, bringing the hordes in tow, they removed it to a glass case at the base of Schiller's bust in the Duchess Anna Amalia Library in Weimar.[62] The skull was interred with solemn Ehrenpflicht, and in a speech given at the ceremony, August von Goethe (the poet's son, speaking on his father's behalf) sought to "[safeguard] these sacred remains"—that is, to "protect a precious treasure from moldy decay," like the urn at the foot of a tomb, and to make the skull accessible only to those "whose paths are guided not by curiosity, but rather by the perception and awareness of what a great man was afforded to Germany, to Europe, indeed to the entire cultivated world."[63] Later that year, when the skull was taken out for cranioscopic evaluation, the elder Goethe kept it on a velvet pillow in his house, just as Soemmerring had done for his friends. Goethe allowed no one but his friend Wilhelm von Humboldt to view it, and (according to Humboldt) he contemplated his own mortality and mausoleum, where Schiller might be interred by his side.

This rictal reminder of mortality compelled Goethe's meditation on death and relics called "On Contemplating Schiller's Skull," a poem that captures the mysticism of phrenology. In a meter reminiscent of Dante, Goethe's narrator stands lost in an ossuary of nameless bones, acting as the "Adept," or expert, tasked with discerning brilliance amid a rubble of shoulder blades. Suddenly, a divine specimen reveals itself, bearing the spark of genius and arousing "such mystic rapt devotion!" that the narrator, unsure if he ought even to touch such an "oracle," frees it from its "musty prison."[64] Goethe's poem captures the covert spirituality of the phrenologist, whose eye is drawn to signs of the divine, and who venerates skulls of genius while the rest are left to mildew in the ossuary of anonymity.

Several features of Goethe's story explain the treatment of composers' skulls in this same period. First, while skull theft could begin at the grass roots, famous heads found their way to cultural arbiters, just as Goethe became the keeper of Schiller's head, and those gatekeeping practices were borrowed from the relic trade of churches. This was the case for the skulls of Haydn and allegedly Mozart, which arrived at music institutions that behaved as sanctuaries. Secondly, the amalgamation of scientific inquiry with Ehrenpflicht explains why Beethoven and Schubert were exhumed not once but twice. The histories of these heads involve wild capers that could belong to the world of fiction, and while they have been told and retold in anecdotal lore, they have not yet been understood as chapters in a history of art-religion.

The tale of Haydn's skull can turn a weak stomach and, in fact, did so for one of its thieves. Shortly after Haydn's death, his head was plundered from his grave by three men who were energized by Gall's theories: Carl Rosenbaum, a former secretary to Prince Esterhàzy and personal acquaintance of Haydn, along with Johann Peter, one of Gall's pupils, and the doctor Leopold Eckhardt. (By 1827, this theft was so infamous that Beethoven's coffin was preemptively encased in brick, which hindered his exhumation decades later.) After close encounters with police, who were deterred in a comical instance when Rosenbaum's wife concealed the skull beneath her mattress, the specimen found its way into an ornamental reliquary that was later donated to the Viennese anatomist Josef Hyrtl.[65] Historians of science know Hyrtl as the most influential and charismatic anatomist of the mid- and late nineteenth century; historians of medicine know him as the author of a leading textbook and a pathbreaker in vascular injection; historians of music know him as a steward of composers' remains. That much is no coincidence, as Hyrtl hailed from a family of musicians: his father was an oboist at the court of Esterhàzy during Haydn's tenure, which explains why Haydn's head was offered to him. Meanwhile, Hyrtl's brother Jacob was a musician who received the skull of Mozart as a gift from an elderly gravedigger. Several nineteenth-century newspapers related the twisted tale of Mozart's skull—now presumed inauthentic—and unlike the better-known story of Haydn's, these histories disagree on the facts. The most cited account hails from none other than Ludwig August Frankl, the advocate of monument fever who introduced chapter 2 of this book, who had become good friends with Hyrtl in medical school. Frankl related to the *Neue Freie Presse* that the gravedigger kept careful track of Mozart's location, having had fond memories of one of his musical works, and several years later when the pauper's grave was emptied out to make space for more bodies, he had rescued the skull and offered it to brother Jacob, who then relayed it to Josef's cabinet.[66]

What happened next shows that Hyrtl, while no admirer of Gall's phrenology, regarded these skulls as relics of genius, and his admiration for them amounted to what some would call desecration. As Hyrtl specialized in the study of hearing organs, it was he who had led the examination of Beethoven's temporal bones after the composer's autopsy in 1827; he stored the organs in a solution at the University of Vienna's Institute of Anatomy, from which they later vanished (rumor had it that the coroner Anton Dotter absconded with the jar to Paris).[67] Later in life, Hyrtl took more liberties with his specimens. He

FIGURE 3.1. Haydn's skull in a decorative case with archivist Hedwig Kraus at the Society for Friends of Music in Vienna, 1954. ÖNB, Image Archives, Lessing LE 54140131A. © akg-images / Erich Lessing.

crowned Mozart's skull with a laurel wreath to differentiate it from those of the prisoners from eastern Europe and Asia that populated his collection. At some stage, we know not when, he inscribed the right temporal bone with a verse by Horace: "Musa vetat mori!"—short for "dignum laude virum Musa vetat mori" or "the Muse forbids the virtuous man to die."[68] Upon his retirement, Hyrtl's anonymous skulls ended up in the Mütter Museum in Philadelphia, a gruesome repository for human specimens that still today grapples with the ethics of display. Hyrtl's heads of composers, in contrast, were bequeathed to houses of culture: Mozart's was given to the museum in Salzburg, where it arrived in 1902 after vanishing for several years, and Haydn's was offered to the Society for Friends of Music in Vienna, where it remained in the archive in a glass vitrine until its grandiose reinterment ceremony in 1952 (fig. 3.1).[69] By the time Mozart made it to Salzburg, he had been inexpertly sawed through and arrived in fragments. As Hyrtl had recently passed away, rumor spread that the aging anatomist had conducted an investigation while nearly blind; the culprit remains a mystery.[70]

The shifting regard for Mozart's skull shows how those who safeguarded composers' remains were torn between piety and curiosity. Those who oversaw autopsies developed a reputation for relic snatching.[71] The most blatant thievery occurred at Gaetano Donizetti's autopsy in 1848, when an Austrian military doctor named Gerolamo Carchen stole the cap of the composer's skull, possibly to compare it with patients at the nearby insane asylum where he served as director. Under mysterious circumstances, the relic made a quick transition from sacred to sacrilegious; by the time the theft was discovered at Donizetti's exhumation in 1875, the skullcap was said, anecdotally, to act as a receptacle for loose change in a store in Bergamo. After its discovery, it was housed in an urn at the Donizetti Museum until it could be reinterred.[72] Likewise, those who stood guard at composers' exhumations could pocket souvenirs. In 1863, Mathias Durst, the director of the orchestra at the Hofburgtheater, protected Beethoven's exhumed body from graverobbers; he took the opportunity to snip a piece of the composer's shroud and encase it in glass, later offering it as a gift alongside "some fine cigars!" signed from "your old and fallible [*schadhafter*] friend."[73]

The violation implied by the word *Reliquienhascherei* does not account for the sense of Ehrenpflicht that attended these little thefts, an extension of doctors' self-image as the guardians of Bildung. Such was the case for the first exhumation of Beethoven and Schubert in 1863, the full import of which is explored shortly. The project itself was an act of piety, replacing wooden coffins with watertight metal that would slow the decay and preserve the contours of genius for posterity. While the remains were aboveground, they were cast and measured by two doctors with a personal connection to the deceased and his Viennese milieu: Gerhard von Breuning, a physician, phrenologist, and son of Beethoven's lifelong friend Stephan; and the medical historian Franz Romeo Seligmann, who can be seen standing center right in Moritz von Schwind's famous panorama of a Schubertiad.[74] Among the two, Breuning had a more visceral relationship with the dead. He communed with Beethoven's skull just as Goethe did with Schiller's: "I kept it by my bedside overnight, and in general proudly watched over that head from whose mouth, in years gone by, I had so often heard the living word!"[75] With his commitment to phrenology, Breuning expressed his hope that the skulls would remain aboveground permanently, housed in a mausoleum of some kind, so that anatomists could continue their study of genius.

Meanwhile, as Seligmann oversaw the casting and measurement of the skulls, he and his colleagues observed that pieces were missing. For many years, it was presumed that the doctors were referring to the temporal bones removed at Beethoven's autopsy in 1827, but a curious discovery raised the possibility that several more fragments had been taken during the exhumation itself. The convoluted journeys of these fragments were charted by William Meredith, among others, in a riveting special issue of the *Beethoven Journal*.[76] In 1968, an unassuming concert review tipped off researchers that a shard of Beethoven's skull was owned by a concert pianist, who received it as a gift from Seligmann's kin. A professor of surgery at UCLA insisted that he would be an ideal guardian "because of my rather unusual involvement in Beethoven's deafness, both as an ear surgeon and as an amateur violinist."[77] Gradually, the family estate located more fragments that were housed in a pear-shaped box labeled with Beethoven's name; Seligmann's son, in his will of 1944, insisted that the handwriting belonged to his father. Meredith suggests that Breuning orchestrated the gift of three skull fragments to Seligmann in thanks for his assistance, and that Seligmann kept quiet about his prized possession, even when the absence of the bones was noted at Beethoven's second exhumation of 1888.[78]

A decade after this masterful act of sleuthing, Meredith issued a surprising corrective. Despite the confident authentication of the fragments by two Viennese doctors, five experts in forensic osteology determined that the fragments' position on the skull had been misidentified. Once correctly positioned, the pieces would need to have been sawed through by the time of the 1827 autopsy, which meant that none of them could possibly belong to Beethoven.[79] Several years later, a forensic doctor offered his corrective to the corrective, proposing an alternative topography and insisting on the pieces' authenticity.[80] It is likely, if still somewhat disputed, that these are relics of Jane Doe.

Thus pathography takes a new path in the twenty-first century: the revelation that the emperor has no clothes, or more specifically, the relic has no aura. In 2023, an international media frenzy was spurred by the sequencing of Beethoven's genome from his locks of hair, a death mask for the twenty-first century.[81] Thirty-three researchers, collaborating across disciplines, proved that Beethoven had a predisposition to liver failure, which absolves him of alcoholism, and uncovered a wealth of biographical details that ranged from lactose intolerance to an extramarital affair among his ancestors that disrupted the family line. A key revelation was the unmasking of a forgery. After one of Beethoven's locks had been sold at auction in 1994 to a urologist who sought Beethoven's cause of death; after the publication of a book that traced the journey of this lock from Beethoven's head to Sotheby's, which claimed in its cocksure subtitle "a scientific mystery solved"; after the book was made a documentary in which lab-coated scientists pry open the brooch for an audience of enraptured onlookers—after all this, the lock in question was not Beethoven's hair, but that of an Ashkenazi Jewish woman. (Recently, when Beethoven's locks were authenticated, testing showed that he was indeed poisoned with lead; the scientific mystery is now solved.)[82] If forgers act as tricksters, as Frederick Reece suggests, then the shock of their deception turns the mirror on our own values and expectations.[83] Here DNA exposed the curiosity cabinet for what it really was: a competition for remains in a market flooded with fakes, fueled by a desire not only to understand genius, but to possess and hoard it.

As I write these words, the quest takes its latest turn. In June 2023, the contested skull fragments were donated by Seligmann's heir Paul Kaufmann to the Josephinium, the museum affiliated with the Medical University of Vienna, which has long been a steward of relic-specimens. There the fragments can be tested against Beethoven's genome to identify them once and for all. In a telephone interview with CNN, Kaufmann confessed that "it is extremely emotional to me to return the fragments where they belong, back to where Beethoven is buried"; in the official proceedings, the Josephinium's rector affirmed that "our collections at the Josephinium are the right place for this," as the institution seeks "the right balance between comprehensible public interest and respect for a dead person."[84] But this news is not entirely new: the donation follows the path of a pendulum that has swung for some time, as relics shift hands from those private to those public, from cultural to medical, and back again. Over a century ago, similar words were spoken about Beethoven's hearing trumpets, his rudimentary aids designed by metronome inventor Johann Nepomuk Maelzel. Since the trumpets were considered artifacts of a history of medicine, they had been displayed in a glass case in the Josephinium. But in 1904, the Prussian Ministry of Spiritual, Educational, and Medical Affairs determined that the objects should be relics, not specimens, and insisted that they join Beethoven's assorted locks of hair, walking sticks, and coffin shards at the Society for Friends of Music. When the society communicated its gratitude, it boasted that it was "the designated site for the relics of great composers" and steward of "pious feelings."[85] If these were held to be instructive objects, their lessons now diverged: where they had once taught medical students about the history of the profession (and humanized their patients), the trumpets were now symbols of suffering and transcendence, relics of the saintly sage.

An Ambivalent Science

In European history, bodies have tended to peregrinate. Saints' bones have moved from church to church, urban expansion has crowded remains into ossuaries, and nationalist upheavals have compelled famous names to be either repatriated to or expelled from mausoleums.[86] In the late nineteenth century, a slew of composer exhumations became the dank underground of the *Denkmalwut* in its quest for ever-taller tombs. After Beethoven and Schubert in 1863, Robert Schumann was the next to be exhumed, in 1879, when his tomb was upgraded with a statue that immortalized his relationship with Clara. The anatomist Hermann Schaaffhausen, who spoke some words during the unveiling ceremony, had obtained the widow's permission to exhume and cast the composer's skull; while he was at it, he took a hair relic for his collection.[87] Beethoven and Schubert found themselves aboveground again in 1888, this time for more symbolic reasons. Along with a small army of famous citizens, they were relocated to the new Zentralfriedhof to occupy a composers' grove.[88] That collection of musical tombs symbolized the eternal conviviality of Walhalla (as in fig. 2.9) while serving, more practically, as a one-stop shop for musical pilgrims. At the graveside, members of the Society for Friends of Music marveled at how brusquely Vienna's Anthropological Society handled the skulls, with no regard for their sanctity. Other exhumations served to authenticate and repatriate. Rossini was returned to Florence from Paris's Père Lachaise in 1887, and in 1894, several contenders for Bach's bones were unearthed in Leipzig.[89] These exhumations served as a literal meeting place for the disciplines, where family members, music writers, art patrons, doctors, and anatomists all bent over the same specimens.

The timing of the exhumations was complex, as they coincided with an uncertain period in phrenology. Since the mid-nineteenth century, craniometry had waned in Germany, despite its continued popularity across the globe. Some anatomists such as Gustav Struve and Carl Gustav Carus continued to work on skulls, but they no longer adhered strictly to Gall's tenets.[90] This shift did not reflect a shaken faith in genius but a heated period in German science when the materialism inherited from Enlightenment could not account for the new frontier of the psyche.[91] In this unstable moment, it was unclear how composers' remains should be studied. Those who leapt at the chance were not the most prominent scientists of their day but relatively lesser-known anatomists whose ambivalence led to a dutiful sort of librarianship, an archival accounting of the body's dimensions.

The traditionalists at these exhumations retained their investment in Gall's theories, this time enriched by the "characterology" of Ludwig Klages, who sought intuitive readings of character through vestiges like handwriting.[92] Exhumations drew a variety of music writers, biographers, and personal acquaintances who sought holistic links between composers' skulls, personalities, and musical character.[93] At the first exhumation in 1863, Heinrich Kreissle von Hellborn, a Schubert biographer and bureaucrat, claimed that many were amazed by the "delicate, almost feminine organization" of Schubert's skull in contrast to Beethoven's.[94] Gerhard von Breuning recalled that "it was extremely interesting physiologically to compare the compact thickness of Beethoven's skull and the fine, almost feminine thinness of Schubert's, and to relate them, almost directly, to the character of their music."[95] Shortly thereafter, the Society for Friends of Music published a detailed report that included measurements of the composers' left temples and a table that compared the bones of Beethoven and Schubert side by side (fig. 3.2). Meanwhile, at the University of Vienna, another outlet for the holistic gaze was underway: in 1885, Guido Adler proposed the systematic discipline of musicology as a science of style criticism (*Stilkritik*), which combined the holistic approach to character with traces of cultural Darwinism that charted the "evolution" of musical styles grounded in racial groups and climates.[96] While Adler insisted he did not elevate famous names in a vacuum, and he defined style as widespread convention, his comparative approach to Schubert, Beethoven, Mozart, and Haydn nonetheless looks like a cabinet of specimens (fig. 3.3).[97]

Not all anatomists were equally convinced by this holistic approach. A plain look at the measurements showed Schubert's cranium to be quite hefty, unlike the dainty specimen Kreissle von Hellborn imagined, just as Bach's skull was later found to have a sloped forehead that departed from Lavater's highbrow ideal. Theodor von Frimmel, an art historian trained in medicine, blamed popular literature and inept portraiture for the myth that Beethoven's forehead was "high, mighty, or imposing."[98] Doubts deepened with the second exhumation of Beethoven, when inconsistent measurements from 1863 were updated, and the result was a compelling debate recorded in the minutes of the Anthropological Society in Vienna. Here three prominent anatomists—Carl Langer von Edenberg, Theodor Meynert, and Hermann Schaaffhausen—disagreed about the validity of phrenology as a science of meaningful comparison.[99] Their debate shows how composers' skulls transitioned from relics to archaeological specimens that shook the foundations of anthropometry.

Langer, a traditionalist, echoed both Gall and Adler's style criticism. He describes the skull of Haydn with admiration, noting its pleasing attributes, "well-proportioned" balance, and large left temple.[100] The mighty and massive skull of Beethoven deviates from the perfect specimen of his teacher. But the largest skull of all is Schubert's, and as if to explain this discrepancy with the composer's graceful melodies, Langer notes that the proportions were distorted by childhood illness. Even this adherent to Gall expressed his skepticism, noting that talented people do not always conform to measurable norms. Schaaffhausen, in contrast, dismissed phrenology in favor of a

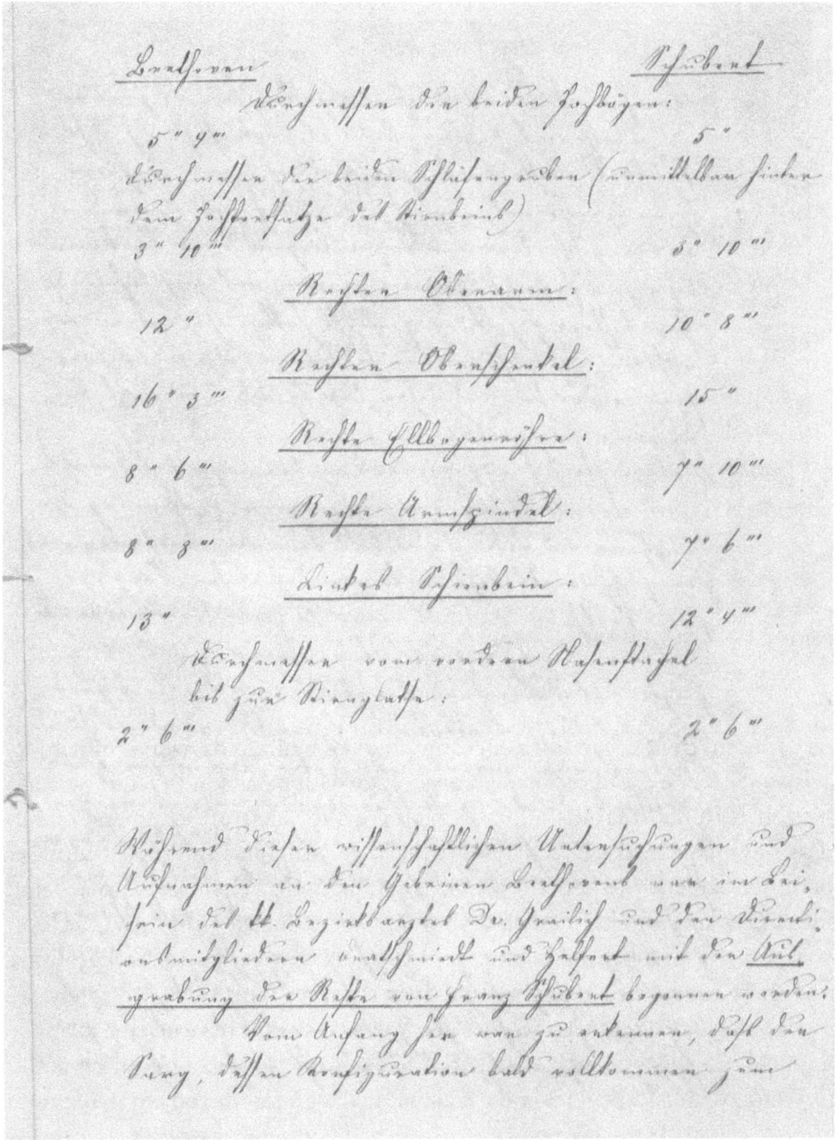

FIGURE 3.2. The comparative approach to composers. On the left, Beethoven's bone measurements (right upper arm, right thigh, right ulna, and so on). On the right, a direct comparison with those of Schubert. Protocols for the Exhumation of Beethoven and Schubert, 1863, p. 9. © Archive of the Society for Friends of Music in Vienna.

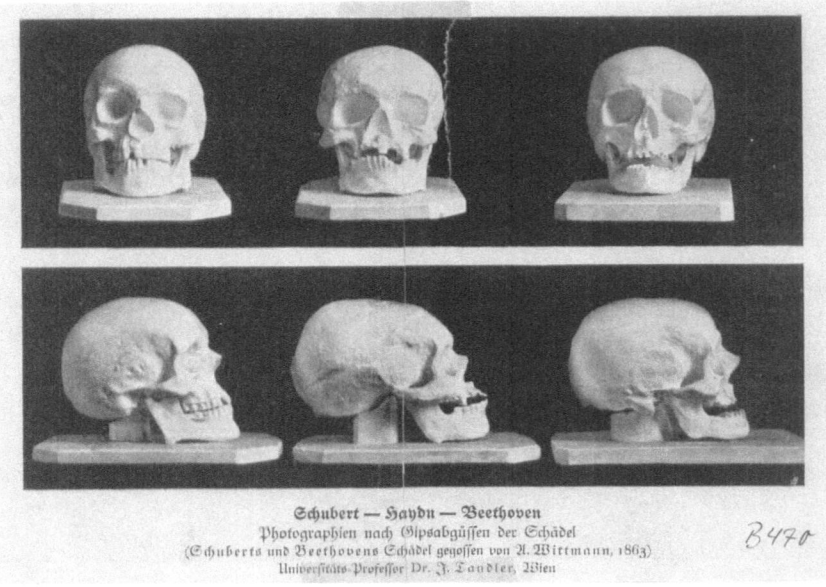

FIGURE 3.3. Photographic reproduction created for Julius Tandler that shows the casts of the skulls of Schubert, Haydn, and Beethoven; Munich, 1913. © Beethoven-Haus, Bonn, B470.

more recent trend in neuroscience: the study of brain folds, whose mutability across a lifetime could reconcile morphology with behavior.[101] While Schaaffhausen regretted that his effort to pinpoint the cause of Schumann's madness was inconclusive, he did observe dense cortical folds and robust auditory bones far superior to those found in an orangutan. When asked to account for discrepancies in his data on brain folds, Schaaffhausen's reply was a clever compromise: in Beethoven's time, geniuses hailed from modest backgrounds, and their convoluted brains reflect their strivings, but today, we rarely see morphological differences because educational reforms have leveled the playing field. All this was flatly disputed by Theodor Meynert, a physician active in Viennese musical circles and a collaborator with Hyrtl. Too often, Meynert argued, scientists mistake one end of the bell curve for the other, explaining pathological abnormalities as signs of genius, when in fact no spiritual activity leaves its mark in matter. He concluded with a sobering thought: what if skulls are just skulls?[102]

If Meynert had wanted to dismiss cranioscopy altogether, he would not have drawn up a detailed set of tables of Beethoven's skull after this second exhumation, as if to capture each contour in amber as Frimmel had done.[103] I suspect that this scrupulous accounting was a sort of Ehrenpflicht, a duty

FROM RELIC TO SPECIMEN

to preserve that we likewise find in Wilhelm His's study of Johann Sebastian Bach's skull in 1895.[104] His's project was fraught with inconsistencies from the start. The aim was to determine precisely how Bach looked, and just as Goethe had selected Schiller's skull from a mess of unremarkables, His was asked to identify Bach's remains from among a wide assortment exhumed from the site. Just one of the two bodies buried in oak caskets was intact, and His's conclusion that this *must* be Bach struck some at the time as a little too convenient.[105] In his published study of the skull, His followed Gall by emphasizing brain capacity, and by suggesting that future studies should compare the heads of Bach and Beethoven. David Yearsley has argued that his project had a more insidious, if unsurprising, agenda to paint Bach in the image of racial Teutonism.[106] That agenda is scarcely veiled in passages that measure allegedly "German" traits—a manly brow ridge, bold nose, and "immensely powerful expression"—and especially in a table that compares Bach's skull with measurements from neighboring German regions to create a topography of Teutonic superiority.[107]

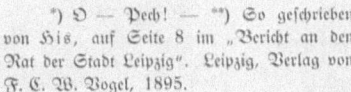

FIGURE 3.4. "The sculptor kisses the scholarly table with devotion." In D.V. (Ernst Klotz), *Das Fragwürdige Todtenbein von Leipzig: Satire auf die tieftraurige Historie vom Leben, Sterben, und Ausgraben der Gebeine J. S. Bach's* (Leipzig: Paul de Wit, 1906), 11. Leipzig University Library.

Yet a close look at this report shows a subtle tone of ambivalence, as if His were not fully convinced of the task at hand. First he admits that Bach's forehead was surprisingly low. Then he denigrates sentimental anatomists who venerate skulls as relics such that they hesitate to saw them to pieces. Ultimately, His admits that the current state of anatomical science can't support a conclusive identification of Bach's skull, nor its bearing on Bach's aptitudes. Instead, he hopes that scientists in the future will have more sophisticated insights into how the "shape of the head reflects the aptitude of people."[108] This explains why his two reports comprise largely numerical data and a detailed record of how the cast was created. His left it to others to make sense of that data.[109]

Amid this flurry of interest in composers' skulls, a satire by Ernst Klotz in 1906 hit the nail on the head. Not only was Klotz critical of the circular logic of Wilhelm His, who "discovered" traits in a specimen selected for those very traits, but he noted how anthropometrists and portraitists share an art-religious adoration of numerical tables. In a damning image, a sculptor falls to his knees before His, who is shown as a Christ figure complete with halo and hand raised in blessing, extending a scroll of measurements of Bach's skull (fig. 3.4).[110] This gets to the heart of the matter: anthropometry wore its scientism as a halo and replaced the allure of bodily relics with a fetishism of measurement.

Calcium and Keratin

Klotz's satire resonates beyond its time. It is easy to imagine substituting the latest methods, as a new generation takes a knee to kiss the sacred genome. With each technological development, we reevaluate relics in the hopes of ever more certainty, of deflating the speculative approaches that have dogged pathography and anthropometry. But what is the value of certainty when that which scientists seek is unquantifiable? These projects are fueled by a sense of wonder at genius, which makes Beethoven's shards into relics, while Jane Doe's are just calcium, and the hair of an Ashkenazi woman is the map of a life few care to know. There is a sense of embarrassment, too, in these bodies lain bare, mined for intimate knowledge and thereby naked. The anecdote-hungry publics of Lavater and Gall felt entitled to the novelty of secrets. We are likewise eager to put flesh back on those bones.

Across these pages, this portrait of the musical doctor as eclectic dilettante fails to account for those who have rebuilt medical-historical writing with scholarly integrity. Jonathan Noble issued a corrective to the inaccurate anecdotes on which composers' pathographies have been based, and Linda Hutcheon and Michael Hutcheon collaborated on a sensitive reading of composers' idiosyncratic responses to declining health.[111] Musicology and pathography need not chide each other for armchair medicine and armchair musicology with mutual mistrust; a fruitful exchange about the gaps between disciplines is precisely what creates disciplinarity to begin with.[112] At its core, pathography seeks to understand how musical sound is shaped by embodied experience, an inquiry it shares with disability studies, embodied cognition, and musicology's corporeal turn.

All the same, a longer history of pathography reveals a genre dogged by its unsavory, and largely unspoken, past. Discredited pseudosciences tend to persist in cultures of monumentality and Ehrenpflicht, as the faces of national heroes are molded to serve a political vision. When early twentieth-century anatomists made claims about skull capacity and character despite their own ambivalence about these methods, they destabilized the border between science and politics in ways that later appealed to German and Italian fascists. As Dante's bones were uncovered during the restoration of his tomb in 1865, a team of doctors analyzed his skull for traits of genius even while they dismissed Gall's phrenology; decades later, in fascist Italy under Mussolini, the project was revived in a new study that pronounced Dante's cranium masculine and weighty, an exemplar of the virile superiority of the Mediterranean race.[113] Those who continue to study artists' remains must reckon with this past, and contemplate the ethics of such concentrated scrutiny of the bodies of canonical European men.

If we line up pathographical diagnoses like specimens in jars, they evoke what Richard Dyer identifies as the grisly aestheticization of the dead white body. Across Western art and its cultural products—horror films, Holocaust imagery, vampires and zombies, Victorian death cults, and even Jesus on the Cross—"the dead white body has often been a sight of veneration, an object of beauty." As Dyer explains, that beauty exposes a peculiar anxiety about white absence and erasure that transforms the white body into death incarnate. If the study of genius has historically been a branch of white supremacist thought, then the impulse to preserve the white body may originate with that deep fear that Dyer calls "whiteness as non-existence."[114] At the turn of the twentieth century, no object was so emblematic of the beauty of white death as Beethoven's plaster mask, which occupies the next chapter. On the surface, this mask acted as a symbol of sacred suffering, but beneath its placid reception lay a deeper fear of anonymity, of erasure and the leveling of death that consumes all in equal measure.

It makes sense to conclude this chapter with the pale faces of the living dead. In 1863, Breuning beheld Beethoven's skull by his bedside with a holistic gaze, as a thing of beauty, its mouth alive with imaginary chatter. He anticipated the ventriloquism that made dead bodies political symbols in the twentieth century. Katherine Verdery notes that bodies "don't talk much on their own (though they did once). Words can be put into their mouths.... It is thus easier to rewrite history with dead people than with other kinds of symbols that are speechless."[115] The politics of reanimation have been a recent focal point in musicology: from the castrato's body as the ghost of an irrecoverable operatic sound, to the "intermundane" processes of "necro-sonic" collaboration in the twentieth-century recording studio, a form of dialogue with the dead that makes late capitalism very late indeed.[116] These Frankensteinian metaphors stress the agency of the dead, their surprisingly vital force in the politics of memory, nationhood, and labor. They also highlight how that vitality is tempered by an asymmetrical ethics, which Verdery articulates more simply as an obligation to give our ancestors a good death. Throughout this chapter, what have appeared as violent acts of disruption and extraction—exhuming, dissecting, mining for intimate secrets—could be charitably understood as a divergent view of what a good death should look like. With great fondness for their objects of study, doctors have tried to make the indignities of death worthwhile, and to give remnants back their voice. The risk is speaking through them.

4

Beethoven's Masks and the Beautiful Death

Many are familiar with Lionello Balestrieri's *Beethoven*, a touching scene of bohemian music making that was unveiled to great acclaim at the Paris World Exhibition in 1900 (fig. 4.1). The image is cluttered and introspective, a study in faces and backs concealing each other, absorbed in contemplation. The musicians are likewise concealed: the hunched form of a violinist catches the light while a pianist's hand darts along an upright piano laden with scores. This makes the wide eyes of the female figure even more arresting, but in a subtle play on convention, she looks just past us as if too immersed in music to break the fourth wall. The only character who faces us directly is the one that can't look at all: a pale cast of Beethoven's face, suspended next to the piano and cascading with dried flowers. Balestrieri's inclusion of this effigy, along with its foliage, is more realistic than symbolic. At the turn of the century, Beethoven's mask hung beside the piano in countless music rooms across Europe, mass-produced with a laurel wreath and a mounting wire (fig. 4.2).

The face on everyone's walls was technically not a death mask at all, but a life mask (fig. 4.3).[1] This was the cast made in 1812 by Franz Klein, the collaborator of Gall who appeared in the previous chapter, and it came in variations that could signal one's status: affordable white plaster, artful glazed ceramic, or more costly bronze. Death masks and effigies of all kinds were desirable in this period, which explains why the term "death mask" became the norm despite that neither portraitists nor consumers found anything pleasing about the actual death mask of Beethoven, whose face had been badly disfigured by autopsy; an advertisement from the Gebrüder Micheli company explained that it was "very deformed and only of scientific value" (fig. 4.4).[2]

Those sunken contours were captured in the days after Beethoven's death by Viennese painter and sculptor Josef Danhauser, along with his brother

FIGURE 4.1. Lionello Balestrieri, *Beethoven (Sonate à Kreutzer)* (1900). Photographic archive of the Revoltella Museum, Gallery of Modern Art, Trieste, Italy, Inv. 139.

Carl, who rushed to the scene to cut two locks of hair as "souvenirs of that venerable head," to sketch the composer's face and hands, and to buy memorabilia such as the razor used for his last shave.[3] Carl's account claims that the death mask was intended as a model for brother Josef's bust of Beethoven, which is a sensible reason to cast this visage. But when Josef Danhauser turned his attention to the project in later years, it was clearly the life mask that inspired his Napoleonic bust, which he advertised in his famed painting *Liszt Fantasizing at the Piano*.[4] The death mask itself was not so much a physiognomic specimen, but a relic and gift; Danhauser offered Liszt a plaster copy of the deformed visage to display in his home in Weimar.

Both of Beethoven's masks are complex objects with an unusual relationship to memory and material presence. Like the funereal photography that later came to replace the casting practice, these were snapshots of the "beautiful death," a nineteenth-century paradigm in which the body undergoes a heavenly transfiguration. They also served as celebrity icons, mass-produced relics, and physiognomies that seemed to gaze out of musical sound. While the Klein mask had mundane origins, it was nonetheless absorbed into the cult of death as a memento mori, a carefully staged symbol of death's inevitability and art's immortality. Caught between living and dead (as all masks are, to an extent), this cast appeared in popular literature as a talisman that resonated with sound or magic. The Micheli Brothers advertisement for the mask noted that the Seventh Symphony was performed at the very moment Beethoven sat in Klein's chair, as if to imply that the mask solidified its historical moment, not solely the contours of the face.[5] Beethoven's mask inverted the traditional object of memory, the relic that connects loved ones with the

FIGURE 4.2. C. V. Muttich, *Beethovens Sonate*. Postcard, ca. 1900. © Archive of the Society for Friends of Music in Vienna.

FIGURE 4.3. (*Left*) Beethoven's life mask, cast by Franz Klein in 1812, photograph by Carl Simon. (*Right*) The death mask cast by Carl and Josef Danhauser the morning after Beethoven's death in 1827, photograph by Dietrich Andrear. © Beethoven-Haus Bonn, B 438 and B 2539.

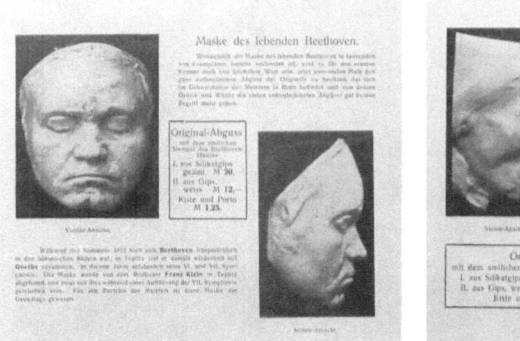 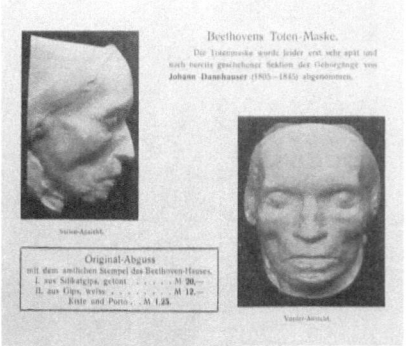

FIGURE 4.4. Pages from a Gebrüder Micheli product catalog advertising the sale of Beethoven's masks, ca. 1900. © Beethoven-Haus, Bonn.

dead. On the one hand, when mounted to the wall, it invited its owner to remember Beethoven as they would a deceased relative, to conjure up fantastical affections. Yet in this casual statement by Micheli Brothers, Beethoven's mask is also an object that remembers. In short stories and essays, it was frequently endowed with a ghostly agency that inserted Beethoven back into the world of the living.

A mask that haunts bourgeois walls, that invites half-dreamed memories and strange stirrings—this will sound familiar to those who know its cousin, the *Inconnue de la Seine*, a colloquial name for the mask of an anonymous drowned woman of Paris. The mystery of the Inconnue, whose philosophical and literary depths have been plumbed by Anne-Gaëlle Saliot, will resurface later in this chapter.[6] The Inconnue established what Beethoven's literary hauntings should look like, and his mask shadows hers on a circuitous route through belles lettres and visual culture. Literary historians Nathan Waddell and Angelika Corbineau-Hoffmann examine how Beethoven's face lurked in English and German literature, cultural historian Richard Leppert suggests that Beethoven's contrasting masks mapped onto his divergent personae in later reception, and Alessandra Comini has traced the iconic scowl through the vagaries of portraiture and the visual arts.[7] These studies offer enticing readings of select literary and visual works, but they do not thoroughly situate the mask in a broader cult of death, nor in a network of ephemeral artifacts such as postcards, ex libris plates, and the copious literary fiction and poetry that populated the papers. Without this wide lens, Beethoven's mask functions as a site of historiography, a linear chronicle of escalating "mythos" or "mythmaking," as Comini calls it. Yet for every artifact that made Beethoven transparently heroic, masculine, superhuman, proudly German, or whatever his mythic traits were held to be, there were countless others that muddied the waters. His image could be erotic or saintly, Christ-like in its suffering or pagan in its dark magic, celestially transfigured, comically Chaplinesque, or a dead thing stripped of its powers altogether. A granular account of Beethoven's masks reveals the strange underbelly of the "Beethoven syndrome" identified by Mark Evan Bonds, a mode of listening in which critics mapped composers' biographies onto their artworks. Bonds points out that listening could be markedly visual, informed by a visage that looms and scowls.[8] A closer look at visual culture of the time, beyond a few monumental examples, shows how the meanings attached to Beethoven's face were far more diffuse than the transcendent sublimity implied by the word "mythos."

When material from this chapter appeared in the journal *Nineteenth-Century Music* in 2020, Beethoven's masks were still directed inward toward his historiography, a barometer for the internal contradictions in the discourse on "late style"—that is, the premise that late and last artworks have special properties felt across the arts, and across artists, regardless of their idiosyncrasies.[9] Since then, the study of lateness has itself become somewhat late. Its interlocutors have dwindled, and its central preoccupations find new life in studies of disability and aging. The relative decline of this discourse in musicology follows an arc set in motion by Gordon McMullan, who redefined late style as a cultural construct rather than an aesthetic universal. Atomizing the topic made it all the more fertile, as lateness could resonate with other autumnal processes: the sunset of Viennese liberalism, the awareness of aging, the incomprehensibility of irony, the flicker of meaning against the grain.[10] The question that remains is how to approach a composer's last years when the foundations of that discourse have been enmeshed in the nineteenth-century cult of death and its attendant fantasies. The aim of this chapter, in marked divergence from my earlier article, is neither to debunk nor to salvage lateness, but to understand how Beethoven's deathbed lore and last visage circulated among shifting paradigms of death. That broader scope reveals lateness to be just one of myriad responses to the problem of posterity in art.

Even as I occasionally turn the mask to face my own discipline by reflecting on its role in Beethoven's historiography, I ultimately maintain that this misses the point. The meanings that accrued on this object were not always, or even frequently, about Beethoven. When the mask first rose to popularity, it evoked the Romantic concept of the "beautiful death," which made it a metonym for the immortality of Art with a capital A. When mounted to the wall, the mask substituted for the domestic crucifix above the bed, an index of secularity. Its popularity persisted through a radical shift in print culture, from death masks to photography, which made it an anachronism that haunted literature. Meanwhile, the mask was implicated in a growing distaste for the (petit) bourgeoisie, which undermined its prestige as metonym for art and rendered it a kitsch object, a desecrated relic. Those politics caught the attention of the Berlin Dada, whose fascination with masks of all kinds was both playful and provocatively antiauthoritarian. By the interwar period, in the wake of so much death, the mask looked grim. Positioned next to the Inconnue on plaster shop walls, Beethoven's face invited ruminations on the anonymity of death, which undermined its status as icon just when the composer's face appeared in somber superfluity for the one-hundredth anniversary of his death. Ultimately, his mask was absorbed into a period of reckoning when death itself, not Beethoven and not Art, became the only real universal.

At the Deathbed

Weather records for Vienna on March 26, 1827, confirm what has now become the stuff of legend: when Beethoven died, thundersnow raged outside his window. In the weeks that followed, poetic tributes marked this as a divine event; later it became a battle against the elements and fate.[11] Beethoven's turbulent death mirrored the supernatural events of antiquity and Christian scripture.

Christ's crucifixion was accompanied by darkness and an earthquake, the prophet Elijah was drawn into heaven by a windstorm, Romulus's apotheosis was attended by wind and lightning, and saints could miraculously command the elements.[12] Beethoven was one of many artists whose deathbeds were retold in ever more mystical ways. In a four-volume biographical novel, the ailing composer finds miraculous inspiration on his stormy deathbed, the "last wingbeat" of the eagle of imagination, while the anthropomorphic weather cries out about his Viennese neglect.[13] In a 1933 story of love between Beethoven and a fictional landlord's daughter, Beethoven calls to God in a thunderstorm, and music flows through him as his love interest eavesdrops in the next room; later, in prayer, she calls him "you little saint."[14] In a Mozart novel, the dying composer exclaims that God was the inspiration for all his works, sacred or secular, and "the earth is my church."[15] Schubert-themed novels revolve primarily around a Bohemian lifestyle, but they never fail to link melancholy with creativity; one compared his suffering to that of Job.[16] It was common for colorful works of biofiction to mask the pain of death with apotheosis in a new martyrology.

The appeal of these stories stems from the same mystery that enchants death masks. These objects are haunting because of what Marcia Pointon, citing Hanneke Grootenboer, calls the withdrawal of the gaze, and even life masks can emulate the moment when the spark of life vanishes from the eye.[17] While deathbed stories share a desire to arrest the moment of withdrawal, they can vary widely in ways that reflect divergent paradigms of death. In a landmark study of 1977, Philippe Ariès chronicled several stages in the Western conception of death through a tapestry of sources that range from bureaucratic records to catacombs to hair jewelry. He argues that early medieval Europeans were ready for death, and approached it with more melancholy than fear. In the late Middle Ages, the deathbed was a site of drama where the dying wrestled with demons and called saints to their aid. The Renaissance memento mori reminded one daily that death awaits, which encouraged living well. In the eighteenth century, deeper attachments to individuals encouraged both elaborate practices of memory and a new terror of death. By the nineteenth century, a diminished belief in hell led death to "[hide] under the mask of beauty" and encouraged the bereaved to find solace in the immortality of love and friendship.[18] Although Ariès presents these paradigms in a historical chronology, the deathbed lore of luminaries combined various approaches to death. Nineteenth-century accounts ranged from dramas akin to the late medieval battle with demons, to the pious submission in the Christian tradition of *ars moriendi* (the art of dying well, confessing sins, and making peace), to the transfiguration that merges a saint's death with the "beautiful death" more contemporary to the period.[19] Even eyewitness accounts of artists' final hours constitute a literary subgenre of sorts, as witnesses interpreted the event through dramatic paradigms that made for a great story.

Visitors to Beethoven's deathbed, for instance, agreed that the storm introduced an element of the miraculous but differed as to whether he fought demons or succumbed in sacred suffering. His final month was one of painful decline, and the artist Josef Teltscher visited several times to sketch what he thought might be the deathbed hour. The result are three pencil sketches that show an unconscious composer nestled deep in his bed, with a surprisingly youthful face that looks at peace.[20] This dignified death aligns with an account by the composer's physician, Andreas Wawruch, who recalled a frail Beethoven in a state of transcendence: Beethoven "performed his devotions with meek submission.... The 26th of March was stormy and dull; towards six in the afternoon a snowstorm began, accompanied by thunder and lightning.—Beethoven died—What would a Roman augur have concluded about his apotheosis from the fortuitous unrest of the elements?"[21] Decades after Beethoven's death, in the wake of the Austro-Prussian War, meek submission was supplanted by a more martial version of events. In his belated eyewitness account, Anselm Hüttenbrenner updated the late medieval paradigm of a heavenly battle with Napoleonic flair:

> There was suddenly a loud clap of thunder accompanied by a bolt of lightning which illuminated the death chamber with a harsh light.... After this unexpected natural phenomenon, which had shaken me greatly, Beethoven opened his eyes, raised his right hand and, his fist clenched, looked upwards for several seconds with a grave, threatening countenance, as though to say, "I defy you, powers of evil! Away! God is with me." It also seemed as though he were calling like a valiant commander [*Feldherr*] to his faint-hearted troops: "Courage, men! Forward! Trust in me! The victory is ours!"
>
> As he let his hand sink down onto the bed again, his eyes closed half-way. My right hand lay under his head, my left hand rested on his breast. There was no more breathing, no more heartbeat! The great composer's spirit fled from this world of deception into the kingdom of truth. I shut the half-open eyes of the deceased, kissed them, and then his forehead, mouth and hands. At my request Frau van Beethoven cut a lock of hair and gave it to me as a sacred relic of Beethoven's last hour.[22]

More striking even than the heroism in Hüttenbrenner's account, which has by now become a truism, is the odd turn from transcendence to materiality when he kisses the body and receives his sacred relic. There are layers of meaning here. The first is of course Christian: Hüttenbrenner positions himself as a saint's disciple and tends to the dead just as Veronica wiped the face of Jesus with her cloth. That act of care, too, culminated in a relic that imprinted the withdrawal of the gaze (the resonance between Veronica's cloth and Beethoven's mask is addressed in due course). The second layer is considerably more secular and hearkens again to Ariès's insights about the "beautiful death." In a provocative passage, Ariès explains that nineteenth-century deaths were understood as a new paradise, "not so much the heavenly home as the earthly home saved from the menace of time"—that is, a kind of secular afterlife imbued with private memory, where friendships continue eternally and can be revived in the here and now through objects, images, letters, and other sites of recollection.[23] This explains why death, and the dead body, are beautiful rather than disturbing—why Hüttenbrenner, in other words, would kiss the face and hands of the deceased, and why funeral photography arose decades later to preserve the body at peace, laden with flowers. Attending the nineteenth-century deathbed, Ariès notes, was "an opportunity to witness a spectacle that is both comforting and exalting. A visit to the house in which someone has died is a little like a visit to a museum. How beautiful he is! . . . Death is concealing itself under a mask of beauty."[24]

This passage explains multitudes. Today it may seem odd how many visitors beheld Beethoven's body in the days after he died, at once paying respects and begging the caretaker for a lock of hair. Ariès helps us understand why this was unsettling for contemporaries at the time, as strangers who never knew Beethoven adopted the intimate visit usually reserved for family and friends. The memoir of Schubert's friend Franz Hartmann betrays a sense of self-importance when he derided other relic seekers despite that he himself had never met the composer. His account of a visit to Beethoven's body is worth quoting at length:

> Thus I saw his magnificent face, which unfortunately I had never had the chance to see in life. There was such a celestial dignity about him, despite the transfiguration he is said to have suffered, that I could not look at him long enough. I went away deeply moved. Once below, I could have wept that I did not ask the old man to cut a few hairs for me. Ferdinand Sauter, whom I had planned to meet there but had missed, came at that moment and I returned to the room with him, telling him of my plan. The old man showed him to us again, uncovering the breast too, which was already completely blue, as was the badly swollen stomach. There was already a very strong cadaverous smell. We gave the old man a tip and begged him for a few of Beethoven's hairs. He shook his head and motioned us to be silent. We were going slowly and sadly

down the steps when the old man called softly from the balustrade above that we should wait at the gate until three fops had left, who had stood tapping their swagger-sticks on their pantaloons while looking at the dead man. Then we went up once more; the old man came out of the door, his finger on his lips, and handed us the hair in a bit of paper. We left with a feeling of mournful joy.[25]

Hartmann's account affirms Ariès's observation that the deathbed was a little like a museum. His derision for the three onlookers resembles the competitive piety of birth-house visitors in chapter 1; unlike these superficial pilgrims to Beethoven, Hartmann sees his own piety as real, and he departs "deeply moved" with "mournful joy." But there also a surprising, and rather honest, shift in this passage from celestial dignity to a disfigured body and odor. Beautiful death was an affectation at best, at least until late nineteenth-century advances in the chemistry of embalming.

It made more sense to see the deathbed as beautiful when a composer's illness was a source of genius, as chronicled in the previous chapter. Here a crucial example is not Beethoven but Chopin, whose tubercular appearance during his life, in a period when fevered flush was fashionable, led him to be regarded as a pale saint in death. In a colorful biography, the abbot who administered Chopin's last rites was said to proclaim that "rays of divine light, flames of divine fire, streamed, I might say, visibly from the figure of the crucified Saviour, and at once illumined the soul and kindled the heart of Chopin. . . . His faith was once more revived, and with unspeakable fervor he made his confession and received the Holy Supper. . . . From this hour he was a saint."[26] It would be a shame to witness such an event without keeping a relic, so the abbot was said to have pocketed the water glass Chopin used in his last rites, which retained a ghostly impression of his lips. Well before this hyperbolic biography, we find touches of a saint's death in Franz Liszt's *Life of Chopin*: his description of the weeping disciples and the angelic robes of Countess Delphine Potocka make this the likely source for the most widely disseminated deathbed scene of the era, a print by Felix Joseph Barrias that was featured as the frontispiece of several lives-of-the-composers collections.[27] But the saint's death quickly cedes to the museum of beauty: as Liszt tells it, when Chopin's body was fringed with flowers, his face was transformed to youthful loveliness, as in a dreamless sleep in a garden of roses. Even the death mask, hand cast, and sketch by Auguste Clésinger "reproduced the delicate traits to which death had rendered their early beauty."[28]

For those who have confronted death masks and locks of hair in museums, this blissful image can seem discordant. Chopin's face is far from serene in the pained photograph taken in his death year, with its sunken gaze; the casts of his face and hands are thrilling yet haunting; the preservation of his heart in a tomb in Warsaw is sacred to some but unnerving to many. There is a fundamental rift here that makes the nineteenth-century cult of death seem foreign. In her rich study of Victorian death relics in literature, Deborah Lutz explains how bodily mementos (hair and tooth jewelry, bits of bone, masks, and much else) participated in a broader set of technologies for remembering people.[29] One could remember with words (handwriting), sight (portraits), and of course touch (relics)—and remembering, or literally re-membering a disassembled body and its lost world, was understood as an act of friendship. This is why casts of body parts were sentimental synecdoches instead of severed limbs, not just for the Victorians but for Victoria herself, who kept copies of the arms of her infant children on plush pillows, anticipating the familiar practice of the bronzed baby shoe.[30] It was this warmth of feeling that compelled the sculptor Harriet Hosmer to capture love in bronze. When she met the poets Robert and Elizabeth Barrett Browning in 1853, she cast their clasped hands, a simple yet potent gesture that elicits a phantom sense of touch.[31] So too does a plaster cast of Chopin's hand. The object invites its owner not just to look but to hold, to press fingertips (veiled as pianistic comparison, perhaps, if the intimacy of the gesture feels embarrassing). Objects that speak so loudly of the absent are literary by nature, Lutz argues, which explains why they pervade fiction and verse.

Masks were just one of a constellation of objects that capture shifting relationships with death, fragility, and the afterlife. It is striking how much variety one finds in Elizabeth Hallam and Jenny Hockey's study of death's material culture: marble effigies mask the effects of time while relics shrivel beneath them; the Renaissance memento mori (some tender like fading flowers, others brazen like worm-ridden skulls) prepare us for a moral death; hair and tooth jewelry are not only relics but micromuseums, weaving an entire family line into a keepsake; funereal scenes made of hair revive the Renaissance memento mori in tombstones and weeping willows.[32] When these objects fascinate, their appeal lies not only in the drama of presence and absence, but in the arguments they make about death that challenge the sensibilities of our time.

Last Thoughts

Deathbed accounts are of further interest to musicology because they encouraged late and last works to become deathbed music, and a reading of the work as an imprint of an event. Last works could be heard either as riveting dramas that enacted the final hours, following the medieval model of death, or as relics of celestial beauty like a lock of hair or a death mask. Artifacts of death, like Beethoven's mask, circulated in an economy that tried to market the mystery of last works as a consumer product.

In some ways this market was created by Mozart's *Requiem*, which was the first deathbed story to be popularized across national borders and media. It is a testament to the enduring influence of this story that it remains well known through Peter Shaffer's 1979 play *Amadeus* and the film that followed, which were rooted not only in the 1830 play by Alexander Pushkin, but also in a tapestry of conspiracy theories, crime fictions, poems, plays, and operettas that span from the 1820s to the early twentieth century.[33] Forgotten iterations of the same falsehoods, like the masked messenger and Salieri's vengeful plot, continue to surface in archives. Even Franz Liszt's personal library of manuscripts contains a handwritten play by "Wermonty"—likely the Polish author Alfons Wermonty—which wove Mozart's compositions into a biofictional rhapsody on the composer's divine genius sacrificed at the hands of human vice. (In a rare twist, the murderous Salieri receives his punishment through the suicide of his daughter, an avid Mozart admirer, by the same poison; perhaps Wermonty was thinking of Victor Hugo's *Le roi s'amuse*.) Mozart's last monologue captures the self-awareness of death that was a convention of the genre:

> Oh, this Requiem, you my dears,
> Should be the final stone in my earthly walls!
> That was the *messenger of death*! Do you see him?
> This Requiem should be my grave song [*Grabgesang*],
> And when I die, it should sound
> At my coffin. It is truly very dear to me.[34]

Setting aside the foreknowledge of death, which has become somewhat of a truism, there is an interesting ambiguity in this passage. On the one hand, this fictional Mozart seems to compose as a form of solace, grieving himself in advance. Yet his tone conveys urgency; he is not going gentle. That ambiguity aligns with a tension in the visual and critical reception of the *Requiem*. Paintings of Mozart's deathbed—there were at least five renditions in the nineteenth century alone—depicted an ever more brooding and tortured soul, which invited listeners to hear a dialogue between Gothic horror and acceptance, as if reviving the medieval battle between angels and demons.[35] Those who admired the large deathbed scene by Mihály Munkácsy, unveiled at a Paris gallery in 1886, might well have experienced horror when enjoying what could be called the first biopic before the invention of cinema: a small ensemble was set up behind the canvas to perform passages from the

Requiem for audiences seated in rows.³⁶ It is thanks to Mozart's *Requiem* that other composers' deathbeds and last works, notably those of Beethoven, were heard to contain both rage and acceptance.³⁷

Mozart's death legends had a more material dimension: the *Requiem* was thought to imprint the dying moment like a cast, which is why the early Romantics referred to it as transfigured (*verklärt*), a word commonly used to describe the heavenly beauty of the face of the deceased.³⁸ The *Requiem* helped to establish a market for enigmatic and poetic fragments: large-scale unfinished works earned the statuesque label of a "torso" or the prestige of a "swan song," while small-scale fragments were more likely to be called "relics," such as Schubert's last sonata (the so-called *Reliquie*), which trails off mid-phrase.³⁹ These were the labels of publishers, of course, who tried to market souvenirs of the beautiful death. Jeffrey Kallberg has noted how dozens of piano miniatures titled "Last Thoughts," "Relic," or "Last Moments" encouraged an undue fixation on lastness in Chopin's oeuvre.⁴⁰ What is strangest about these miniatures is their lighthearted tone, quite at odds with the drama of the *Requiem* or the graceful architecture of Schubert's last symphony. A jaunty waltz by Carl Reißiger, which was marketed as Carl Maria von Weber's "last musical thought," was reprinted, arranged, and lathered into variations for every imaginable instrument from 1835 until 1900. The success of that piece prompted Anton Diabelli to present his piano arrangement of a string quartet sketch by Beethoven as his "last musical thoughts," and Adolf Martin Schlesinger offered his own set of last thoughts by Bellini, Donizetti, Chopin, and Beethoven (this time represented by his Bagatelle in B-flat Major, WoO 60).⁴¹ These souvenirs of domestic sheet music occupied the same print market as curated collections of famous last words.⁴²

When we revisit Beethoven's deathbed as it was refracted through memoirs, biographies, and reactions to his last works, we find a constellation of approaches that span the sentimental keepsake and the final drama. His deathbed was a meeting place for two personae that surface repeatedly in his critical reception: Napoleon and Christ. The heroic persona, apparent in Hüttenbrenner's martial flourishes, was held to triumph over suffering, while the Christ persona, apparent in the accounts of both Hüttenbrenner and Wawruch, met death with grace. Two soteriological words appear frequently in writings about Beethoven's final days: *Versöhnung* (reconciliation) and *Erlöser* (redeemer), which allude to the belief that Christ healed the rift between God and humankind and redeemed the world through suffering.⁴³ Nor were the personae always opposed: in his 1897 study of genius, Hermann Türck linked the figures of Napoleon and Christ through their shared "striving after the highest eternal state of being," or what the Romantic generation frequently called the genius's "redemptive power" (*Erlösungskraft*), and during the First World War the idea resurfaced in a nationalist biography called *Beethoven the German*: "Beethoven's development went from world dominator to world transcender, from Napoleon to Christ, and toward every unity of antiquity and Christianity that the Tenth Symphony would have brought."[44] We find this approach in J. W. N. Sullivan's influential psychobiography of Beethoven's spirit from 1927, the anniversary of his death year. Sullivan held that Beethoven's music navigates two essential drives, the "capacity for suffering" and the "power of self-assertion" that enabled him to overcome it, and the last compositions were an arena where his personae found reconciliation.[45]

An extensive reception history of Beethoven's late and last works lies outside the scope of this chapter; suffice it to say that the material culture of memory, the visitation practices of the deathbed, the drama of final hours, and the nineteenth-century beautiful death have simmered at the margins of the scholarly discourse on lateness, and that the boundary between popular culture and music criticism is more porous than often acknowledged.[46] It is largely due to Beethoven's puzzling oeuvre, combined with his dual faces of death, that the very idea of lateness captivated doctors and humanists for decades. The motivations of the doctors, or some of them at least, are illuminated in the previous chapter; the humanists can be harder to pin down. For those tantalized by the promise of rich hermeneutics, late works present an exciting enigma. Produced at the end of a career, this music can be more experimental than juvenilia, or curiously naive in ways that contradict that expectation. A newcomer to the vast array of articles and books on late style, some of which date to the nineteenth century, might be puzzled that the presumed characteristics of lateness driving this discourse have changed little. The ossification of this discourse may be a token of deference to Edward Said, who quite literally had the last word: his influential swan song on lateness, left unfinished at his death, shares the spotlight with two essays by Theodor W. Adorno, and together these authors codified an assertion that late works are "catastrophes" that run "against the grain" of their era.[47] The resulting conversation has tended to pivot around a surprisingly stable set of dichotomies: complex versus naive, innovative versus conventional, personal versus impersonal, individual versus epochal, serene versus irascible.[48] Lurking amid those dualisms is of course Beethoven, who was defined as the composer of extremes already in his lifetime (or as Franz Grillparzer put it in his funeral oration, "from the cooing of the dove to the thunder's roll") and whose late bagatelles and *Missa Solemnis* perplexed Adorno enough to spur an intergenerational debate.[49]

This is one reason Beethoven's face matters. In the same period that Beethoven's life mask rose to popularity, conversations about his late works had shifted from a broad range of critical discussions to a sentimental sweet tooth for "last thoughts." Adorno found this irritating, which explains why his essay on late style from 1937 returns a dimension of decay and distortion to a conversation dominated by platitudes of transcendence. His distaste for the bourgeois cult of "touching relics," as he called them, signals not only the musical trinkets marketed by publishers, but also the deterioration of critical writings about Beethoven's last works as mirrors of the beautiful death.[50]

One artifact, and its author more generally, typifies the sort of thing Adorno might have found repellant. It is the so-called *Grabgesang* (literally "grave song") that concluded a book by Ludwig Nohl, whose popular writings offer a much richer picture of nineteenth-century taste than the more insulated debates of his contemporaries. If sacralizing Beethoven was a key project of the New German School and especially Hans von Bülow, as Karen Leistra-Jones has argued, then Ludwig Nohl was a central (if today unacknowledged) organ of that mission.[51] Nohl was quite simply the most prolific and widely read public intellectual whose works lauded Wagner and Liszt as heirs of Beethoven. With the exception of his biography of Beethoven, most of his books have fallen into obscurity, which some of his contemporaries would have applauded given his reputation for grandstanding.[52] Most significantly for this study, Nohl regularly Christianized Beethoven's suffering: he amplified Wagner's notion that deafness made Beethoven an anchorite; he called Beethoven the *Musikgottessohn*, the musical son of God; and his biographical writings made Beethoven's deathbed a site of Christ-like transfiguration.[53]

Only someone fixated on last works could turn Beethoven's bachelorhood into a mark of martyrdom. To conclude his 1875 monograph on Beethoven's love life, Nohl wrote that the dying Beethoven sublimated his unrequited love for women into a love for humankind—a love that was likewise unrequited, Nohl claimed, because his contemporaries so cruelly misunderstood his late works. To illustrate this unusual argument, Nohl affixed his own art-song arrangement of the "Lento assai" from Beethoven's last String Quartet, op. 135.[54] Nohl's text, which opens with the incipit of a well-known folksong, reenacts Beethoven's last thoughts on the deathbed, as a second Christ:

> I am tired, I soon shall rest,
> I have lived and suffered enough.
> Wound in my heart, wound in the aches
> That life has given to me.
>
> Youthful vigor, great and bold
> Mature ambition, serious and true:

I sought to embrace the entire world,
To lovingly bring it contentment.

But who can grasp life?
Stormily it drove me from the tracks
And nothing was left of all my loves
Except my song.

For even in my aching death,
I will sing the song of love.
For love is life,
Love, the only human joy.

Remedy to all earthly affliction,
All-embracing, merciful to all!
Love is truth, copious light,
Blessed life, eternal being!⁵⁵

(*Müde bin ich, geh' zur Ruh, / Hab' genug gelebt, gelitten. / Wund im Herzen, wund in Schmerzen, / Die das Leben mir gebracht. // Hoch und kühn der Jugendmuth, / Ernst und treu des Mannes Streben: / Wollt' die ganze Welt umfassen, / Liebend bringen ihr das Glück. // Doch wer mag dem Leben nah'n? / Stürmend war's mich aus dem Gleise. / Und von alle meinem Lieben / Blieb mir nichts als mein Gesang. // Doch ob auch zum Tode wund, / Will der Liebe Lied ich singen. / Denn die Liebe ist das Leben, / Liebe, einzig Menschenglück. // Lösung aller Erdennoth, / Allumschlingen, Allerbarmen! / Lieb' ist Wahrheit, Lichtes Fülle, / Sel'ges Leben, ew'ges Sein!*)

Naturally, Nohl's arrangement serves not only as a poignant conclusion, but as an advertisement for a *Hausmusik* side-hustle: at the close of these tender words, he mentions that his song is available as sheet music from Breitkopf and Härtel.⁵⁶ Nohl stages Beethoven's death as a kind of Passion play, as Beethoven's last words from the cross, and ultimately he produces a touching relic. Beethoven's mask circulated in this established market of deathbed souvenirs.

The Cult of the Face

If one flips through recent guestbooks of the Beethoven-Haus Museum in Bonn, one finds rough sketches of the composer's portrait on nearly every page. This visual abundance is distinctive to Beethoven, and it hails from a moment in the 1890s when representations of the Klein mask exploded across visual culture, surfacing on ex libris plates, postcards, allegorical art nouveau prints, and home decor. This mask was more than an "icon" in the simple sense of a ubiquitous image. As an imprint of the face, it assumed the prestige of a *vera icon*, a "true image" that emulated the face relics of Christ and solidified the idea that Beethoven's suffering was sacred.

In a Christian context, the vera icon refers to various images of the Holy Face of Jesus that walk a fine line between icon, imprint, and relic. These miraculous images on cloth—known as *acheiropoeitos*, "not made by hand"—first appeared in early Christian Byzantium in a period when the Old Testament ban on graven images rendered stone a taboo medium. There the Mandylion of Edessa, among the earliest of such cloth relics, was the miracle that persuaded a king's conversion; by the Middle Ages, similar icons were associated with St. Veronica, who gently wiped the face of Christ on his way to the Crucifixion and retained a holy imprint on her veil. In Western Europe, cloth relics called "Veronicas" were considered the domain of female mystics and nuns, which led cloisters across Europe to commission altarpieces that captured women's encounters with Christ's face and, more interestingly, to manufacture Veronicas for private devotion.[57] Even a relic "not made by hand" could retain some of its power when copied, which meant that a single page could bear four or eight faces of Christ; likewise, copies of the Mandylion in thirteenth-century Byzantium were thought to retain their power when the original relic was used as a signet.[58] In her meditation on the drives that underlie representations of the disembodied head, Julia Kristeva argues that a vera icon like the Mandylion "is not viewed, it is absorbed, it is experienced: it translates an invisible world into its visible lines."[59] Face relics, like other severed heads, represent the invisible (death, God, the transcendence of art) through the loss of the visible, the skull or the face floating alone in the void.

At the turn of the twentieth century, Beethoven's mask and Christ's vera icon were cut from the same cloth. There are historical grounds for that connection: Beethoven's mask rose to popularity during a period of renewed attention to Christ's face, which began with the exhibition of the Shroud of Turin in 1862 and its more cosmopolitan dissemination as a haunting photo-negative in 1898.[60] Casts of the Klein mask, unlike other masks of the period, were typically adorned with a laurel wreath, a visual pun that hinted at the poetic pairing of laurels with thorns that we saw in chapter 1. For many, Beethoven's mask and Christ's cloths were understood as the "true" face in a sea of imperfect images because their life-sized dimensions made them powerfully real; in the late Middle Ages, for instance, the actual measure of Christ's body could be a relic in its own right.[61] Unlike other kinds of relics, even a simulacrum of the vera icon retains its Benjaminian aura, which makes it possible to mass-produce a material presence. Casts of Beethoven's face participated in a cult of pietà that could be coded female, like Veronica's veil; later in this chapter, we will see how fictional admirers of the mask were primarily women, which went beyond a simple correlation of parlor decoration with femininity.

Granted, Beethoven's mask was not the only cast of its kind, nor did it spring from nowhere like a true acheiropoeitos. Portraitists implied that the life mask was a vera icon when they made his firm-pressed lips an iconic gesture of resolve, as Comini has shown; already in 1922, a skeptic admonished readers for forgetting that the mask was made by human hands and explained that Beethoven's expression arose from the discomfort of the procedure.[62] In the nineteenth century, the profession of the *formatore*, the maker of molds, gained prestige when museums and design schools sought replicas of classical statuary and ancient artifacts. Their activities help to explain the variety of decorative objects one finds alongside Beethoven's mask in the catalogs of plaster companies.[63] In London, the Atelier Brucciani specialized in the Greek and Roman classics and cast death masks only when commissioned, but the Parisian Atelier Lorenzi supplemented their catalog of antiquities with the masks of luminaries beloved in France, such as Verlaine, Géricault, Beethoven, Chopin, Liszt, and Wagner. In Berlin, Micheli Brothers offered a Germanocentric catalog that reflected the aims of Bildung. Their busts of pedagogical reformers graced school foyers, while home libraries might boast an Enlightenment thinker in miniature or a shelf-sized Greek statue. This company offered the widest assortment of masks, the most popular being those of Beethoven, Goethe, Wagner, Liszt, and Shakespeare.[64]

One finds a different landscape in the inventory of Goldscheider in Vienna, a ceramics manufacturer whose glazed terra-cottas were more artful than white plaster. Here the face of Beethoven treads the line between the fine arts and decoration that characterized the *Jugendstil*—that is, the German and Austrian offshoot of the art nouveau in which nymphs, muses, and stylized flowers intermingle in surreal allegories. Among Goldscheider's inventory, buyers could take their pick of an artist's rendition in natural red terra-cotta or in crackled white glaze with a turquoise laurel wreath; one might even opt for a neckless head of Beethoven with exaggerated hairstyle balanced on a pedestal. More interesting, though, are the pieces that surround this Beethoveniana in the 1890s and early 1900s: busts or masks not only of figures like Goethe and Schiller, Voltaire, Gluck, and the all-time favorite Queen Louise, but also of fictional characters such as Robinson Crusoe and anonymous exotics such as a Bulgarian woman, an African man, a laughing monk, or a turbaned Arab. Here Goldscheider capitalized on the character heads (*Charakterköpfe*, or *têtes d'expression*) that were models for sculpture as well as the ethnographic casts commissioned by anthropology museums. By interspersing European "geniuses" with these racial "types," the Goldscheider catalog made home decor an extension of the phrenologist's cabinet.[65]

One item of Beethoven-themed decor stands out because it represents a much broader fixation on this mask among the Jugendstil: a Goldscheider vase from 1897 titled "Music," in which a seminude muse reaches across the strings of a lyre while Beethoven's mask floats high above, emerging from a textured backdrop of laurel leaves.[66] The floating Klein mask was in vogue at this time: ex libris plates and prints show it rising into dark skies above windblown landscapes or emerging as a pale apparition in shadowy voids. Equally common were depictions of the mask hovering over the shoulder of a performer, as if music could summon the composer's spirit (figs. 4.5 and 4.6).[67] When Beethoven's mask took to the air, it exuded the power of a Veronica or a Mandylion. Franz von Stuck, a prominent Munich secessionist who painted a number of Christological scenes, kept this mask in his studio and rendered it in ghostly brushstrokes that play with the dichotomy of laurels and thorns (fig. 4.7).[68]

Why should Beethoven appear dreamlike, haunting, and talismanic in the visual arts of this period? A clue lies not only in Beethoven's Christology, but in the culture of his companion, the Inconnue of the Seine. In reality, she was a life mask of unknown provenance; in popular opinion, she was the death mask of a mysterious suicide pulled from the river, which was feasible given

FIGURE 4.5. C. W. Bergmüller, *Beethoven-Sonate*. Reprint of oil painting (n.d.), *Illustrirte Zeitung*, March 24, 1927, 407; later *Etude Magazine* 48, no. 3 (March 1930): 151. Author's collection.

FIGURE 4.6. *Beethovenkopf über nächtlichem Gewässer (Head of Beethoven over Nocturnal Waters)*. Reprint by Manfred Maly of an etching by Georg Wimmer (n.d.). © Beethoven-Haus, Bonn, B 2439.

that the Paris morgue regularly made casts of criminal faces and put corpses on display. The Mona Lisa smile of the Inconnue made this the single most popular mask of the period. In Paris, the masks of Beethoven and the Inconnue could be seen paired in the shop windows of *formatori*, if we are to believe Rainer Maria Rilke in his novel of 1910: "The face of the young drowned woman, which was cast in the Morgue, because it was beautiful, because it smiled, smiled so deceptively, as though it knew. And beneath it, the face that did know. That hard knot of senses drawn tense; that unrelenting concentration of a music continually seeking to escape."[69] Rilke suggests that this coupling was not arbitrary, but that these objects spoke to each other in a language of the dead; what exactly they both "knew" remains a mystery to the living.

In Anne-Gaëlle Saliot's wide-ranging exploration of the Inconnue, we find an object on the move. The mask traverses class and artistic register, from cheap souvenir to recurring dream in interwar surrealist art and literature. Saliot argues that the Inconnue indexes the modern city in ways that echo Walter Benjamin's writings, and that it interfaced with products of its time: the cult of suicide, the detective novel, the crime section of the paper, and the new woman. For art critics of the period, the Inconnue posed a challenge, as plasterworks and other mimetic print technologies were disparaged as too indexical (carrying a material trace of the original) and insufficiently iconic (resembling the original, as would a painting). As a result, the object lent itself to the burgeoning medium of photography, a print of a print. Whether it hung on a parlor wall or peered through the pages of literature, this mask

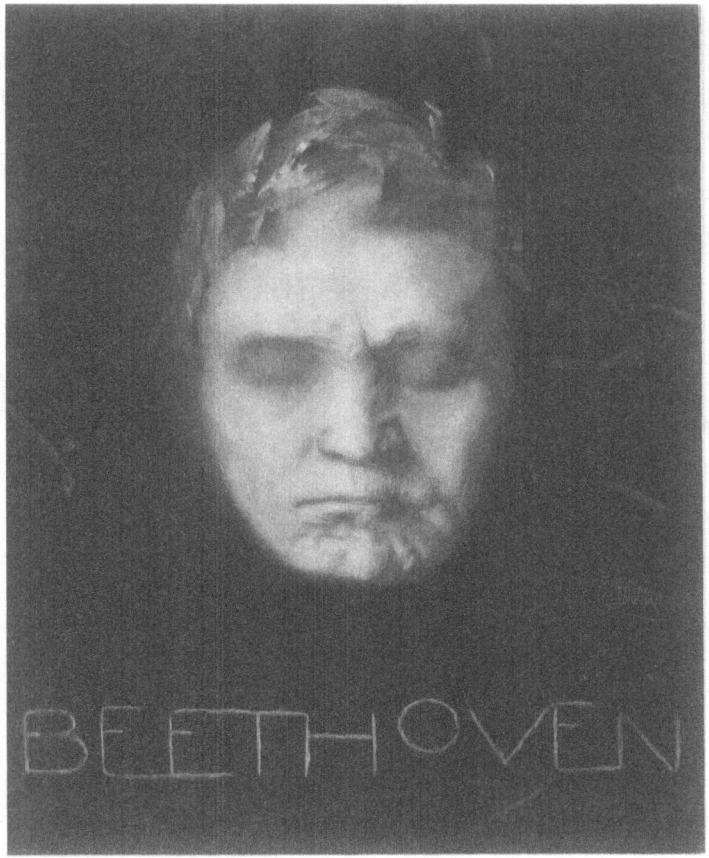

FIGURE 4.7. Franz von Stuck, *Beethovenmaske mit Lorbeerkranz* (*Beethoven Mask with Laurel Wreath*). Reproduction by the Franz Hanfstaengl Verlag (1900) from painting (1896). © Beethoven-Haus Bonn, B 1389.

confronted urban dwellers with an uncomfortable reality: that death, in modern times, represents an anonymity that is inevitable.[70]

Beethoven's mask appears only in passing in Saliot's study, and the Inconnue appears nowhere in Comini's account of the mask, but I suggest that these two objects circulated in quiet dialogue. Both casts look simultaneously living and dead, and when mass-produced, they become not just copies but simulacra (that is, copies of a copy) that lose the grain of the original, with smooth and dreamlike surfaces that invite enigma. As a result, both Beethoven and the Inconnue functioned as broader symbols of death, not only indices or icons, which explains why these masks were frequently photographed in folds of heavy cloth or veils in the manner of funeral photography.[71] When Beethoven's death house was demolished in Vienna in 1903, a commemorative etching showed his mask floating below the doomed house along with the incipit of the funeral march from the *Eroica*; here it is not so much a phantasm, like in Stuck, but a flat emblem of death affixed to the foreground.[72] A Viennese admirer made that connection more literal when he salvaged stones from the wreckage and commissioned an artist to carve the life mask into the rock, as addressed in chapter 1.[73] By far the most common appearance of this symbol was a variety of still-life scenes that positioned the Klein mask on the wall like a crucifix, surrounded by flowers, scores, sketches, or musical instruments (fig. 4.8). These are melancholy yet redemptive memento mori, reminders that human life is ephemeral but art is immortal.

FIGURE 4.8. (*Left*) A photograph from the Atelier Brüser. (*Right*) A still life that was widely sold as a lithograph by Wilhelm Menzler, *Beethoven* or *Polyhymnia*, 1919. © Beethoven-Haus, Bonn, NE 81, vol. 7, no. 128 and B 1251.

FIGURE 4.9. "My songs will live on when I myself am gone." In etching by Walther Rath, 1916. © Beethoven-Haus, Bonn, B 552, K 4/8.

There is a naive optimism here that makes these shrines into touching relics of death like those disparaged by Adorno. One example makes this especially clear. In 1916, artist Walther Rath made an etching that he later gifted to the Beethoven-Haus in Bonn, in which the Klein mask, hanging from a wall, overlooks an open notebook and a skull against an almost electric aura of linework (fig. 4.9). In a corner Rath etched an epigraph: "My songs will live on when I myself am gone." Here Rath paraphrases the poet Annette von Droste-Hülshoff, whose 1851 collection *The Spiritual Year* used the Catholic devotional calendar as a platform to work through her doubts as she questioned God and faith. Droste-Hülshoff's poem for the fifth day of Lent was written from a dual perspective: we hear the thoughts of Jesus on the Cross (marked as such by an explicit reference to the Gospel of John), but at the same time, this *lyrisches ich* stands in for anyone and everyone taking their dying breath. Quite unlike the usual picture of Jesus, this narrator questions God's motives, interrogates death, and remains acutely conscious of each failing nerve fiber as the body shuts down. In the end, as his brain fails, the narrator realizes that even God's word is not spirit but matter, a confused tangle of clots and veins. There is a tone of desperation here, as an uncertain Christ cries in anguish, "Behold, the songs may live—but I am gone!"[74]

When read beside the original poem, Rath's excerpt rings hollow. Droste-Hülshoff suggests that apotheosis happens only in hindsight, and that eternal life is little consolation when faced with the material horror of death. But in Rath's etching we find simple solace. It could be argued that this heartfelt oversimplification turns this image, and other Beethoven-themed memento mori, into kitsch. The complexity of kitsch as an aesthetic category cannot be understated, and the next chapter disentangles this concept more systematically. Here it suffices to recall its key definition as an adamantly sentimental expression of a platitude, an aspiration toward Romanticism that dead-ends in a cliché.

Some of the earliest discourse on kitsch took Beethoven's mask as an example. In an essay of 1922, unpublished for decades, the Bauhaus designer Oskar Schlemmer distinguished between honest and dishonest kitsch: whereas folk festivals and circus sideshows have an authenticity that makes them "part of the world's rich tapestry," the commercialization of high culture has produced a more insidious form of kitsch "that reduces the great, the perfectly formed, and the aesthetically beautiful into tiny, cute bric-a-brac, such as the metallic, electroformed death mask of Beethoven, a paper weight shaped like the Milan Cathedral or light bulbs shaped like grapes! . . . Perhaps people need this glitz, even if it's spurious; perhaps people need something they regard as beautiful that is cheap."[75] Schlemmer might have observed not only the mass production of the mask itself, but also its use as an emblem on postcards, stamps, chocolate cards, teacups, and pipes; by the 1920s, the object was more a predecessor to Hello Kitty than a phantasm of the art nouveau.[76] In 1908, as if to recover some of the mask's former prestige, Micheli Brothers partnered with the Beethoven-Haus in Bonn to brand it with a museum crest to increase its value.[77] When the company contacted the museum, their representative noted that the mask had entered a mainstream market: "The Beethoven mask is already very widely disseminated; indeed it has become the fashion to use it as a item of decor, even in circles that have neither a relationship with nor a passion for it."[78] A decade later, the Viennese wit Alfred Polgar confirmed that Beethoven's face "is an international requisite for brightening up a room." Polgar griped (with typical *Wiener Schmäh*, or curmudgeonly humor) that dilettantes have cheapened Beethoven's physiognomy with their parlor musicking: "Above how many hundreds of thousands of pianos from one end of the globe to another broods this mask, abandoned to the *Sonate Pathétique* (don't be sloppy, Melanie!) and the four-handed onslaught that makes symphonies groan and overtures tremble. The most extraordinary countenance that God's pencil ever drew—surrendered to housewives and popularity!"[79] One finds echoes of Melanie's sloppy pianism in other literature of the period. In his study of Wyndham Lewis's novel *Tarr* (1928), Nathan Waddell points out how Beethoven's mask embodies the naive tastes of Bertha, an aspiring bourgeois who hampers the artistic sensibilities of a male protagonist. In her home, the mask becomes a kitschy replica of high art: "there was the plaster cast of Beethoven (some people who have frequented artistic circles get to dislike this face extremely), brass jars from Normandy, a photograph of Mona Lisa (Tarr could not look upon the Mona Lisa without a sinking feeling)."[80] If, by the 1920s, Beethoven's mask was associated with the saccharine tastes of middle-class women, most especially those of the petit bourgeoisie who emulated the upper middle, then this marks a gendered divergence from the Inconnue. Both began their lives as decor that could be called kitsch, but by the 1930s, the Inconnue drew the male gaze of the surrealists while Beethoven's face was, as Polgar put it, "surrendered to housewives and popularity!"[81]

The misogyny in these texts is entirely characteristic of turn-of-the-century attitudes toward kitsch. Andreas Huyssen has shown how early twentieth-century male writers on modernism and its antithesis tended to feminize mass culture, with critiques that ranged from subtly gendered language to fears that insipid women threatened art, politics, and autonomy.[82] If kitsch was coded female, and Beethoven's mask was coded kitsch, then it follows that the Dadaists would thumb their nose both at Beethoven's masculinity and at the misogynist panic of the modernists. For the cover of the *Dada Almanach*, published just after the first international Dada fair of 1920, Otto Schmalhausen refashioned Beethoven's mask as the face of Charlie Chaplin, which included a mustache, an odd tuft of forehead hair, and wide-open blue eyes with heavy eyeliner, which was understood as both kitschy and effeminate.[83] In this image, the interwar Berlin Dada reappropriated kitsch as a medium for art, with a series of projects that anticipated Benjamin's essay on mechanical reproduction by fifteen years, as argued by Sherwin Simmons. For the Dadaists, Chaplin was a champion of modern mass media; Beethoven's image, on the other hand, had become a paradigm of the staid and stodgy. For Wieland Herzfelde, who annotated the work in the *Dada Almanach*, Schmalhausen's mustachioed mask rescued Beethoven from the oppression of bourgeois respectability (and, implicitly, from housewives).[84]

These are all distinctly male perspectives on Beethoven's mask as a woman's trinket. It remains an open question whether women did indeed have a special relationship with this mask just as nuns kept relics of the Holy Face. Certainly Irene Wild's panegyric to Beethoven's face and form, discussed at length in chapter 1, has a tone that is both spiritual and erotic, as Wild "feels her way" to a looming visage whose eyes alight when her cosmic presence is nigh. Her poem appears to respond to two contemporary currents: a literary fixation on Beethoven's lost loves and a visual eroticization of the muse, who substitutes for a lover once Beethoven renounced earthly attachments for art.⁸⁵ It lies outside the scope of this chapter to ascertain just how common Wild's sentiments were among Beethoven's female admirers. It seems reasonable to speculate that music-loving women were impacted by a visual culture that showed the Klein mask embraced by female forms, which would render Beethoven's closed eyes more sensual than deathly. Nude female figures often graced the allegories of the Jugendstil; the artist and life reformer Hugo Höppener (or "Fidus") took this a step further in allegorical images that argue for a utopian return to nature, which explains why Beethoven's mask-as-temple looms above both female nudes and an ideal man, woman, and child.⁸⁶ In biographical fiction, a blind woman was said to be the living muse for Beethoven's "Moonlight" Sonata, and in a 1904 allegorical representation of the work by Franz Stassen, she floats behind the mask with hair down to her ankles. A print by Ludovic Alleaume from the 1920s is even more intimate: a ghostly woman, possibly a muse, whispers against Beethoven's cheek while her hands cradle his head and hair. The intimacy of hair stands out even more strongly in Alois Kolb's allegory from 1906: here Beethoven's hair, rising above a face modeled on the mask, is a soft bed for a naked couple's embrace. (In 1968, artist Michael Mathias Prechtl offered a smart satire called *Beethovens Erotica*: a Rubinesque muse cradles a cello atop Beethoven's mane, with a defiant patch of armpit hair that taunts the male gaze.)⁸⁷ It is worth pondering whether these cultural products invited female Beethoven lovers to imagine themselves as the muse, which would introduce a dimension of eroticism to this mask just as the Inconnue enticed male surrealists.

Typically, to pronounce an object kitschy is an aggressive move unless one emulates Walter Benjamin, who encountered the potpourri of mass culture with curiosity as a series of enigmatic dreams. On the one hand, Beethoven's mask invited artists to dream like Benjamin and give this mass-produced object new life as an apparition; on the other hand, the critical derision toward kitsch deflated the mask's symbolism and revealed it to be yet another useless thing. I suspect that Beethoven's mask was reconceived as kitsch amid a growing sense of alienation from the nineteenth century, whose artifacts still cluttered the material culture of a modernizing world. It was Beethoven's emergence as an *anachronism*, then, that burst the bubble of transcendent immortality. The concept of anachronism guides Saliot's study of the Inconnue, an object that was "a symptomatic expression of a modern world haunted by the earlier modernity of the nineteenth century."⁸⁸ Beethoven was likewise a token of that earlier modernity, a figure said to press ahead of his epoch in his own time, but whose image in the wake of World War I represented the residues of a nineteenth-century bourgeoisie that was now covered (as Benjamin put it) in a layer of dust.⁸⁹

Animating the Anachronism

It makes sense that Beethoven's mask would become an anachronism: not only were casts an obsolete technology in the era of photography, but anachronism is also the logical outcome of the eternal, of living too long and overstaying one's welcome. The extended afterlife of Beethoven's face invited writers and artists to animate his mask, to pry open the closed eyes and resurrect his body in a modern world, at times to comical effect. In the visual arts, that animation could be quite literal. Franz von Stuck opened Beethoven's eyes in a series of plaster reliefs, as if animating his earlier painting, and many of his contemporaries added hair, eyes, or the outlines of a body in renditions that were still unmistakably mask-like.[90]

Among the most eerie of these anachronisms was a work by the Munich-based artist Max Klinger. The piece in question is not, in fact, the famously confounding monument he created for the Viennese Secession exhibit in 1902, but rather a painting from over a decade earlier called *Pietà*, which he completed along with other Catholic-inflected works after a stint in Rome (fig. 4.10).[91] The anachronism in this painting is indescribably strange, and its significance for the aesthetics of kitsch will be more fully explored in the

FIGURE 4.10. Max Klinger, *Pietà: Maria und Johannes trauernd am Leichnam Christi* (*Pietà: Mary and John Mourning at the Body of Christ*). Oil on canvas. Rome, 1890. Dresden, Staatliche Kunstsammlungen, Inv. no. 2460. © SLUB Dresden / Deutsche Fotothek / Unknown photographer.

next chapter. Like countless other depictions of the Pietà, the Virgin Mary looks in anguish upon the body of Christ while St. John the Evangelist consoles her, but here, John's face is the Klein mask of Beethoven made flesh. The face of Beethoven has an unnaturally sharp jawline that hints at the medium of collage, as if the mask were superimposed onto a painting by the Nazarenes. The result is that Beethoven, being out of place and somehow *wrong*, draws the eye, while the face of Christ looks quite ordinary and disappears into the periphery.

The common explanation for this image, by Comini among others, is that the artist frames Beethoven as a modern redeemer for a secular epoch.[92] This explanation misses something of the original context of Klinger's religious paintings, reading them (anachronistically, I might add) through the lens of his later monument. That monument did indeed make Beethoven into a demigod who unifies Hellenism with Christianity; the jumble of symbols that adorn this piece of "colossal kitsch," as Comini calls it, include Prometheus, the Crucifixion, the birth of Venus, and an odd conflation of John the Evangelist with John the Baptist.[93] This monument and its controversial reception are ferociously complicated; its utopian aspirations toward a Third Kingdom have been explored by Kevin Karnes, its tangled political reception by Anna Harwell Celenza.[94] But if we set aside the monument for a moment and focus on Klinger's other works from the 1890s, his *Pietà* emerges in a new light. In this decade, he produced numerous etchings that betray deep religious ambivalence, portraying a human Christ who stands off to the side in dynamic scenes, not yet a messiah. He also created several series on death that infuse the nineteenth-century beautiful death with Gothic and medieval grotesquerie.[95] In *Dead Mother*, a young woman lies pale in a ruined vestibule before a forest, flowers strewn about her hair and pillow; her infant child perches cat-like on her chest. In *Death* from his series *A Love*, the ghost-white corpse of a woman looks serene while her mourners are faceless terrors in the shadows. And in a striking commentary on Hans Holbein's *The Body of the Dead Christ in the Tomb*, Klinger's *Death as Savior* shows a male corpse entombed in a three-dimensional frame with his face turned away—an aniconic reversal of Christ's cult of the face—while the altarpiece above shows a man prostrating himself before a white angel as dark beings wrestle away his wife and child.[96] All these images predate Klinger's *Pietà*. What they share in common is a critique of death's alleged beauty that blurs the line between iconicity and anonymity. Klinger poses the bodies of the anonymous to look sacred, like saints or Christ, while Christ himself is shown in the midst of an ordinary day in the life of a prophet. It is not so much straight-faced adulation for Beethoven that we find in Klinger's *Pietà*, then, but rather an extension of this religious ambivalence, in which both Beethoven and Christ occupy the uneasy space between redeemer and man.

Against this backdrop, Klinger's monument in 1902 is not the benchmark for his work, but rather a marked departure. His various monument projects (first Beethoven, then Brahms, then unfinished plans for Wagner and Liszt) sought to maximize the medium, laden with symbol upon symbol. Klinger's association with the symbolist movement is doubly significant. It meant, on the one hand, that his love of music found a compelling visual outlet in proto-surrealism, as Walter Frisch has shown in his study of Klinger's *Brahms Fantasy* series.[97] On the other hand, symbolism implicated Klinger in heated debates about class and taste. In 1905, art critic Julius Maier-Graefe rebuked the popularity of the symbolist painter Arnold Böcklin, whom Klinger openly admired, as a symptom of the stuffy philistinism of the German middle class; those same accusations came to haunt Klinger's monuments and overshadowed the rich nuances of his earlier prints.[98] As a result, Klinger is remembered not for questioning Beethoven's transcendence, as he questioned that of Christ, but for hastening Beethoven's *Verkitschung*, his cheapening in a trite cult of transcendence. Klinger's art, and the cultural politics in which he was enmeshed, drew attention to another layer of anachronism: that of the bourgeoisie itself, whose taste in art was increasingly out of step with the times.

In the next chapter, I show how this larger anachronism worried a generation of critics who feared the decline of art and thought. But icons of the decaying bourgeoisie, like Beethoven's mask, were not always such a serious matter. We would be missing something if we ignored the simple fact that anachronisms, like all misfits, can be delightfully *funny*. The many whimsical takes on Beethoven's animation show how the middle classes could laugh at themselves, especially in Vienna, where droll humor reigned. Short stories and essays called feuilletons, named after the newspaper section devoted to light fiction, were especially keen to mock the talismanic aura of the mask. A journalist in 1926 griped, with exaggerated dismay, at Beethoven's mask displayed as a glasses model in a Viennese shop window.[99] In a story of 1933, the Klein mask mounted above the piano sets off a series of false fire alarms.[100] In another, Beethoven's mask on the wall thoroughly terrifies a dog, who only gradually comes to recognize that the thing is dead.[101] In yet another, an amateur pianist hangs the mask over her instrument only to find herself mysteriously alienated from Beethoven's music, tormented by his "curse." She attributes Beethoven's "revenge" to her scrimping purchase of the plaster mask rather than a bronze alternative that might better honor the composer.[102] These paranormal tales reflect the concurrent popularity of parlor séances and stage mediums. In 1895, a London medium channeled

Beethoven onstage and notated, before the audience's eyes, his Tenth Symphony and a violin sonata in C; she described her process with the pronoun "we."[103] No wonder Beethoven's face hovers in visual culture as if raised from the dead.

One of the charms of the feuilleton genre is that it treads a fine line between earnest and facetious; it is hard to tell whether these stories contain a kernel of longing for Beethoven's presence, or chuckle at that desire, or both. In a story by Desider Kosztolányi, published in 1914 in the edgy literary magazine *Der Brenner*, Beethoven returns to earth not as a messiah but as an ordinary traveler weary from the road.[104] What follows is a retelling of E. T. A. Hoffmann's "Ritter Gluck" of 1809, in which the dead Gluck returns to guide the performance practice of his own works. In Kosztolányi's tale, a Czech hornist named Bolcek dedicates his life to dual missions: defending art's sanctity against the impiety of uncouth audiences and consuming a nightly excess of wine. One evening, from the concert stage, he spots a chilling sight: Beethoven himself is seated among the chattering audience to hear his own symphony. The incredulous Bolcek, wondering at first if he hallucinates from drink, recalls that Beethoven is indeed dead: at home, the musician has the death mask hanging above his bed, and he has seen the facsimile of Beethoven's last notes along with a ghastly sketch of him in his last hour. Yet, lo and behold, Beethoven's face is close enough to touch, to see his open eyes, to speak and hear an answer. Granted, he looks a little pale, as if tired from a long journey; his right elbow is smudged with dirt, likely (Bolcek supposes) from the walls of the grave. Around him the audience, ever badly behaved, fails to recognize who sits in their midst. At the end of the concert, Bolcek, though rattled, builds up the courage to greet Beethoven, who proves to be as affable as a dear friend. The composer regrets that he must depart the next morning on the train, never to return, but he invites the musician to call on him at his inn later that evening. When Bolcek knocks, he finds the room empty, candles unlit, with "Beethoven" scrawled sloppily in the guestbook.

There is much of interest in this little ghost story. First, it is a rebuke of the undisciplined listeners who fail to recognize the very visage that hangs on their walls. There is a subtle indication, too, that the last bastion of art is not the erudite but the folk, like this small-town brass player. Above all, the story captures the uncanny thrill of the mask come alive, juxtaposed against the perhaps disappointing *normalcy* of the resurrected Beethoven, who sits among the masses and has a train to catch in the morning, a striking predecessor to the commuting composer in Kagel's *Ludwig van*. Kosztolányi seems to ask, through this affectionate spook, whether old music has any value in the modern world unless it is periodically visited by its creator.

In all these artifacts of animation, a desire to recover Beethoven's gaze was stymied by the Romantic irony of the memento mori, the sentiment that an artist's songs live on. The *aliveness* of Beethoven's music, its humanity and gaze, were hard to reconcile with a visual culture saturated with images of his flat *deadness*. This helps to explain the strain of discourse illuminated by Daniel K. L. Chua in his study of Beethoven's humanism: Adorno invested Beethoven's music with "the gift of sight," while Benjamin broadly defined the experience of aura as art's "ability to look back at us."[105] Benjamin's remarks reflect his conflicted response to photography, the only mechanically reproduced medium that can retain its aura as it captures "the fleeting expression of a human face" in a "cult of remembrance of dead or absent loved ones"; it is no wonder that, a half century later, Barthes compared photography with the art of embalming.[106] Animating Beethoven's mask in the early twentieth century was, in a sense, a kind of photographic manipulation that enfolded Beethoven into the most contemporary practices of remembrance. But the picture is not quite so simple. Chua grapples with a sticking point in Beethoven's humanism: that his music's "gaze"—which is felt most strongly in the embodied and painful stammering of his Cavatina from op. 130, or the "Crucifixus" of his *Missa Solemnis*—gestures to not only an ordinary humanism, but an "Other humanism" encapsulated by Christ. "Could this gaze," Chua asks, "be the face of God in Christ, reflected in the countenance of the Other?" While embodied moments in Beethoven's music are by no means a "direct vision of Christ incarnate," they beckon to a face of "human vulnerability" that is "similarly defenseless and exposed, a face subject to public humiliation and condemned because it failed to account for itself."[107] Even as Beethoven's mask, with its laurel crown reminiscent of Christ's crown of thorns, makes no direct appearance in Chua's text, the material culture of Adorno's time seems to haunt Chua's philosophy.

These remarks about the human vulnerability of Christ suggest, if inadvertently, a deeper level of meaning in Beethoven's masks. At first glance, these masks quicken the blood because of their immediacy, pressed against the composer's own flesh, pores, and scars. But in the end, Beethoven became oddly ancillary to the afterlife of his own face, which ran rampant beyond his control. His helplessness reflects the brutal vulnerability of the dead body, subject to others' whims. When feuilletonists noted the bespectacled mask in a shop window, or when they poked fun at petit bourgeois superstitions about its dark powers, they hinted at the desecration of a corpse. Out of the constellation of Beethoven's reception we find the first signs of empathy for that vulnerability, for the deadness of his body, which contains a potent echo of the Pietà.

The Leveling of Faces by Death

This chapter has charted how Beethoven's face fell from the heights of sentimental symbolism and the relics of Christ, down into kitsch and finally into anonymity. The beautiful death masked a deeper fear of erasure and nonexistence, of death as secular abyss. By ossifying flesh, the mask concealed the decay of the memento mori to imply a sublime and eternal life through art. But when the dead face became anonymous rather than a vera icon, the practices that attended the deathbed—affectionate care and stewardship of the embalmed, eternal friendships, flowers—looked like anachronisms.

In 1927, the centenary of Beethoven's death, the Klein mask surged in popularity for the last time. One year prior, Ernst Benkard had published *The Undying Face*, an annotated gallery of death masks that captured the imagination of modernists and surrealists alongside the general public.[108] In the wake of Benkard's book, the press came alive with analyses of famous faces and waxed poetical about the expressivity of Beethoven's features to commemorate his death year.[109] Yet his two masks, like the others in Benkard's catalogue, had taken on new meanings. Death masks were irresistible to artists who were drawn to primitivism and ethnology, to the uncanny double and the mechanical man, and who had seen the horrors of Great War with its gas masks and surgical reconstructions of broken faces.[110] In Benkard's book, dignified luminaries appeared with painfully distorted features that revived fresh traumas. The collection marked a twentieth-century turning point when the celestial beauty of nineteenth-century death rang hollow, and the indignity of death's visage became the memento mori for modern times.

At the height of those conversations, the novelist and essayist Alfred Döblin ruminated on Benkard's masks.[111] Döblin encountered a paradox as he flipped through Benkard's book: here were the faces of famous individuals, of recognizable icons, rendered into a "vast, peculiar moonscape" of silent objects, exactly like the mystery of the nameless Inconnue. Döblin called this "the leveling of faces and images by death," in which death erases personal striving and leaves only "stones that have been rolled around and polished by the action of the sea for decades."[112] This provocative metaphor makes masks into fossils that line the collector's shelf like any other organic matter. In Döblin we find the secular antithesis of celebrity that signals a broader moment of reckoning: a turning point when the longing to preserve the beautiful dead is exposed as a farce, when efforts to embalm the artist appear extravagant or even desperate. In Benkard's book, Beethoven and other German "geniuses" follow the Inconnue of the Seine to lose their mythos and find anonymity for the first time. In a sense, they were finally laid to rest.

5

Art-Religion *Verkitscht*

In 1934, a folksy humor paper from Austria's former empire published an anecdote that captures a turning point in concert life. In a coffeehouse, a band plays Beethoven's Second Symphony. The aptly named "Herr Bildung" strides up to the conductor and asks, "lovely piece, who's it by?" to which the conductor replies, after rifling through his binder's volume, "music piece number 197."[1] The butt of the joke is admittedly varied: both listener and conductor are equally incompetent, the casual venue does an injustice to art, and the social class lampooned could be the same petit bourgeois readership to whom this newspaper catered. What is abundantly clear is a shift in the tenor of the word Bildung, a once respectable form of moral self-cultivation through masterworks. Here erudition is nothing but an empty gesture.

It seems appropriate to begin a chapter on art-religion *verkitscht*, or cheapened, with this cheap joke. The bumbling figure of "Herr Bildung" reflects a growing interwar cynicism that marks a turning point in composer veneration. As canonical figures were made accessible through commodities, the reputation of the *Bildungsbürgertum*, the educated middle classes, grew tenuous. When enrichment was bought and sold through the economies of piety shown in this book, the products associated with moral uplift came to resemble "kitsch," a derogatory term that circulated among evangelists of traditional Bildung. Art-religion persisted in this environment, but it did not age well. For critics of these practices, who often hailed from the very social classes they rebuked, the fixation on composers' bodies, lives, and afterlives resembled the manias of mass culture. In dark humor and satire, secular religion looked newly threatening in the same period that saw the rise of fascism.

In musicology, conversations about mass culture often begin with Theodor Adorno's essays of the 1920s and 1930s, which unearthed the dangers of the recording, radio, and film industries. Shortly after the war, Adorno's more systematic exploration of this subject informed the Anglophone polemics against the middlebrow, notably Dwight Macdonald's concept of the "midcult," the condescending marketing of literary classics.[2] Adorno seems to respond to that discourse in his 1959 theory of *Halbbildung*, which some translate as "pseudo-culture."[3] Here he laments that art has been reduced to an empty shell, packaged in products like biographical novels and listening guides that caption symphonic themes, and that those who claim to have Bildung are spiritually impoverished masses huddling in shared delusion. Adorno was one of many voices in this discourse. His concerns found precedent among a variety of critics who feared art's demise, such as the architect Adolf Loos, the literary modernist Hermann Broch, and the journalist Adolf Weissmann.[4] To read around Adorno, we must attend not only to formal criticism by his contemporaries, but to the many whimsical reactions to mass culture that crop up in satires and feuilletons. A wide range of texts reveal how Bildung was increasingly disparaged as an industry that wore piety as its false front.

This chapter explores a simple question: how, and in the opinion of whom, was art-religion cheapened? A similar question lies latent in Alexander Rehding's *Music and Monumentality*, which periodically invokes the fragile line between sublimity and kitsch, but does not probe exactly why and for whom the sublime lost its magic.[5] Granted, tracing the disintegration of art-religion is an elusive task. It is far simpler to chart how a practice consolidates, with additive emulation and reinforcement, than to locate erratic moments of failure. I invoke the word *verkitscht*, then, to describe a piecemeal process that predates the formal use of the word "kitsch." In the vernacular, kitsch has come to denote a cheap copy of art made in poor taste, a moralizing rebuke that is more judgmental than curious. Recently, inventive studies have thickened the concept to include the experience of outdated material worlds. In musical life, that could mean laying wreaths at monuments, a nostalgia for the Biedermeier in composer-themed goods, or bodily relics that persisted at the sunset of nineteenth-century funeral culture. As the practices discussed throughout this book were held to perform ever more anachronistic functions in modern life, satires of those practices came to resemble critiques of kitsch and mass culture.

The historiography of declinism is admittedly complex, especially in Austria. To invoke a crisis of Bildung treads on the well-worn path of Carl Schorske, whose thesis revolves around the failure of Vienna's liberal bourgeoisie, for whom the temple of art offered a retreat from political engagement just as modernists retreated into psychological states and the decadence of ornament. Among historians, the chief critique of Schorske has been his conflation of politically disparate groups under a unifying theory.[6] That critique is borne out by the ambivalent object of satire in the writings of the feuilletonists: for socialists on the left, the problem lay in the consumer clutter of music culture, while for those on the right, Bildung and its cheapened counterparts could be associated with educated Jewry. (Readers of the joke that opened this chapter might have assumed that "Herr Bildung" was Jewish.) It was unclear who, exactly, was thought to genuflect at the temple of art. Was it the assimilated Jew acting as a replica of a German, whom the pan-Germanists, and later the National Socialists, saw as degrading their culture? Or was it the populist followers of antisemitic and anti-intellectual movements who enacted nineteenth-century rituals at the feet of marble busts in Walhalla? In the earliest discourses on mass culture, the identity of the deluded masses is nebulous, which in turn muddies the conceptual history of kitsch.

The first task of this chapter, then, is to disentangle that history and posit an alternative reading of kitsch that explains why art-religion collapsed under its own weight. I then offer three vignettes that epitomize the failure of art-religion. In the 1880s, the first glimmers of doubt hailed from critics of liberal Vienna, who poked fun at the veiled commercial intent of *imitatio*, the imitation of Christ and the saints that allowed composers to promote themselves through grandiose homage. Satirists who laughed at the overblown monumentality of Wagner, Liszt, and the little-known August Bungert noted the anachronism of this practice: when composers treat the dead like the living, their devotees respond by lauding the living as if they were already dead. At the turn of the twentieth century, Bildung was more explicitly characterized as a false religion by writers from divergent political positions, whose critiques of music lovers as deluded masses were entangled with the earliest discourses on kitsch. By the rise of National Socialism, piety was pushed to a breaking point. To understand this final turn, I look to Anton Bruckner's embalmed remains, which were admired as a saint's relic in a series of candlelight rituals before, during, and shortly after the Third Reich. Bruckner's interment cultivated the illusion of a beautiful death and a deep medieval past that floats outside of time. But as his body became an organ for shifting politics of commemoration, and as the National Socialists animated the body to speak on their behalf, the tomb could not keep its sanctity. Piety reached its point of no return.

Kitsch and Anachronism

As we saw in the previous chapter, the Romantic meanings that accrued on Beethoven's mask were undercut by two properties of modernity: kitsch and anachronism. The mask's superfluity and sentimental symbolism caught the eye of the earliest kitsch critics in the interwar period. Bauhaus designer Oskar Schlemmer listed Beethoven's mask among other "tasteless, mass-produced articles" that "populate the dwellings of the workers and the bourgeoisie," two quite different social classes who, in Schlemmer's view, shared a naive delight in insipid things that aspire to greatness.[7] Meanwhile, the object became a misfit in literature and especially the symbolist paintings of Max Klinger, whose *Pietà* of 1890 shows Beethoven's face plastered, with visible stylistic rupture, onto a scene from the Passion (see fig. 4.10).[8] Klinger makes it clear that Beethoven does not belong.

This painting captures the experience of anachronism, a concept that is considerably more nuanced than the mere survival of artifacts into the present. While the term has a range of existing meanings, from the creative reenactments of history enthusiasts to the temporal superimpositions of art criticism, I define the anachronism in simpler terms: a friction that arises from the conflation of disjunct material worlds, which draws attention to the temporal misfit and complicates the boundary between sincerity and farce.[9] Not only does the anachronism look backward in time, like Beethoven's face plastered onto St. John's body, but it arises when an archaic, deep past aspires to continuity with the present, yet fails. The result is an object lost in time, familiar yet jarring, a temporal experience of the uncanny. While some anachronisms can be flagrantly silly, others capture the pointed stare of the deadpan or the nonchalant ambivalence of pop art. This quality of whimsical sincerity is one of many properties that the anachronism shares with kitsch, and to explore this fertile intersection is to participate in a broader turn that expands the concept of kitsch beyond bad taste.

Any investigation of kitsch requires a close look at its layered historical and polemical meanings, as the term has been a moving target. While the word's etymology remains speculative, its various origin stories hail from cynicism about a value void. "Kitsch" may stem from the archaic verb *kitschen*, which implies the muddy smears that arise from rubbish collection; among art dealers of the 1870s, it became shorthand for knock-offs popular among foreign tourists at the market, which explains the alternative theory that "kitsch" was a mispronunciation of "sketch," a lithographic replica.[10] The term was limited to the art world until the interwar period, when it took on new life in polemics about aesthetics in an age of mass reproduction. The ingredients for kitsch discourse circulated for several decades before the word emulsified them. It was common in the late nineteenth century to debate whether the sensory enjoyment of art could be pathological and immoral, or whether crowds act as a hive devoid of reason. That irrationality was coded feminine, as Andreas

Huyssen has observed, and this implicit misogyny directed critics' attention to domestic decor, pretty things, and the dandyism of the flaneur.[11]

By the time "kitsch" was the word of choice for those concerns in the 1920s and 1930s, it had taken on three new dimensions. First, it became a means for designers who preferred a spare aesthetic to distinguish themselves from rival movements like the Jugendstil or Werkstätte, notably in Vienna, where ornament became politically charged. This is the dimension that has engaged historians of design. Second, kitsch signaled a new breed of unthinking human (or what Hermann Broch later called the "kitsch-man"), who degraded or falsified Bildung and stymied the formation of the rational, individual self.[12] This is the aspect that informed Clement Greenberg's influential essay of 1939, which defines kitsch as a clumsy imitation of art among a status-hungry petit bourgeoisie; less often cited, but equally important, is the generation of interwar critics discussed later in this chapter who saw frictions among middle-class strata.[13] Third, kitsch invited a philosophy of everyday life. As Walter Benjamin strolled through the Parisian Arcades, he ruminated in notebook after notebook; his oneiric philosophies of objects, called *Denkbilder* (or thought images), show how trivial goods contain multitudes.[14] It is tempting to wield Benjamin's influential essay on mass reproduction, in which art's ritual "aura" dissipates as it is copied, to support those (like Adorno) who condemned reproduction and standardization to recover a more rigorous mode of Bildung. But that approach caricatures Benjamin by neglecting his more curious encounters with commodities as artifacts of memory, time, and the unconscious.[15]

These three developments—modernists' opposition to frivolity, the degradation of Bildung, and the emergence of a philosophy of objects—account for much of the twentieth century's discourse on kitsch. The word represents not an aesthetic, but an accumulation of political thought whose sedimented layers ossified at midcentury. That ossification is most visible in *Kitsch: The World of Bad Taste*, an influential 1969 volume edited by art critic Gillo Dorfles. This emporium of hand-wringing shows how infamy can also be good press: the authors seem both repulsed and fascinated by kitsch, which results in a primitivist ethnography of the Western self as other.[16] The volume echoes, in a sense, the 1909 undertaking of art historian Gustav Pazaurek, who outfitted Stuttgart's craft museum with an exhibition of bad taste that flattered these "abominations" with attention.[17] Yet Dorfles's volume, while combative, is also expansive in its purview. Kitsch in this collection encompasses bronze monuments in city squares, religious icons and statuary, and elements of myth that informed the media image of celebrities. In this volume lies the first trace of a critical discourse on art-religion verkitscht, a process in which pillars of

Western society that have become second nature are understood anew as curious, exotic, or barbaric.[18]

Throughout its turbulent history, the concept of kitsch has exposed the politics of materiality. In an era of mass reproduction, it reflected anxieties about art's status; in an era of crowds, it reflected fears of the decline of Bildung and thereby reason; and later in the twentieth century, it signaled life under late capitalism through the superabundant cuteness of the knickknack.[19] There is a hidden dimension of kitsch that crisscrosses through these layers, and that advances us beyond the oversimplified trope of "bad art": kitsch has a precarious relationship with historical time. While kitsch is the product of modern technologies, it prefers to retreat from modernity into the warmth of memory. For Sam Binkley, the repetition and conventionality of kitsch reinscribe what sociologists call "embeddedness," or the reassuring habits of premodern life that create a sense of security in a volatile and otherwise "disembedded" world.[20] Binkley's insights explain the nostalgia for Biedermeier coziness in early twentieth-century commodities marketed with the composer's image, or what Gustav Pazaurek called *Aktualitätskitsch* as early as 1909.[21] Household brands like Stollwerk and Gartmann chocolate, Wagner margarine, Liebig's soup cubes, and Eckstein tobacco ran collectible card series focused on figures of Bildung (or, as the margarine brand put it, the "German thought and achievement" collection). Inside a wrapper one might find scenes from Mozart's life and operas, caricatures of Beethoven glowering at the inn, or Schubert accompanying a jolly Schubertiade.[22] The appeal of these products lay in their warm retreat into outmoded domesticity.

In response, interwar critics framed kitsch as a medium of memory. For Adorno, writing in exile just before the outbreak of World War II, kitsch prompts the gullible to misremember the past with phantom nostalgia.[23] A decade earlier, in his meditation on surrealism, Benjamin defined kitsch as the process by which things draw the human into their interior, turning people into containers for a lost nineteenth-century milieu; yet ultimately these objects fall victim to a layer of dust in the parlor, held close yet distanced by the patina of time. Kitsch, then, explains the experience of objects that exist in an uncanny zone, close yet far, living and dead.[24] In 1935, Ernst Bloch published aphoristic musings, in a collection called "Dust," which present kitsch as one of many false fronts of a petit bourgeoisie that has grown stale. His unflattering portrait shows the middle class bored on gossip, paging through magazine pictures and kitsch literature—that is, the "sayable stuff" that lies about the charms of "our little yesterday" while nonetheless "rummaging around in the mustiness."[25] There is a sensory dimension to this critique, conjuring up the smell and feel of antiquarian worlds.

This sensory and temporal facet of kitsch prompted Celeste Olalquiaga's expansive, if rather enigmatic, meditation on the "kitsch experience."[26] In a belated counterpoint with Benjamin, Olalquiaga defines kitsch as a utopian conviction that replication will not diminish an object's aura if that aura stems not from authenticity, but from the glow of memory. If kitsch lets us bask in the past, it confronts us with outmoded materialities like saints' relics, curiosity cabinets, cluttered parlors, and Victorian funeralia that merge the living with the dead, and these, rather than "bad art," are the focus of her study. When viewed through this frame, residues of composers' houses and lives are kitsch: the frozen interior of Berg's apartment with half a cigarette in the ashtray, the death masks in vitrines, the plaster replicas of Beethoven's desk items on sale at the gift shop, and especially hair jewelry such as the landscape from Beethoven's hair or the pearled bouquet from twisted locks of Liszt.[27] These are a distinctive form of kitsch that replicates memory in an obsessive surplus.

If kitsch can be defined as an immersive experience of past material worlds, this explains why art-religion, a relic of the nineteenth century, was gradually verkitscht. Art-religion enacted nineteenth-century concepts like Bildung, Ehrenpflicht, the beautiful death, and a belief that genius resides in the bones. The twentieth century saw a shift in the tactile relationship with the dead. Family members were less likely to keep hair jewelry or casts of hands and limbs, and when surrogate traces such as desktop busts, masks, postcards, or recordings filled the void left by relics, they miniaturized reverence into consumer goods.[28] In this period of uncertainty about how to remember the dead, satirists characterized an overabundance of material traces as a chaotic swamp, concert audiences as animal herds, and pilgrims as manic cults. In short: when relic culture became an anachronism, its material residues turned into kitsch. In the vignettes that follow, I chart how Bildung traveled from pious duty to mass culture in three stages: the mistrust of the cheap replica, the fear of deluded masses, and the breaking point of ritual enchantment when relics served the volatile politics of the twentieth century.

Imitatio and the Cult of the Living

In Richard Wagner's final years, two of his most impassioned critics, Daniel Spitzer and Eduard Hanslick, penned some of the earliest critiques of a cult of the living. Spitzer's 1880 novella *Wagnerians in Love* mocks the middle class through a fictional band of Wagnerians, with characters like the music-loving doctor who mimes the "Rákóczy" March while taking a patient's pulse, or the tasteless matron of the arts whose foyer is a forest of busts.[29] While all the characters are exaggerated devotees, the most extreme figure is the aspiring composer Goldschein (possibly a caricature of the conductor Hermann Levi), who has converted from Judaism to Wagnerism and now dons plush attire while drafting his opera *Schwanhilde*. When Goldschein meets Wagner, he begs for the composer's ratty toothbrush; at first he plans to brush his own teeth with it in an annual ritual, but instead he commissions a watch chain to be made from its bristles, a relic that is later donned and kissed in concert by a famed pupil of Liszt.[30]

This and other foibles make Spitzer's satire an arresting parody not only of cults, but of a distinctly Christian concept: *imitatio Christi*, or the devout emulation of the lives of saints or Christ.[31] It is not the toothbrush relic itself, but Goldschein's intention to *use* it, that makes his reverence absurd. *Imitatio* is precisely what irritated Hanslick two years later in "The Cult of Wagner," an essay that rebukes the Christian and sermonizing tone of the *Bayreuther Blätter*, the organ for Wagnerites and their dogmas. Hanslick notes that Wagnerians adopt the composer's lifestyle and views, from Schopenhauer to vegetarianism, and fold like-minded individuals into their circle as second-order saints, or *Nebenheiligen*. What Hanslick finds most puzzling are those who emulate of the living—quite unlike admirers of Mozart and Beethoven, who are more invested in biographical insights than cultish reenactment. How illuminating these Bayreuth pages will be, Hanslick laments, to the cultural historian of the future who charts the decay of the educated class.[32] In these early critiques of art-religion, both emanating from liberal Vienna, we find the seeds of skepticism toward parasocial behavior.

Granted, a mistrust of Wagnerism was nothing new. Already in 1877, the composer Wilhelm Tappert had published a lexicon of critical abuse that was expanded into numerous editions.[33] Aside from a few jabs at Pope Wagner and at Bayreuth, the idol temple, most of the slights in this volume pertained not to Wagner's cult but to the sensory effect of his music.[34] Tappert's collection reveals the first strains of a debate about musical brainwashing that was amplified after Max Nordau's 1892 book *Degeneracy*: not only were critics bemused by Wagner's own sensual preferences, but his endless melodies were held to infect listeners like a contagion.[35] By attributing Wagner's success to delirium, critics directed attention away from the agency and behaviors of Wagnerites, which made this a discourse very different from that of Spitzer and Hanslick, who unearthed the apparatus of art-religion that propped up these intoxicating sounds. Wagner's was the first in a series of extravagant cults, and to understand how art-religion was verkitscht, we must illuminate the moment when devotees to Wagner, Liszt, and forgotten figures like August Bungert began to appear ludicrous.

Essential to that project are the insights of Nicholas Vazsonyi, who exposes the mechanics of self-promotion that were embedded in Wagner's rhetoric.[36] Vazsonyi argues that Wagner was no ordinary megalomaniac but rather a strategic master of marketing, and central to that claim is the concept of *Selbstinszenierung*, the process by which luminaries stage themselves like theatrical directors. Wagner's methods of self-staging were expansive, and one move in particular helps to explain the critique of his followers: his strategy of "marketing martyrdom," or renouncing material gain just as Christ or the saints embraced poverty. This sacrificial stance anticipates the antimarket pose theorized by Pierre Bourdieu, which I addressed in chapter 1: only by claiming *l'art pour l'art*, art for art's sake, does art gain material value in this upside-down economy.[37] *Imitatio*, while not named as such by Vazsonyi, amplified the antimarket pose. It is well known that Wagner and Liszt positioned themselves as heirs to Beethoven: the famed anecdote of the so-called *Weihekuss*, Beethoven's kiss of consecration on the young Liszt's brow, compelled Wagner to engineer a kiss of his own in his semiautobiographical novella *A Pilgrimage to Beethoven*.[38] In this literary act of self-staging, a German composer (or pseudo-Wagner) is disgusted by the materialism of his English companion as they journey to visit Beethoven in Vienna; only the sober German, of course, gets the master's mark of approval, the Weihekuss. At the core of Wagner's historicism—his teleological sense of self-importance, as he carries the torch of German art from ancient Greece into the future—lies a distinctly Christian chain of imitation and pilgrimage. Each disciple becomes a new saint.

In a sense, *imitatio* added a halo to strategies that had long been active in the music-publishing industry. Emily H. Green has shown how, starting in the late eighteenth century, frontispiece dedications on musical scores functioned as careerism veiled in humility and pious deference. It is no coincidence that Green, like Vazsonyi, invokes the symbolic economy of art theorized by Bourdieu; few models more directly explain the hidden transactions of the public gift, the prestige of deference, and the hidden economy of piety. By dedicating a work to Haydn, for example, an aspiring composer could ride on famous coattails and insinuate forms of mentorship that were almost as fantastical as Wagner's pilgrimage.[39] But Wagner's feigned humility surpassed these earlier models, and he drew attention to the hypocritical act of marketing piety. This explains why Spitzer's Wagnerians are all social climbers and the most ardent disciple bears the name Goldschein, which evokes not only the sparkle of the Rhine gold but the word for "money bill," an antisemitic insinuation. While *imitatio* remained an appealing strategy for those who borrowed Wagner's marketing strategies—after all, the very concept justified the act of emulating Wagner—it became a risky maneuver that could appear conspicuously self-serving.

This was complicated terrain for Franz Liszt, whose own approach to Selbstinszenierung crisscrossed with Wagner's in a web of fraught homage. Liszt's career demonstrates the full scope of *imitatio*, which found expression in many domains: as a stance in self-promotional writings, a behavior that enacts deference, a musical homage or arrangement, and the building of monuments, both material and symbolic, through concert programming and ritual events. Where the emulation of Beethoven was concerned, Liszt was an earlier architect than Wagner. Wagner's pilgrimage novella was drafted in 1841, and his polemical production of the Ninth Symphony followed five years later, but already in the late 1830s, Liszt promoted himself through all-Beethoven concert programs and symphonic transcriptions that made his own hand uniquely visible.[40] His leading role in the Beethoven celebrations of 1845 and 1870—a monument unveiling in Bonn and a centennial in Weimar, respectively—was more prestigious than Wagner's polemics in the press. For both festivals, Liszt composed celebratory cantatas that fused his own music with quotations from the master and inserted himself into history. (Generally speaking, most festival choruses that touted timelessness have fallen through the cracks of music history, not least because rousing finales like "Heil Beethoven" were more tasteful in 1845 than they were after 1945.) For Ryan Minor, Liszt's two cantatas betray a gradual shift in Beethoven's status from living memory to ossified heritage, and for Alexander Rehding, they show how Liszt monumentalized himself at Beethoven's feet: "While Liszt *appeared* to let Beethoven speak for himself, he in fact used Beethoven as a ventriloquist's dummy, and let him speak for Liszt."[41] In his later years, Liszt took minor orders that made his *imitatio Christi* an obligation of his faith, but even in these late years, the strategy could be self-serving.[42] Shortly after Wagner's death, Liszt composed an elegy called "At Wagner's Graveside," which merged a motif from *Parsifal* with one from Liszt's own cantata, *The Bells of Strasbourg*. What sounds at first like a tender ode to friendship contains layers of *imitatio*: Liszt's inscription insists that his own cantata came first, blurring the line between master and epigone. Wagner was of course helpless to disagree.[43]

These activities raised a question already during Liszt's lifetime: when does *imitatio* lose its piety and cheapen the imitator? The most vocal critics of Liszt's automonumentalization were members of what Laurie McManus calls the "priesthood of art"—Clara and Robert Schumann, Joseph Joachim, and others—whose criteria for priesthood favored sobriety over self-staging.[44] But a focus on the warring ideologies of a few well-known actors misses a key point. While Liszt engineered many aspects of his celebrity—the most novel, perhaps, being the circulation of his likeness through modern print media such as photography—it was not Liszt but his devotees who took things too far. Late in his career, Liszt's admirers adopted the monumental practices he himself had designed for Beethoven, Schiller, and Goethe, lauding him in ways usually reserved for the dead.

Liszt's living cult took three main forms: poetic, musical, and ritual. It was common for friends to send poems as gifts in the tradition of keepsake albums. (In one intriguing instance, an acquaintance inscribed a long ode to Liszt's performance of his *Dante* Sonata in one of Liszt's own notebooks.) Intimate expressions like these typically diverged from formal encomia read aloud during special occasions, a gesture similar to being presented with a medallion.[45] Liszt's effects contain ample evidence of both categories, which is not surprising in light of his celebrity. But what was *unusual* for the period was the conflation of these two categories when admirers mailed poetic gifts to Liszt that emulated monumental odes. In a handwritten poem sent to Liszt for his sixty-eighth birthday, choirs jubilate in Apollo's fields, a temple is bedecked, wreaths are offered at the altar, and all greet and hail Liszt, the Orpheus of modern times.[46] It appears that Liszt's admirers absorbed the language of festivals held in his honor, which became a regular occurrence across European cities. During an 1860 festival in Weimar, organizers retrofitted Liszt's own festival piece for Schiller from the previous year by replacing the name "Schiller" with "Liszt." In the poetic ode declaimed there, those gathered were said to consecrate "a monument that you yourself build."[47]

One of these festivals marked a turning point for art-religion verkitscht. This was the crown jewel of Lisztiana: the Budapest celebration of 1873, an event so grand that it took even Liszt by surprise. Ostensibly, the occasion was Liszt's fiftieth year in the public eye, but the covert aim of the jubilee was to compete with Weimar to reclaim Liszt as a son of the Austro-Hungarian Empire.[48] To understand the ostentatious glitz of this event, I turn to the operetta librettist Hugo Wittmann, who wrote prolifically for the feuilleton of Vienna's *Neue Freie Presse*. It is no coincidence that his cynical remarks about the proceedings appeared in the same liberal paper that took anticlerical and secularist positions in political discourse, and that served as the organ for both Spitzer and Hanslick to critique Wagner and Bruckner. When the *Neue Freie Presse* deployed Wittmann to Budapest as a foreign correspondent, he witnessed grandiose events ripe for parody.[49]

The jubilee lasted three days and dominated the dailies for a week. Granted, it was common for papers to reprint speeches and odes in great detail, as attendees in the era before amplification could scarcely hear a word, but the intensity of the reports was unusual even for its time. One paper proclaimed that it was every Hungarian's moral duty to laud Liszt for the fatherland. Another noted that the gods of antiquity reclaimed their favorite mortals in youth, but in Liszt's case, a mortal has become a God himself and may be celebrated as such.[50] And so he was: upon arrival, Liszt was presented with gold and silver laurels by Hungarian nobles, and that evening, a crowd gathered below his apartments as military bands serenaded him with the march he himself wrote for the Goethe jubilee. To Wittmann in the square below, Liszt's silhouette in the window drifted past like a strange ghost. The next day saw a performance of Liszt's oratorio *Christus* conducted by Hans Richter. The oratorio was prefaced not only by a series of speeches, but by a festival cantata by the Hungarian composer Heinrich Gobbi. While the full score has been lost, a piano arrangement of the prologue gives us a taste. It opens with a mysterious bass tremolo and ambiguous harmonic language in a style study of Liszt's tone poems. The mystery cedes to a martial fanfare and concludes with a Hungarian march dotted with Scottish snaps.[51] The choral sections echo Liszt's own Beethoven cantata from 1845: "Hail him! Hail!" intones a soloist, and the choir later responds, "behold as heavenly dreams arise tenderly in the master's soul; songs, sweet, like vesper knells, to spread deep devotion, nearing the heights of heaven to which his spirit wanders forth. Only there he lives, only there he stirs, hearkening the choir of Seraphim."[52] Liszt lies so close to heaven in these passages that it is a wonder he is not already dead.

Wittmann's report was honest about these overblown proportions. He confessed that he was no music critic, and the best he could offer was a pathological assessment of the concert's "Lisztomania" ("the more people faint or suffer from cramps," he wrote, "the more beautiful the composition; if some die of stroke, so deeply moved are they, then one may telegraph out the news of an incomparable success").[53] Yet Wittmann found much to remark on besides the concert proper. He was amused by the odd "tangle" of dignitaries gathered around Liszt in his priestly robes ("countesses, princesses, an archbishop, a German professor, Hungarian counts and ministers, Jesuits and gypsies").[54] Much of his report caricatures the bloviating Ludwig Nohl, the music writer we encountered in earlier chapters. Nohl, Wittmann explained, is a gruesome advocate of the Liszt-Wagner artistic vision whose writings so inflate and convolute his heroes that he renders them dead as a doornail. Nohl's extravagant toast at this event is a platform for a boast about his own Beethoven research (or, as transcribed by Wittmann, "meine

Beeeethovven-Fooorrschungen!"), laced with a dash of Romani music to help things cohere, and at the moment he might finally say something meaningful, a gaze to the heavens with the words "I will say nothing more."[55] Regarding Liszt himself Wittmann has nothing but kind words. He admits he can not blame this great man for the ostentation of his jubilee and holds no grudge against this modest silhouette in priests' robes, seated among royalty and barraged with odes. His satire shows how a cult of the living could cheapen an otherwise respectable person. It is among the first portraits of a composer verkitscht.

The most surprising thing about Wittmann's text is its place among Liszt's personal papers and clippings, which suggests that the composer could laugh at the excesses of his own fame. Despite his early reputation for tossing ladies his cigar butts, the elderly Liszt approached his material afterlife with genteel humility. Shortly after the jubilee, when the city of Budapest converted his apartments into a museum, Liszt felt obligated to will them just enough "relics" to populate their exhibits, but "more would be too much!"[56] And when asked about his ideal funeral, he expressed to his partner Carolyne that he wanted a burial whose modesty befitted a priest: quite unlike Rossini's procession, his ceremony should be "as simple and spare as possible," and he should be lain to rest under the cover of night exactly where he died, with neither music nor entombment in a church.[57] In an ironic twist, those very wishes drew even more attention to Liszt's remains. After his death and discreet burial in Bayreuth, a contest erupted between representatives in Budapest and Weimar who sought to rescue him from Wagner's shadow by exhuming and reburying him with proper pomp. In life, Liszt was a cosmopolitan who occasionally staged himself as a national figure; in death, those allegiances collided and supercharged a sense of competitive Ehrenpflicht.[58] A heated debate erupted in the chambers of Hungary's government, where some demanded that an ongoing petition to exhume Prince Ferenc Rákóczy II from Turkey should be higher national priority than a cosmopolitan artist who did not even speak Hungarian. (That project later found legs in 1906, when Rákóczy was exhumed and reburied in a nationalist spectacle.) In response, a satirical paper issued a cartoon in which both Liszt and Rákóczy rise from their graves with arms outstretched like Christ resurrected, crushing members of parliament under their tombstones; the caption reads, "In vain you complain; they shall rise again anyhow!" (fig. 5.1).[59] Spoofed here is not only an excessive attention to the dead, but the strategy of *imitatio Christi* that Liszt, like Wagner, had mastered perhaps too well.

These satires point out a strange eagerness for the composer's death and the commemoration it might afford. In 1805, for instance, a false rumor of

FIGURE 5.1. Liszt and Rákóczy rising from their graves in *Bolond Istók*, March 6, 1887. Handwritten translations into German by Lina Ramann. © Klassik Stiftung Weimar, Goethe- und Schiller-Archiv, 59/372, 1.

Haydn's death in Paris initiated the commemoration machine, with a funeral hymn already underway when the news broke that he still lived.[60] Seventy years later, death was not even a necessary prerequisite to be festooned. Esteeming the living in this way shows a generation that was self-consciously aware of what Joseph Roach called the "body cinematic," the afterimage of "the body natural" that is both immortal and tantalizingly incomplete.[61] In a culture of incessant lauding, composers had to imagine their own afterlives in ways that surpassed the usual will and testament. But when lesser-known composers tried to influence their body cinematic, they could be accused of megalomania. By the turn of the twentieth century, canonicity had become a speculative enterprise where critics tried to sniff out counterfeits in a market saturated with art-religion. When imitators failed, they made sacred *imitatio* into a cheap replica that laid bare the mechanisms of self-promotion.

This is precisely what happened in the curious case of August Bungert, a now-forgotten composer who was known in his day, somewhat disparagingly, as the North German Wagner. We know vastly more about Bungert than one might expect thanks to a comprehensive volume by Christoph Hust, which reconstructs the composer's virtuosic self-staging as a German patriot, a Nietzsche exegete, a prophet, a Wagnerian–turned–Wagner adversary, and above all a businessman who exploited his professional network.[62] Bungert's emulation of Wagner's cult caught the attention of his contemporaries, despite his insistence on the integrity of his image. With a Wagner-sized ego apparent in every word he wrote, Bungert assembled a circle of disciples led by the devoted Max Chop, a music critic who founded a newsletter that emulated Wagner's *Bayreuther Blätter*. There a ragtag group of German nationalists sought to promote Bungert's best-known composition: his operatic tetralogy *The Homeric World*, a cycle of over one thousand pages based on Homer's Odyssey that was composed between 1880 and 1896. While Bungert's compositional style was quite different from Wagner's—among other things, he favored diatonicism over chromaticism to portray the purity of the Hellenic world—the scope of his ambition looked remarkably like the *Ring* cycle, and Chop made matters worse by circulating a guide to Bungert's leitmotifs.

It did not help Bungert's case that, one year after his cycle was complete, he petitioned the mayor of Godesberg to erect a festival house for the exclusive performance of his Homeric masterpiece. His vision was an international hub of summer tourism on the Rhine, modeled quite consciously on Bayreuth and funded by royalty. He had indeed secured some support from his advocate Carmen Sylva, the erudite queen of Romania who authored poetic texts for Bungert's Lieder, and who served as the protectorate for the Bungert Society.[63] When he wrote this letter in 1897, Bungert was riding high after the success of *Odysseus's Return*, the last opera of the cycle, in Dresden, but in Berlin the following year, that same opera was so thoroughly lambasted that interest in his Godesberg festival dissipated. Almost overnight, Bungert went from an artist on the rise to a laughingstock fringed by a zealous coterie. In Berlin, the feuilletonist Josef Schrattenholz spoofed Bungert's ego in a series of puffed-up poems, while Hamburg's Ferdinand Pohl accused him of plagiarism, calling his opera a "colorful hubbub [*Brimborium*]" and a "mythical variety show."[64] One Berlin review made it plain: "Wagner has been out-Wagnered [*überwagnert*]; his Nibelung Trilogy is repeated, or will be repeated, just as soon as August Bungert has finished his music-dramatic hexalogy *The Homeric World* and—tremble, Bayreuth!—has realized his plans for a Bungert-Festspielhaus in Godesberg."[65]

Pohl's review offers another insight: he considers Bungert the epitome of the "chaste Bildung Philistines [*dem keuschen Bildungsphilister*]," those who wanted their bookshelves of Great Works to come alive in shallow spectacle.[66] It was not only that Bungert copied Wagner's oversized proportions; he embodied a fanaticism for the classics that was seen as commercial and insincere. One paper noted that, if others imitate Bungert as he did Wagner, we will soon drown in "Dantean Worlds" and "Shakespearean Worlds" and no end of twelve-opera cycles that turn all of classical literature into a musical fresco.[67] Bungert appears here as the musical version of symbolist painter Arnold Böcklin, whose Hellenistic canvases were so explosively popular that they were later framed as kitsch for the middle classes.[68]

It is admittedly amusing, in this light, to read the pages of the monthly *Bungert Bund*, the newsletter that circulated among Bungert's fans from 1911 to 1914. To read this journal in its entirety is to trace the arc of an artist forgotten. At first, there is excitement in the air, a call to arms rife with declinism. The political stance of the Bungerites is clear: a conservative counterweight to modernism, cosmopolitanism, and decadence, an assertion of rural North German pride, and a belief in Bungert as the ideal vehicle to transmit the values of ancient Greece to the populace in an act of Bildung. Each issue contains a section called "Defense against Attacks," where Chop and his compatriots spilled considerably more ink than the critics whose blows they parried. As Bungert's music drifted into obsolescence, that combative energy found a broader outlet in disdain for the "hypermodernism" of urban musical life and its so-called cliques (who, exactly, belongs to these cliques is never clear). With an ever dwindling authorship, and obituary after obituary for its members, the journal became a strange vacuum for a lone Max Chop shouting into the void, punctuated by occasional submissions from Bungert himself. In 1914, the last year of his life, Bungert offered a poem to the *Bund* that compared modern music with the sounds of cats and dogs.[69] Ever defiant, he stood by and watched as his music became a bourgeois anachronism.

Imagine the backlash when, in 1912, a Darmstadt professor named Willibald Nagel published a rebuke of Bungert's disciples titled with nothing but a question mark.[70] Nagel hailed from the same small Rhineland town as Bungert, and while he had few qualms about this music, he suspected some chicanery of the *Bund* in an age when art for art's sake had become a business. To erect a hall in competition with Bayreuth is folly, Nagel claims, because Bungert's music, while innocuous on its own, is a phony imitation of Wagner's. Would it not be strange, he adds, if every Alpine mountain were topped with a festival hall with its own funicular—a Lehmann-Bund, a Schulze-Bund, a Müller-Bund, ad infinitum? Everyone wants to idolize something, he explains, even if it's a five-penny idol like the golden calf. At these words, the Bungerites were incensed. Posterity has proven them wrong.

Spats like these show a crucial turn in the skepticism toward art-religion. Long before Walter Benjamin formalized his concept of aura and its degradation through mechanical reproduction, critics were already suspicious of cheap imitations, and especially *imitatio* with its aura of the sacred that masked a commercial function. When Nagel imagined a world filled with Bayreuths and miniature religions, he seems almost to anticipate, by seventy years, Jean Baudrillard's dystopian portrait of our world populated by simulacra, by copies of copies that are ever more distanced from their originals.[71] In Christian practice, *imitatio* is not inherently suspect. But in an era of commercial interests, hyperbolic ceremonies, and Wagnerian marketing, *imitatio* lost its aura. It was the impulse not only to replicate, but to cloak the replica in bourgeois respectability, that irritated critics. Their concerns prefigure by several decades what came to be called kitsch.

Mass Delusions

The mistrust of *imitatio* coincided with another population of gullible followers. The 1870s saw the first public debates about spiritualism and occult movements, and by the turn of the twentieth century, the response from administrative entities like the scientific establishment, the church, and state governments had grown more heated. High-profile trials of mediums, such as Anna Rothe's conviction of fraud in 1903, persuaded the public that charismatic leaders of alternative religions are a threat to society; satirists followed close behind, lampooning spiritualism and its naive followers.[72] That impression was compounded by a concurrent discourse on crowd behavior. Gustave Le Bon's alarmist treatise of 1895 was introduced to German readers by Georg Simmel, sociologist of the mass and individual who feared how urban modernity could desensitize the human, and theories of mass culture soon pervaded German and Austrian sociology, philosophy, and art criticism.[73]

In this period of skepticism about alternative religion, the veneration of composers took on the character of a cult. In a compelling coincidence, two authors, within a span of a few months, wrote nightmarish satires that mocked art music's collapse into kitsch. In December of 1912, the Viennese sociologist Edgar Zilsel published his "Mozart and Time" in the same magazine that later presented Kosztolányi's ghost story. In February 1913, in the journal *Die Musik*, the musicologist Alfred Heuss published "A Day at the Monastery of St. Ludwig."[74] The political orientations of these authors were quite different: Zilsel was a Marxist, while Heuss later became infamous for his conservatism. The resonance between these essays, which invoked dreamlike allegory to critique music lovers as mindless masses, shows how mass culture was politically slippery, targeted by all political orientations in turn.

Both satires are cut from characteristically Viennese cloth: they exhibit what Edward Timms has called the "symbolic decoding of trivia," a characteristic trait of Viennese discourse in which "objects of everyday experience [are] emblems of ideological positions"⁷⁵—that is, a mode in which any single newspaper article, performance, coffee shop, or fashionable thing could prompt an interrogation of the cultural moment. It follows that kitsch discourse originated here as a reaction to Viennese *Gschnas*, a lively word from Austrian dialect that refers to costume parties, tasteless frippery, and an indigestible hodgepodge of foods.⁷⁶ From 1938 to 1951, writing in American exile, the writer Hermann Broch remembered Vienna as the "metropolis of kitsch," a "culture void," where the impulse to preserve the city as a museum left behind the overdecorated shell of a declining empire.⁷⁷ In 1908, in this city of frilly facades, the architect Adolf Loos channeled his aversion to surface ornament into an incendiary rant against "uncivilized" artisans and racial others, invoking Beethoven and Wagner as spiritual balm for the educated.⁷⁸ Despite his claims to modernity, Loos was a nineteenth-century gentleman at heart, designing cozy spaces of contemplation that would revive the interiority of Bildung and shelter the mind from urban chaos.⁷⁹ His essay and its backlash exposed a rift between those who took refuge in older forms of Bildung and those who embraced the frivolity of Gschnas.

In this climate, a new literary form was popularized by the satirist Karl Kraus: the "didactic fantasy," in which surreal dreams illustrate a point. Kraus's periodical *Die Fackel*—widely read during its four-decade run—sought to expose the facades of Austria's liberal Bildungsbürgertum, or what he called the "Janus-faced" bourgeoisie whose progressive politics masked their conservatism and backwardness.⁸⁰ (This demographic was also Kraus's readership, which gave the critique its bite.) Music, too, could serve as a mask for brainwashing, a theme to which Kraus returned across his career. He noted the hidden politics of jaunty Viennese popular tunes, the effervescence of operetta, the propagandistic potential of men's choirs who lather up pan-Germanism on composers' anniversaries, and the sacralization of Bayreuth.⁸¹ For Kraus, Bildung was a false front that veiled imminent collapse. Kraus has been dubbed the "apocalyptic satirist" for a reason: his fantasies often culminate in nightmarish disasters that evoke the earthquakes of the book of Revelation, as the bourgeoisie tears itself apart.⁸²

Enter Edgar Zilsel. Today, he is best known for his Marxist intervention in the sociology of science; in 1912, he was a young philosopher of logical empiricism, an aspiring sociologist, an avid concertgoer, and an apparent admirer of Kraus. His first publication was an anticapitalist fantasy in *Der Brenner*, a German periodical modeled on *Die Fackel*.⁸³ Through a surreal dream, Zilsel ruminates on music's false claims to timelessness that veil a material marketplace. The story begins as dilettantish listeners—all of whom, we quickly learn, are well-dressed animals—vacate the Vienna Staatsoper and stampede about in conversation, dispersing to feed in their stalls. When the narrator looks up, he sees the greedy spider of time, in whose web the world decays into a fog of mildew. He finds himself suspended in a cloud where pure tones arrange themselves like crystals, while the "slurry of earthly feelings" flows below.[84] Here in the heavens, the narrator feels music with a visceral intensity, his hair singed by Brunnhilde's magic fire and cooled by the crystals of Mozart's sublime sounds. As he yearns for Mozart's presence, reaching out to the eternal tones for traces of their creator, he falls into an eighteenth-century tableau of Mozart's workaday realities that jolts him from the sublime to the ridiculous. The narrator, as a sort of time traveler, encounters Mozart as timeless essence, only to be slapped in the face by Mozart the anachronism.

If Zilsel's allegory ended here, we might think that he leveled his critique at ignorant operagoers, or at the disappointing banality of Mozart's daily life, which interfere with an immersion in musical sound. But timelessness, for Zilsel, is neither attainable nor desirable. The story ends as the spider of time dissolves human civilization into a swamp, in whose depths lurk a threatening version of the Ouroboros, the symbolic snake of eternity that eats its own tail:

> Streams of mud trickled from Mozart's grave and merged with an ocean of scum, which, diluted by new floods of rain, swelled slowly, and burying everything beneath it, only lifted sluggish foliage aloft to its firmament. The leaves expanded into the immeasurable and from the printed pages swelled toads with their slimy gleam. A gramophone circled, raven hinged, and cawed out the newest operetta. But it swarmed in the slurry; a cluster of dithering Wagnerians and other titanic clerks had worked their way up from under the newspaper pages and clambered onto that saving raft, but they tugged the others down by the hair; sticky maggots, fattened on the dead rococo and patriotic artists [*Heimatskünstler*]. In the depths spun the fabled snake that eats its own tail: the progress that progresses toward progress. It exhaled its vapor, I bolted upright, yet I smelled the stale mist of sulfur and fell into the morass.[85]

Here the beauty of music drowns in *stuff*, in sheet music, gramophones, and newspapers churning out the latest polemics. Zilsel's disillusionment with the technologies of reproduction and dissemination prefigure the "culture industry" of Horkheimer and Adorno—that is, an industry of sameness that churns from the radio to mark the dead end of bourgeois liberal ambition.[86] Zilsel's allegory shows the failure of the mythologies of progress that had fueled Wagnerian rhetoric by showing us an anachronistic and nonlinear model of time, in which history and the present dissolve into slime. In the end, there is no canon, no pure music, no heaven or hell, but a predatory dustbin of history that devours all.

If this allegory weren't evocative enough, it takes on more meaning when read against a slim manifesto that Zilsel had started to sketch that same year. In 1918, delayed by war, he published *Die Geniereligion* (The cult of genius), which was among the first systematic critiques of the deification of secular figures.[87] Zilsel saw relic collecting and pilgrimage as tools for demagoguery: "with holy awe, as if on pilgrimage to Lourdes, we journey to these genius graves" where "we treasure the relics, autographs, hair locks, quills, and tobacco boxes of our great men just as the Catholic Church treasures the bones, accessories, and clothing of saints."[88] These are only the trappings of religion, he argued, and like the swamp with its Ouroboros they are underpinned by deeper dogmas: a cult of sentimentality that insulates geniuses from rebuke, a conflation of genius with divinity, and a belief in heroic individualism that devalues collective achievement. Zilsel identified the mechanism by which these dogmas spread as the almost automatic "rubbing-off" of ideas from "genius priests"—namely charlatanic public intellectuals and their army of "titanic clerks"—in ways that diminish the rational thinking of the masses.[89]

For Zilsel, with his Marxist sympathies, "the masses" are not the unruly lower-class mob that troubled Weimar intellectuals of the 1920s, but the educated classes who are spoon-fed what Adorno later called *Halbbildung*, or diluted culture. This makes Zilsel unusual among social scientists of mass culture. Stefan Jonsson has shown how the slippery concept of "the masses" was an idée fixe in Austro-German art, literature, sociology, and psychology in a period of political instability, when both Austria and the Weimar Republic tried and failed to establish a democracy. What made "the masses" so nebulous, Jonsson shows, was an indexical function that named and pointed without coming into focus: "It named a savage part of humanity, and it pointed toward the wilderness, territory uncharted by the social scientist."[90] The only way to define that uncharted territory was by contrasting the mass with the individual, and in turn, the individualism of Bildung became a bulwark against the mass.[91] Once Bildung was commodified, as in the "patriotic artists" who drown in the capitalist slurry of Zilsel's fantasy, the educated middle classes were no longer immune to the dangers of mass culture. Nor are Zilsel's masses the haughty bourgeoisie one finds in critiques of decadence, but stony-faced and industrial, the white-collar equivalent of the worker swarm in Fritz Lang's *Metropolis* or the wave of suits pouring into the U-Bahn in Walter Ruttmann's *Berlin: Symphony of a Metropolis*.[92] Zilsel's operatic animals were a familiar metaphor for those bored with the rituals of concert life. In 1921, Alfred Polgar offered a comparable portrait: at the orchestra, hornists sleep during long rests, trombonists read the papers, and a well-dressed listener munches on sugared nuts with a "silly enraptured expression on her face." One day, Polgar notes, she will be reincarnated as a grazing donkey.[93] While the boredom in Polgar's satire echoes Hector Berlioz's *Evenings with the Orchestra* from 1852, its macabre sensibilities mark this as a product of the twentieth century. Here concert life is not only stagnant but decaying, as a memory of a concert past is interrupted by intrusive thoughts that all these people are now dead, worm riddled and feeding the daisies. This was a different face of cultural degeneration: the atavism of empty-headed tedium.

When the word "kitsch" spread after the war, the unthinking followers of Bildung were implicated alongside the petit bourgeoisie. For cynics, art itself was nothing but the "miscellaneous stuff of fashion" (*Modekram*).[94] These were the words of the German playwright Frank Wedekind in 1924 in his sketches for a play entitled *Kitsch*, in which characters in a love triangle represent strata of class and taste: a young artist is a highbrow, a composer and savvy businessman is a lowbrow, and an art historian is a harbinger of the middlebrow, an expression that emerged at midcentury to deride the dull bourgeoisie who cling blindly to Bildung. Peter is a hot-blooded composer (or painter, in later sketches) who locks horns with the art history professor Zugschwert as they compete for the hand of a haughty Swiss version of Lulu. The lowbrow is a low blow: a Grand Opera composer named Nachtigall who is a Black Jewish avatar of Meyerbeer, denoted by a slur that implies his degeneracy.[95] Not only does Nachtigall strive for profit, which earns contempt from Peter, but his anachronistic tastes and aspirations make him an object of scorn. In one scene, he admits to his foolhardy *imitatio*: "at times, in soliloquy, I accidentally call myself Beethoven." Each of Wedekind's three artists finds himself verkitscht in a different way: Nachtigall writes for the masses with naive aspirations, Zugschwert is so stodgy that he fails to recognize the kitsch of his own Bildung, and even the skeptical Peter confesses that his latest portrait of Nachtigall is nothing but "a kitsch."[96]

Wedekind's play marks the earliest use of *kitsch* to describe what Zilsel implicitly critiqued: here the concept is not a jab at the lower classes, but a social environment devoid of rigorous Bildung in which all classes flounder equally. The literary modernist Hermann Broch found himself caught in that quagmire. His long-simmering unease with the dynamics of the mass culminated in an unfinished treatise, *A Theory of Mass Mania*, which sought to explain the populist tendencies that had lately coalesced into fascist demagoguery. Across this ambitious and interdisciplinary text, Broch positions himself as an Enlightenment humanist who seeks to salvage the positive societal effects of religion as a defense of democracy; he intervenes, with some pessimism, in what he perceives as a cultural vacuum that leads society to revert to primitive rituals and totems.[97] Relatedly, Broch wrote several critiques of kitsch. In an essay of 1947, he argued that musical style is continually reborn to evade *Verkitschung*, as if kitsch were nipping at the heels of art.[98] A few years later, he consolidated these ideas into a lecture delivered to Yale's German department, arguing that kitsch arose in the nineteenth century as Romanticism teetered between cosmic heights and triviality. Following Norbert Elias, whose essay of 1935 Broch echoes but does not cite, he suggests that the German middle class tipped Romanticism over the edge in their mimicry of aristocratic taste, a mindless *imitatio*.[99] Regardless of the merits of this theory, Broch's discussion of music looks like a tribute to his audience of Germanists: Bach and Beethoven pass muster, while Chopin and Tchaikovsky are pinnacles of kitsch and Berlioz is "only just bearable."[100] Wagner earns special attention because his bombast and hypersentimentality fit Broch's own definition of kitsch, but to avoid blaspheming the master, he makes Wagner a victim of the culture vacuum whose great achievement was his graceful dance on the razor's edge of kitsch.[101] If Broch here seems neurotically insistent on German genius, he was indeed compensating for something. John A. Hargraves has traced Broch's earliest love of music in his letters, and here we find a young man who embraced operettas and piano duets, relishing Saint-Saëns and Grieg ("I'm terribly fond of him," he wrote). Suddenly and inexplicably, Broch found himself disgusted by his earlier pleasures, souring on Puccini, Tchaikovsky, and even Schubert ("it's all been said already").[102] Broch is compelling not only because his writings linked kitsch with mass mania, and linked mass mania with religion. Latent beneath those ideas was a coming of age, a dynamic process of wrestling with guilty pleasures in a period when Bildung was ever more commodified.

A few months after Zilsel published his essay, another fantastical critique of art-religion appeared in the journal *Die Musik*, which later became a mouthpiece for musical propaganda in the Third Reich. Publications in this journal were a far cry politically from Zilsel's anticapitalism, nor did its contents align with critiques of the bourgeoisie by Kraus, Wedekind, and others who rubbed shoulders with modernists. The author was a young Alfred Heuss, a Bach scholar who later attacked modern music and mass culture in his position as editor of the *Neue Zeitschrift für Musik*. Heuss's mission, as Nicholas Attfield explains, was to recover a distinctly eighteenth-century practice of moral and spiritual listening by focusing on music's architectural form, not its Romantic excesses.[103] In 1923, Heuss issued a scathing review of the young Paul Hindemith, whose foxtrots he found orgiastically depraved; Hindemith's letter to the editor, dripping with acid sarcasm, saluted Heuss as the "Noble

Relic of a Better Century, Guardian of Teutonic Art," whose staid hermeneutics were "kitschy scrawlings"; that is, Heuss's pedantic nostalgia for lost Bildung was an even worse form of kitsch that Hindemith found representative of the National Socialists.[104] But this spat was yet to come. In 1913, a young Heuss aimed his frustrations with mass culture not at foxtrots, dance halls, or sexual depravity, but at the nineteenth-century cult of genius run amok.

Understanding this early essay requires a closer look at the years leading up to World War I, which saw a vogue for designing Beethoven temples. A series of initiatives, some derived from German life reform movements, fashioned a new religion that sprang from Beethoven's music. In 1907, the Munich architect Ernst Haiger sketched a symphonic temple dressed with an allegorical frieze; each performance of the Ninth would be illuminated by fire that streamed from the choir loft.[105] Oddly enough, Haiger's idea gained enough traction that a committee was convened in 1913 to erect the temple in Stuttgart in time for Beethoven's anniversary celebration in 1920.[106] (Like the foiled fêtes of 2020, that plan was aborted in the wake of war and the Spanish flu, which explains why festivities for Beethoven's death day in 1927 were unexpectedly robust.) Meanwhile, in the salons of Paris, the eccentric architect François Garas painted a series of watercolors that included a "temple of thought" dedicated to Beethoven. Inside this mass of buttresses and spires, still under construction and draped in pulleys, Garas envisioned a bust of Beethoven rising from a vast clock made from rows of naked human bodies.[107] Buoyed by these projects, the Dutch musician Willem Hutschenruyter published a pamphlet called *Das Beethovenhaus* in 1908, which was translated into German in 1911.[108] His Beethoven temple would be situated on the sand dunes of the Netherlands, and it served as an antidote to the ills of concert life, rooting out potpourri programs and loose etiquette. In his opening salvo, Hutschenruyter chronicles all the imaginary temples mentioned here. A Beethoven religion is emerging, he claims, which needs state support to flourish.

When Heuss encountered this pamphlet in 1912, he responded with a didactic fantasy that prefigures Mauricio Kagel's *Ludwig van* by a half century. The piece is relentlessly entertaining. It begins when the author drifts into a dream with Hutschenruyter's book in his lap, and he finds himself at the gates of a Grecian-Gothic temple, reenacting Wotan's first glimpse of Walhalla. As a carillon rings out the A-flat theme from the Fifth Symphony, Heuss meets a pilgrim who acts as his tour guide, and they wander a massive complex of splendid concert halls, atria, gardens, and libraries. Heuss satirizes concert culture and academia in turn. The pilgrims pass by several temples, each dedicated to the endless repetition of a single work, and they visit a library that contains one hundred thousand volumes on Beethoven that average one thousand pages each. Like prospective students, they tour the campus of Beethoven University, complete with its institute for the identification of the Immortal Beloved and a scientific laboratory to identify forged documents, a jab at Anton Schindler's falsified entries in the conversation books.[109] (Decades later, Kagel's *Ludwig van* poked fun at this systematic positivism, as a self-important scholar hammers out his tenth article on the topic.)

When Heuss's pilgrim explains how the Beethoven religion began, the story takes a turn toward fascism and totalitarian reeducation. Heuss explains that, long ago, there were a variety of sects like the Bachanten and the Mozartnarren. While the followers of Mozart caved quickly to missionary efforts, the Bach heathens were harder to assimilate. There was once, too, a sect so large that it might be called a party: the Wagnerians. Fortunately, their compatibility with the Beethoven religion allowed them to be absorbed. With some relief, Heuss's pilgrim explains that names like Bach, Handel, and Mozart have been forgotten, and "this specialized knowledge is only cultivated by a few historians."[110] The temple itself has existed since the year 180—that is, 180 years after Beethoven's birth, which has marked the Common Era ever since the "Beethovenian worldview" was declared an official state religion. Now, in the new era, children learn about the life and teachings of St. Ludwig in schools and aspire to make this pilgrimage once in their lives, a rite of passage for the self-actualized. All citizens of the new republic are subjected to grueling state exams: two beats from the middle of a work by Beethoven must be identified and dated to the day and hour of composition. One hears tell of a candidate who misattributed an excerpt to Haydn, and subsequently died of hunger while serving a life sentence in the Florestan Prison.

Like Kraus's apocalyptic fantasies, then, Heuss's utopia is gradually exposed as a dystopian nightmare. In the end, the pilgrims reveal their true nature. Riots break out during a performance of the Ninth, the robed pilgrims annihilate themselves in bloody battle, the temple crashes down around the narrator's ears, and he jolts awake to find that he has wrenched Hutschenruyter's pamphlet in half. These are no erudite devotees, but an unruly mob living in a police state. Their violence resembles that of the relic collectors found in this book, who splintered Beethoven's cradle or sawed Mozart's skull in half. At stake is not the preservation of the past, which Heuss appreciated, but the effacement of music's architectural beauty by a monomaniacal fixation on the composer.

Stepping back from this disarray, we find ourselves a far cry from Daniel Spitzer or Hugo Wittmann, the satirists of Wagner's cult and Liszt's ovations. Late nineteenth-century skeptics could feel disenchanted by art-religion, could observe pomposity with remove, and could cautiously guess at posterity, wondering what new genius might establish the next Bayreuth. But in the early twentieth century, doubts about art-religion were suffused with violence, even as that skepticism emerged from various political positions. Both Zilsel and Heuss yearn to recover a Kantian ideal of musical beauty that has been effaced by a dangerous caricature of religion, by "genius priests" who hide their machinations behind the composer's charisma. In Zilsel lurks a fear of drowning in capitalism, which masks a deeper fear that crystals of sound may be an illusion. For Heuss, the conservative exegete, there is a fear of mass stupidity and the collapse of knowledge. Theirs is a distinctly twentieth-century kind of laughter, deadpan and dark.

The Composer Embalmed

For Zilsel, Heuss, and other critics of kitsch and mass culture, the rituals of Bildung are a sinister anachronism. In these satires, art-religion is not only a replica, but a reanimated corpse, a grandiose ceremony for the dead whose followers chew their cud with a blank stare. The allusions to death and violence in these satires responded, if subtly, to the morbid curiosities of Austrian funeral culture, which coveted the *schöne Leich*, the beautifully staged body. Parading bodies about served to animate politics, as Katherine Verdery has argued, and both Austria and neighboring Hungary, its former empire, were particularly keen to speak through their dead.[111] In chapter 1, we saw Beethoven's spirit freed from the walls of his demolished house by a funereal ceremony. In 1954, Haydn's skull was held aloft to glitter in the flashbulbs before it was reunited with its body. Bartók was exhumed and welcomed back to Budapest in a grand state ceremony. A year later, at the feet of the Iron Curtain, a quarter of a million Hungarians gathered to rebury their revolutionary prime minister, Imre Nagy (whose corpse was lovely in theory; in truth, the skeleton had been cheapened by the addition of giraffe bones from a nearby zoo). This cavalcade of Austro-Hungarian corpses shows a turning point for art-religion verkitscht. With the embalming or exhumation of the composer's body, and the anachronistic rituals that accompanied the coffin, the effort to preserve the "lovely corpse" as an incorruptible relic was disrupted by the inevitable decay of these remains. In the twentieth century, piety veiled much more than commercialism: it masked the moral rot of the National Socialists, who famously reanimated nineteenth-century rituals in an ideological zombie culture.

To argue this point, I turn to an extraordinary artifact: Anton Bruckner's embalmed body. This requires some setting of the scene. From the Upper

Austrian capital of Linz, one can take a bus to St. Florian, a small town appended to an Augustinian monastery. Bruckner was employed there as an organist for several years, and his admirers still flock to hear his organ played in a gilded chapel. On the floor beneath the organ, one treads over Bruckner's name inlaid in marble, bounded by the outline of a tomb. A private tour by request leads one from peaceful stone corridors down into the depths of the crypt, where the air grows damp with a distinctive charnel odor. The space beneath the nave is lined with alcoves where the coffins of former abbots are lain to rest, two by two. In the center of the nave, what look at first like solid walls are niches that house the bodies of distinguished canons in long rows. Deep in the crypt, directly beneath the organ, these narrow corridors open into a spacious, vaulted room. Along the back wall is an ossuary, with its hair-raising wallpaper of eye sockets and femurs.[112] While the canons and abbots are clustered together in cozy nooks, gathered in good company for eternity, this room contains a lone tomb that rises from a pedestal, its metallic surface overgrown with foliage motifs. This is the resting place of Anton Bruckner, just as he wanted it.

Throughout this book, composers appear in bits and pieces that underscore the absence of the celebrity body. Bruckner's tomb is a rare exception. While the organ plays above, sublimating the body into sound, Bruckner remains stubbornly *there*. Like Beethoven's mask in the previous chapter, Bruckner's remains are caught between living and dead, in part because his embalming lent him the air of a nineteenth-century "beautiful death," in which a flower-fringed body lies tranquil on display. The anachronism of this tomb lies not in a return to lost or nostalgic pasts, like kitsch, but in a refusal to acknowledge time at all. Bruckner's last wishes, however they were intended when he drafted them, forced later generations to confront the indignities of decay. In 1929 and again in 1996, the outdated paradigm of the beautiful death met the cold autopsy table of twentieth-century forensics, which turned the relic into an anthropological artifact.

It is an ongoing challenge in Bruckner scholarship to disentangle the composer's own wishes, which were rarely articulated with much effusion, from those of this pupils and admirers after his death. We can say for certain that Bruckner had more precise specifications for his interment than many; those wishes may reflect his own Catholic piety and his relationship with death, rather than a monomaniacal desire for devotees.[113] Posthumous accounts affirm that Bruckner sought out tangible contact with the dead. It would be grossly exaggerated to call this "necrophilia" as it was certainly not sexual, and his behavior was consistent with the Catholic materiality of devout Upper Austrians, or practices derived from the baroque culture of *pietas austriaca*.[114] All we can say for certain is that Bruckner was fond of a photograph of his mother on her deathbed, and while the details are likely exaggerated, he was said to have barged uninvited into the crowd of doctors who examined Beethoven's skull when it was exhumed. Witnesses reported that he laid his hand tenderly on Schubert's brow, then took Beethoven's skull in both hands and spoke to it; two accounts recalled him saying, "you would have let me hold you, dear Beethoven, if you had lived, even as this strange man wants to forbid it." Another claimed he pressed the skull to his heart, at which point a doctor exclaimed, "this man is insane!"[115] At the time, this story splashed across the papers and worsened Bruckner's reputation for provincial eccentricity among his urban detractors. Lesser known was the composer's reaction, decades earlier, to the embalmed body of Maximilian, Archduke of Austria and emperor of Mexico. Several years after Maximilian's execution in Querétaro, his embalmed body was transported back to Austria, and a photograph circulated in the press. Bruckner was fascinated. He wrote to a friend: "I would like, at any cost, to see the body of Maximilian. Please be so good, Weinwurm, as to send a reliable person into the palace, ideally; let it be inquired with the Office of the Court Chamberlain whether the corpse of Maximilian may be *viewed, in the open* in the coffin or through glass, or if only the closed coffin may be viewed."[116] That was in 1868, and in the years that followed, Bruckner was known to socialize with a group of young doctors who gathered regularly at a café. It was here, or so speculates Klaus Heinrich Kohrs, that Bruckner learned about the new science of embalming with formaldehyde. He befriended a pupil of Josef Hyrtl—the anatomist we met in chapter 3, who possessed Mozart's skull for a time—which emboldened Bruckner to compose a letter to Hyrtl asking whether the soul lies in matter or transcends it.[117]

It appears that, when he composed his will and testament, Bruckner was not yet certain of the answer to that question. His body, he declared in 1893, would be preserved by injection and interred, at considerable expense, beneath the organ.[118] It is reasonable to suspect that Bruckner was afraid of being buried alive, a concern that was so common in this period that many caskets were outfitted with a bell. It is also possible that Austria's culture of material piety and Ehrenpflicht led him to question Christian tenets in which the soul is released at death, which made the decay of the underground feel threatening. Whatever the reason, Bruckner insisted that he should not be buried in the churchyard, as would have been standard procedure for civilians who were not members of the clergy. This made him an outlier in the crypt below St. Florian.

A more difficult question is whether Bruckner wanted his body to be viewed, or whether this reflected the wishes of later devotees. For the first several decades of Bruckner's interment, his tomb was outfitted with a full-body

 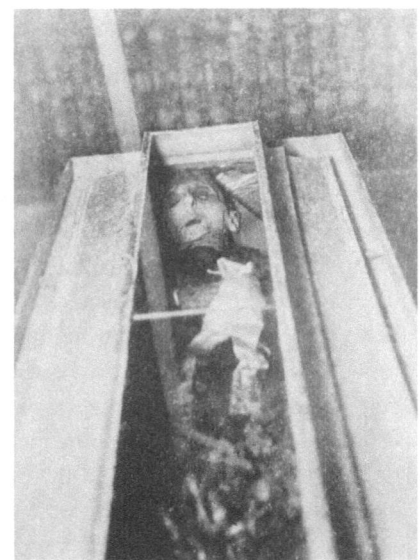

FIGURE 5.2. (*Left*) Photograph of the embalmed body of Emperor Maximilian I of Mexico, by François Aubert, 1867. © Metropolitan Museum of Art. (*Right*) Photograph of the embalmed body of Anton Bruckner taken in 1921. Austrian National Library, Vienna, Music Collection, Misc. 293/186.

window that emulated that of Maximilian or the Habsburgs generally (fig. 5.2). In 1930, Max Auer, the president of the International Bruckner Society and foremost authority on Bruckner, insisted that the plan was purely hygienic and "not so that the curious masses should get a peek."[119] More recently, Andreas Lindner argued that Bruckner was no narcissist, but rather sought out the most sanitary means of transport from Vienna to St. Florian, thinking solely of hygienic ordinances.[120] The fragmented notebooks of Bruckner's pupil and authorized biographer August Göllerich suggest otherwise: "He designed his coffin himself during his lifetime in Vienna, very broad, like his clothing, inlaid with a glass pane along the full length of his person, so that one ~~could~~ can see him [*sic*]. (Auer saw him very well preserved, as he was well embalmed.)"[121] If Bruckner did in fact wish to be seen, perhaps we need not call it narcissism. By offering a window into his eternal sleep, he may have wanted his admirers to experience the same tenderness he felt toward the deathbed photograph of his mother.

Opportunities to commune with Bruckner had an element of the sacred. Lifting the lid was not mundane, but reserved for special anniversaries of the composer's birth and death, when concerts would culminate in a descent to the crypt. A substantial number of these events were noted in the papers, in such a piecemeal fashion that I suspect many visits went unreported, including backstage tours for conductors and important guests.[122] A typical example is the one-hundredth birthday in 1924, when the Vienna Men's Choral Society sang Bruckner's elegy "At the Grave" around the tomb-shaped inlay in the narthex; a sermon-like speech was given; Bruckner's organ was played; and finally, all processed into the crypt to observe the open coffin. Even the most curt reporters could not help but wax poetic: "From the yellow light of a few candles emerged the unforgettable face of Bruckner, looking deeply pale, like a resin mask."[123] The birthday event was modeled on the twenty-fifth anniversary of Bruckner's death a few years prior, which drew large crowds from the surrounding cities, not least a field trip of middle schoolers from Linz.[124] In the chapel was heard Bruckner's Requiem in D Minor and his "Epitaph" (*Nachruf*) for choir and organ, followed by a speech from the canon that emphasized Bruckner's deep Catholic conviction, and a trip to the crypt to admire the master under glass.[125] The late Bruckner's pupil Franz Gräflinger reenacted the event over the radio. He recalled that the crowd, candles and torches in hand, was so large that only a small portion could pack into the vault. There in the gloom, "ghostly glimmers of light flickered onto the rows of stacked skulls in the alcoves. There stood Bruckner's sarcophagus. Today the lid is off. Through the glass one sees the dead master, preserved through embalming, who appears heavily narcotized and wax-pale in color."[126] Those assembled heard the canon give a speech about the depths of Bruckner's faith, and Gräflinger's account of its contents ends in the prayer affirmation *fiat, fiat*. With this resting place beneath the organ, he adds in a later episode, "so became St. Florian the mecca for all Bruckner pilgrims."[127]

It would be too simple to interpret these processions solely as Catholic. These were layered rituals in which Bruckner's body took on anthropological, political, and spiritual meanings. For one thing, encounters with the embalmed were less startling in the period than they might seem today. Visitors to Vienna often made a trip into the crypt beneath the church of St. Stephan to view the city's Habsburg forebears; detailed accounts of what one might find there, along with tours of Austria's lesser-known crypts, were frequent fodder for the illustrated dailies.[128] That interest in domestic remains was made more riveting by photographic technology: when Tutankhamun's tomb was uncovered in 1922, it was publicized with carefully staged images of archaeologists at work.[129] The procession to Bruckner's tomb at his 1924 centennial, so soon after King Tut dominated the papers, might well have been poised between ritual ovation and archaeological curiosity.

Even those in a spiritual mood might have experienced these processions as both Catholic and cosmological. After the First World War, a substantial critical movement endowed Bruckner's symphonies with a mysticism that transcended his Catholic piety, making him both an occult presence and a redeemer of the German national spirit.[130] (It is no surprise that Heuss, a decade after his "Sankt Ludwig" satire, worried that the Bruckner-sect had gone too far. He imagined a cult member's conversation with Bruckner's ghost: "Don't forget, dear Saint Anthony, you have an astral body.")[131] Already in Bruckner's lifetime, his music was enmeshed in Vienna's occult circles in ways that musicologists have only begun to examine. Bruckner's pupil and secretary Frederick Eckstein was an active theosophist and mystic, and even the famed Rudolph Steiner, the father of anthroposophy, studied composition with Bruckner. In later years, Steiner penciled into Göllerich's notebooks a tidy map of Bruckner's usual seat in the Café Griensteidl.[132] There are grounds to speculate that, for select attendees, the Catholic ethos of this site intermingled with more esoteric visions of immortality.

Down in the crypt, visitors found a medievalistic refuge from modernity, an arrested sense of time. These same qualities surface in interwar critical appraisals of Bruckner's symphonies, whose temporal expansiveness made for an archaic listening experience.[133] It is impossible to say whether the suspended time and space in these symphonies was encoded with political meanings, such as the concept of *Lebensraum* that later became a push to control the space of Europe. That argument by Karen Painter looks admittedly less secure in the wake of Pamela Potter's intervention: music is an unreliable carrier of political messages, and efforts to pinpoint the "Nazification" of the arts can be a political project to quarantine art infected with Nazi contagion and shield earlier products of German and Austrian culture.[134] It is difficult, then, to decipher what political meaning this crypt, or Bruckner's music, might have had in the interwar period. What we can say is that this was an unusual temple of art: not the concert hall of urban Bildung, but a medievalistic reenchantment suspended in time.

It is clear enough that the processions to Bruckner's body were art-religious in nature. Unlike the living cults of Wagner, Liszt, and Bungert, these rituals did not garner much skepticism in the press, in part because skeptics did not usually attend. Bruckner ceremonies were more regional than cosmopolitan, such that Gräflinger's radio show dubs all of Upper Austria the "Brucknerland." The tone of these broadcasts made Upper Austrian cultural events a counterweight to liberal Vienna, following Salzburg and Bonn decades earlier. (Two decades later, the Reich Ministry of Culture contemplated a project to dismantle Vienna as "city of music" and to recenter Austrian culture in Linz.)[135] Even in the absence of blatant critique, I would argue that Bruckner's tomb offers insights into art-religion verkitscht. If we consider cheapening to mean the bursting of a bubble, a transcendental moment whose sanctity

ART-RELIGION *VERKITSCHT* 167

collapses on itself, then the integrity of this tomb was threatened by two intruders. The first was mold; the second was Hitler.

During one of St. Florian's festivals in the 1920s, a visitor noticed a big problem. Today Adolf Wenusch is known, if at all, as a doctor who studied the health effects of tobacco; in his spare time, he was a self-proclaimed "veritable Bruckner nut." In 1929, Wenusch led an impassioned campaign with the leadership of St. Florian and the International Bruckner Society concerning the abject state of the remains. There is so much mold, he wrote, that it can be seen sprouting from Bruckner's nose. If this were England, there would be elaborate climate control, but here in Austria the damp conditions are leading a national monument to rot. At the very least, he insists, a cast of the embalmed face should be taken for science.[136] In response, the abbot commissioned a formal report from a professor of hygiene in Vienna, and the result is a deadpan description of the macabre, complete with spider egg sacs and two flies that circle lazily beneath the glass.[137] The indignity of this scene was not lost on the stewards of Bruckner's legacy. Both the abbot and Max Auer thought it "barbaric" to reembalm the body with the latest chemicals, and they denied a full restoration; after all, the abbot noted, Bruckner's body was viewed only on rare occasions.[138] They settled on a thorough cleaning of the coffin's interior, leaving the body undisturbed, until the issue was raised again in 1961. This time, St. Florian sought state support from the Bureau of Historic Preservation for a more hermetic coffin. The insularity of their rituals now worked against the cause, as Bruckner's remains were deemed insufficiently valuable to Austrian heritage (beyond the Brucknerland, that is).[139] Dutiful in its obligations, the abbey fronted the funds for the new coffin, and this time the body was concealed from view. In an act of compassion, the institution placed greater value on protecting Bruckner's dignity than his tissue.

By 1996, the centennial of Bruckner's death, his value to the nation had increased, and the bureau determined that the composer would be honored, not desecrated, by a full restoration. The job was not for the faint of heart, and a specialist was hired to transport the body across state lines to an anthropological institute in Basel. A time-stamped log recounts what can only be described as a madcap tale.[140] First there was the long drive with Bruckner in the back, with a chemical smell so strong that the sunroof had to be opened despite falling snow, all of which caused confusion at the tolls. Then there was the removal of a startling three kilograms of mold, all collected and preserved, which compelled the team to record the state of Bruckner's body in wince-worthy detail, from the accidental loss of fingernails while removing his jacket to the desiccation of his genitals.

It may seem crude to further reproduce that detail here. But we must grasp the indignities of decay and restoration to understand how Bruckner's resting place aboveground has forced those who preserve his legacy, largely out of a love of his music, to grapple with a memento mori. The nakedness of Bruckner, in particular, shows how preservation and desecration are impossible to extricate. In the laboratory, Bruckner's remains cease to be a relic; torn from their atmospheric ossuary, they become a specimen on the table. That transformation reflects the paradox of embalming itself, which seeks chemical immortality in a liminal zone between modern science and archaic ritual. The very technologies of restoration that salvaged Bruckner, and that necessitated this impious gaze on his body, originated with a different exhibitionary complex: the collection and display of Egyptian mummies and Indigenous remains by Western museums whose claims to those bodies are still passionately contested. During Bruckner's restoration, the colonial politics of ethnographic collecting made these remains quite different from those of composers whose stone tombs occupy more familiar landscape of communion. In the lab, Bruckner's body looks foreign and exotic, an artifact of the rural folk traditions and local pieties that have long compelled Europeans to turn an ethnographic lens on their own populations. Out of context, from the outside, the art-religious rituals that preserved this body look like superstition, like false claims to immortality.

The second intruder to desecrate Bruckner's tomb is better known. In 1937, amid the columns of Walhalla (the marble hall of German heroes on the Danube), Adolf Hitler unveiled Bruckner's bust in a ritual spectacle that used Bruckner's own music as its cinematic backdrop.[141] Joseph Goebbels, the Reich's minister of propaganda, gave a lengthy speech that made Bruckner a vehicle of Nazi ideology, and his anticlerical omissions secularized the devoutly Catholic composer.[142] An anonymous editorial, in a last gasp of free speech before the Anschluss, protested the event and insisted that Bruckner was a composer of "deep Christian humility," not the mouthpiece of a "heroic worldview of German humanity."[143] The following year, a squadron of Nazi guards occupied St. Florian in what Auer later called the monastery's "darkest day," the start of a long project to convert the site into a secular Bruckner Society overseen by the Führer and dubbed the "Bruckner Bayreuth." Shortly after the war, Auer's professions of disgust in a local paper mourned the desecration of a temple and rued the day when Hitler himself descended to the crypt and offered Bruckner's body his salute.[144] Whatever Auer's position on the Third Reich, he (like many others) embraced the promise of state support for the arts, notably his International Bruckner Society. Already in 1936, prior to Goebbels's speech at Walhalla, Auer seems to have anticipated the propaganda machine that would soon descend on St. Florian.[145] Two months after the notorious Olympics in Berlin, Auer went on the radio to report the festivities at St. Florian for Bruckner's anniversary, and envisioned a "Bruckner Olympics" where conductors and organists might compete in the chapel of St. Florian to an audience of "Bruckner pilgrims from across the globe."[146]

More surprising than St. Florian's co-optation by the Nazis, which is by now well known to historians, was the return to ritual processions for Bruckner's death day in 1946 and 1956, organized jointly by the abbot of St. Florian and Bruckner's pupils. But the tone of these graveside ovations had shifted. At Austria's *Stunde Null*, its Zero Hour, festivals for the misunderstood master could present the nation as a victim and martyr to the Germans.[147] The return to Bruckner's tomb absolved the composer of his associations with the National Socialists and reconfessionalized him as a Catholic. In 1946, after a requiem service was heard in the church, visitors processed to the crypt to hear a speech by Bruckner's pupil Franz Xaver Müller, which a local paper described as an "expiation of the Master" that was wholly different from his veneration at Walhalla.[148] The medievalism of Bruckner's tomb spoke of a long, proud Austrian past that predated the turmoil of recent memory. In this same postwar period, Bruckner's music was likewise presented as an empowering anachronism that returns its listeners to a premodern state of spiritual transcendence. In a speech given in 1955 at the Upper Austrian Bruckner Days in Linz, the expansive tranquility of Bruckner's symphonies was upheld as an antidote to the industrial-capitalist abuse of time, which has become a commodity, a thing we "have."[149] The following year saw the last public procession to Bruckner's remains. Curiously, its ritual borrowed the cinematic approach that had furnished Goebbels's ceremonies with a soundtrack nearly two decades earlier. Rather than process in silence, the crowd descended while an organist's improvisations resounded through the walls; as flowers were laid on the coffin, Bruckner's "Aequale" sounded from afar. It was a fitting use of the organ that Auer so often called Bruckner's "sounding tomb" and seemed intended to resonate Bruckner's very bones.[150]

For readers on the outside looking in, the ritual treatment of these remains can seem alien. To regional Bruckner lovers, the tomb may still today appear as a natural extension of the composer's Catholic faith.[151] These divergent views reflect a core problem of secular religions: it is unclear whether devotion to a figure like Bruckner begs the same anthropological remove that established religions do, and if so, any critique of these processions would belong in the realm of journalism, alongside the barbs of Wittmann and Nagel.[152] What Bruckner's body demonstrates, regardless of opinions about its sanctity, is how comparable practices of art-religion could serve divergent politics. The National Socialists are well known for "political religion," for hollowing out religious and civic ritual and seeding its marrow with a revised ideology. Established confessions like Catholicism could be overwritten with a pan-confessional *Gottgläubigkeit*, or belief in God, while new religions like Ariosophy and Theosophy were alternately mined and purged.[153] When Hitler and Goebbels sought to overwrite St. Florian as a monastery dedicated to Bruckner, they took advantage of some of the tomb's unique properties: its established rituals for a secular figure, and the enchanting anachronism of the celestial face, which helped to engineer a new confession that felt ancient. The "Bruckner Bayreuth" became the latest copy in a long tradition of *imitatio*. This time, Wagner's template proved useful to a state religion that needed an artistic site of pilgrimage.

Bruckner's tomb offers the most violent example of how Ehrenpflicht leaves the composer exposed: naked on the autopsy table and subject to volatile forces. The fascist adaptation of art-religion explains why much of this book has felt prescient of the Third Reich: the measurement of skulls and masks, the collective effervescence of the civic ritual, the odes declaimed to a temple of art, the forceful obligations of duty (*Pflicht*), and even the interwar exhibitions of bad taste that anticipated the "degenerate art" exhibits two decades later. In the interwar processions to Bruckner's remains, alongside the revival of those rituals by the National Socialists, Bruckner's body was reanimated to speak for the living in convoluted tongues. The result of this ventriloquism is that we can no longer extricate Bruckner's own intentions from the gothic decadence of art-religion. As monumental ambitions toppled into political kitsch, Bruckner's tomb could not help but speak. Today it is a more patient and curious form of questioning. Mutely, it asks: does your century still have a place for my beautiful death?

Coda

> What guides [poetic thinking] is the conviction that although the living is subject to the ruin of time, the process of decay is at the same time a process of crystallization; that in the depth of the sea, into which sinks and is dissolved what was once alive, some things "suffer a sea change" and survive in new crystallized forms and shapes that remain immune to the elements, as though they waited only for the pearl diver who one day will come down to them and bring them up into the world of the living.
> HANNAH ARENDT, *Men in Dark Times*

> Question your teaspoons.
> GEORGES PEREC, "Approaches to What?"

I opened this book with a scene from Mauricio Kagel's *Ludwig van* that has haunted me for a long time. The futile yearning to recover the past, and the ethnographic gaze at alien historical artifacts, invited me to confront the antiquarian tendencies that lured me into the archives and resulted in this book.[1] In 1968, not long before Kagel fished busts from the bathtub, Hannah Arendt described Walter Benjamin as an aesthete who trawled the past for pearls, a mode of "poetic thinking" distinct from collecting. When composers' traces are saturated with music's aura, they are enchanted with poetic thinking that seems to preserve them from the elements. But Kagel warns that Benjamin's pearls can come up transformed, decayed, no longer beautiful in death.

This book ends in a more ominous place than it began. Tracing the arc from piety to kitsch shows how preservation is never far from desecration, how European high culture can resemble exotic hoarding, and how devotion can teeter on the edge of violence.[2] In the anxious German discourse I have outlined, kitsch leaves an aftertaste of self-loathing. The voices of interwar kitsch critics ring with disgust at their empty-headed neighbors who cheapen art, culture, and religion. Their disdain feels familiar to those of us who grapple with "the canon"—with high culture embalmed. Disdain can become abhorrence when we acknowledge that the canon acts as a metonym for the colonial, racist, and classist systems of exclusion that held it aloft. But, as this book has shown, the violence inherent in canon formation was not always unidirectional and not only a product of colonial and missionary zeal. Canons were formed by that very zeal directed inward, just as interwar critics saw kitsch as a foreign intruder into bourgeois life, a monster under the bed.

In 1973, the experimental essayist Georges Perec demanded that we question our teaspoons: "What's needed perhaps is to found our own anthropology, one that will speak about us, will look in ourselves for what for so long we've been pillaging from others. Not the exotic anymore, but the endotic."[3] Perec anticipates the imaginary alien of Bruno Nettl's *Heartland Excursions*, who descends into a midwestern conservatory and observes the rites and demigods of the concert hall.[4] In a sense, *The Composer Embalmed* has excavated beneath the substrate of monumentality, offering a more thorough account of Western piety—one that questions Beethoven's teaspoons. It follows that the devotees in this book are not exotic creatures of the nineteenth century, but the endotic ancestors of concert culture and musicology.

Even so, a rift has lately formed between the musicologist as specialist, who remains detached from veneration with the stern gaze of the kitsch critic, and a wider listening public that takes part in morbid curiosity and parasocial intimacy without deriding these as guilty pleasures. In celebrity culture, the guileless charm of kitsch has been redeemed by camp, its glamorous cousin. When Susan Sontag identified the camp sensibility in 1964, she looked back to Walter Benjamin as a flaneur in the Arcades and ahead to the liberatory politics of queer culture and celebrity artifice, the triumph of excess. There are intriguing resonances between Sontag's rubric and the rituals of art-religion. When bands of pilgrims serenaded Mozart's cottage at the feet of a monument taller than the structure itself, were they not engaged in a "melodramatic excess" that was "dead serious"? When Irene Wild composed her rhapsodic ode to Beethoven's spirit, was this not "a seriousness that fails," which might be seen as camp for its blend of "the exaggerated, the fantastic, the passionate, and the naïve"?[5] If academic expertise carries the guilt of German kitsch, rather than the liberatory potential of camp, it acts as a secularizing project that dampens the lively and immersive experience of the parasocial. The question is how to feel spiritually moved when cognizant of a sleight of hand. Even a modern pilgrimage to Graceland or Dollywood can be deflated by commercialism cloaked in piety, by ethical lessons couched in sound bites that both charm and disappoint.[6] It is clear that devotional practices, pilgrimage, and relic collecting continue into the present; less clear is whether devotees enjoy (or are allowed to enjoy) them as naive camp, or with a sense of ironic detachment that marks art-religion's fall from grace.

Relics are stubborn. As they persist, incorruptible, they become political objects that expose the fears and ambitions of their stewards. In a late nineteenth-century period of civic and national tensions, historic houses froze time not only to invoke the sacred, but to insulate composers against the perceived threats of the afterlife. Anatomists likewise shielded an archaic

CODA   173

definition of genius when they examined skulls with methods that were themselves relics, a science embalmed. Their activities generated cabinets of death masks that acted as a credo to art's immortality. After the First World War, a wave of disenchantment with the beautiful death gave relics of bourgeois sentiment a sour taste. In the embarrassment of the Zero Hour, after the National Socialists had made secular piety their template, a swift return to earlier rituals demarcated those years as an aberration. In the Cold War, while the communist East cultivated its political dead, Kagel displayed cabinets of endotic specimens to question West Germany's teaspoons. Today, relics are made political again in a postmodern push to exonerate kitsch from cantankerous critique. Perhaps this book arose, quite unintentionally, from a late capitalist attraction to the dark underbelly of the curiosities, with their charming surface that conceals violent and desperate tendencies.[7]

But to be honest, and pardon my piety: I felt no trace of gothic violence or fixation when I decoded the word *mizpah* in morse code, nested among the names of Beethoven's pilgrims. It felt like a pearl.

Acknowledgments

O ANNO!
May your speedy search and broad controls
Be ever bless'd
One must invest
Mere moments to fill research holes.

No *Fraktur* script nor query strange
Can quell one's quest
To dredge the best
Of hist'ry in a given range.

My acknowledgments must begin with a paean to archivists and librarians. At the Beethoven-Haus in Bonn, I was assisted by Silke Bettermann, Dorothea Geffert, Stephanie Kuban, and Maria Rößner-Richarz. At the birth-house archive of the Internationale Stiftung Mozarteum in Salzburg, Sabine Greger and Gabriele Ramsauer shared rare materials that bear no accession numbers, and Christoph Großpietsch offered many insights. At the Goethe- und Schiller-Archiv in Weimar, Stefanie Harnisch and Evelyn Liepsch assisted with my study of Franz Liszt's estate. In Vienna, I learned a great deal from Otto Biba at the Society for Friends of Music. In both Vienna and St. Florian, my work on Anton Bruckner was informed by conversations with Andrea Harrandt and Friedrich Buchmayr; Dr. Buchmayr offered an insightful private tour of St. Florian's crypt and historic rooms. In Toblach, Saskia Santer and Herbert Santer shared many insights and allowed me to visit Gustav Mahler's cottage. At these institutions and others, a great many associates assisted with image scans and permissions. Broadly speaking, those who develop databases like ANNO (AustriaN Newspapers Online), the virtual newspaper archive of the Austrian National Library, or extensive digital collections and catalogues like those of the Beethoven-Haus, the Internationale Stiftung Mozarteum, the Salzburg Museum, and the Klassik-Stiftung, open up new cultural-historical questions that would otherwise be impossible to pursue.

Funding from several sources supported the research and writing of this book. The Oregon Humanities Center at the University of Oregon furnished a precious ten weeks of writing time and a publication subvention; the Provost's Office of the University of Oregon offered a Faculty Research Award that funded an archival research trip, while their Humanities and Creative Arts Summer Stipend supported two months of writing; the University of Hawai'i at Mānoa's Endowment for the Humanities offered a summer grant that funded an archival research trip; and the DAAD (Deutsche akademische Austauschdienst) and Social Science Research Council funded a year of archival work. For a subvention that covered the cost of image permissions, I thank the American Musicological Society—specifically, the AMS 75 PAYS Fund and the General Fund of the AMS, supported in part by the National Endowment for the Humanities and the Andrew W. Mellon Foundation. At the University of Oregon's Office for the Vice President for Research and Innovation, Catherine Jarmin Miller offered valuable feedback on funding proposals.

I am grateful to those who saw this book through review, revision, and editing. Marta Tonegutti, the acquisitions editor in music for the University of Chicago Press, series editors Nicholas Mathew and James Q. Davies, and editorial assistant Kristin Rawlings all offered generous guidance to a first-time author. I am grateful for the incisive feedback of three anonymous reviewers who read the manuscript closely and thoughtfully. My sincere thanks to copyeditor Kathleen Kageff and production editor Elizabeth Ellingboe for their sharp eyes on the manuscript and proofs. Michelle Sulaiman assisted with the bibliography. A portion of the material in chapter 4 appeared in an article in the journal *Nineteenth-Century Music*; my thanks to the University of California Press for permission to republish this content.

This book has benefited from mentors far and wide. My earliest interests in musicology, cultural studies, and materiality were sparked by Emily I. Dolan, Eric Jarosinski, Jeffrey Kallberg, and Gary Tomlinson. At the University of Chicago, my doctoral adviser Berthold Hoeckner modeled what strong writing and thinking should look like. I owe a further thanks to faculty at that institution who shaped my ideas: Thomas Christensen, Martha Feldman, Robert Kendrick, Steven Rings, Anne Walters Robertson, and Lawrence Zbikowski. At the University of Hawai'i at Mānoa, I had many stimulating conversations about this book, and research generally, with my colleague Katherine McQuiston, whose mentorship still serves me today.

I am grateful to my colleagues in musicology at the University of Oregon for building a vibrant intellectual community. My thanks to Habib Iddrisu, Lori Kruckenberg, Jesús Ramos-Kittrell, Lindsey Rodgers, Zachary Wallmark, Larry Wayte, and Juan Eduardo Wolf, along with my retired colleagues Marian Smith and Marc Vanscheeuwijk. Among these, some deserve special mention: Zach for his publishing advice and his remarks on the introduction and coda, Lori for her kind encouragement and her comments on my last chapter, and Marian for her wisdom about archival work and for the library she passed down to me. I appreciate colleagues outside musicology who have supported my work in many ways, notably my music theory colleagues Jack Boss, Drew Nobile, and Stephen Rodgers; Steve deserves mention for his feedback on a conference paper related to this project. I have benefited from research support offered by Sabrina Madison-Cannon, the dean of the School of Music and Dance, who helped to fund a productivity training program from the National Center for Faculty Diversity and Development. I am indebted to art historian Joyce Cheng, with whom I spent many evenings discussing kitsch; my thanks as well to the thoughtful readers in the interdisciplinary draft workshop that Joyce founded. For their consultation on specific facets of this project, ranging from translation to prospectus to image permissions, my thanks to Stacy Alaimo, Diana Garvin, Martin Klebes, Dorothee Ostmeier, Kate Petcosky-Kulkarni, and Kate Thornhill. I am grateful as well to Lara Bovilsky, architect of the Writing Circles program, and Mike Murashige, writing consultant and developmental editor, both affiliated with the University of Oregon's Center on Diversity and Community administered through the Division of Equity and Inclusion.

Several colleagues at other institutions generously shared their expertise. For their comments on chapter drafts, I thank James Q. Davies, Linda Hutcheon and Michael Hutcheon, Sarah Iker, Benjamin Korstvedt, Emily Richmond Pollock, Frederick Reece, Claudio Vellutini, and Lindsay Wright. Others acted as a sounding board over the phone: Jessica Peritz, Marcelle Pierson, and Martha Sprigge. On my most recent research trip, Christoph Hust offered many insights that informed my last chapter. I owe a long-standing debt of gratitude to Martin Eybl, Christian Thorau, Jutta Toelle, and Melanie Unseld for the warm hospitality they have shown during several research visits to Austria and Germany. I look forward to our continued exchange as we build bridges between our scholarly communities.

An unusual opportunity impacted this book in a profound way. The Faculty External Mentor Program from the University of Oregon's Division of Equity and Inclusion funded a three-year mentoring relationship with Francesca Brittan. Francesca read and commented on every chapter in this book with brilliant insight. This same program allowed me to invite Joy H. Calico and Emily I. Dolan to critique the manuscript in a workshop. Their annotations were an embarrassment of riches and strengthened the book beyond measure.

I do not quite know where to start when it comes to thanking friends in Chicago, Honolulu, and Eugene, as these cities have gifted me more extraordinary people than I can name here. A special thanks to friends who reminded me of the vital aspects of ourselves that lie beyond our chosen careers. The sense of humor in this book will look familiar to the Lillys, Horovitzes, and Fines from whom I inherited it. Among them, special thanks are due to Alan Fine and Elizabeth Lilly, who have not only supported but actively participated in my curiosity. I dedicate this book to the memory of all four grandparents, who were avid readers with a genuine love of the arts. I would have enjoyed sharing this book with them. It seems fitting to close with the text of a postcard I received in Berlin in 2015, which explains, in better words than I could, why they are much missed: "Have you found the source of Wagner belly button hair now flooding West Germany? All jokes and just for fun. I love you and think of you often. Hope the research is going well. Love, Grandpa."

Archives and Abbreviations

Abbreviations: Frequently Cited Institutions and Collections

B H B : Library and Archive of the Beethoven-Haus, Bonn
G S A : Klassik Stiftung Weimar, Goethe- und Schiller- Archiv
I M S : Internationale Mozart-Stiftung (now the Internationale Stiftung Mozarteum), Salzburg
M G A : Mozart Geburtshaus Archiv, a collection of the Internationale Stiftung Mozarteum, Salzburg
Ö N B : Österreichische Nationalbibliothek, Vienna
V B H : Verein Beethoven-Haus, Bonn
W G M : Library and Archive of the Gesellschaft der Musikfreunde, Vienna

Other Collections Consulted

Josephinium Archive, Vienna
Landesbibliothek, Coburg
Robert-Schumann-Haus, Zwickau
Staatsbibliothek, Berlin
Wienbibliothek im Rathaus, Vienna
Wien Museum Archive, Vienna

All translations are the author's unless otherwise noted.

Notes

Preface

1. Ariès, *Hour of Our Death*, 471–74.
2. Kutschke, "Celebration of Beethoven's Bicentennial."
3. Gibbon, "Beethoven Returns to Bonn."
4. Kagel, *Tamtam*, 80.
5. Ludwig van Beethoven, "In questa tomba oscura," WoO 133, 1806–7; poetic text by Giuseppe Carpani. See also Stavlas, "Reconstructing Beethoven," 59.

Introduction

1. Letter from Julius Franz Borgias Schneller to Baroness Gleichenstein, Freiburg im Breisgau, October 15, 1832, in Schneller, *Julius Schneller's Lebensumriss*, 258. This passage was copied into a provenance letter that attended Beethoven's cane, spoon, and hair, which had made their way to the collections of Anton Graf Prokesch von Osten; later, his son donated the relics to the WGM, October 27, 1906: WGM Archive, ER7 and ER10. Beethoven had multiple walking sticks, another of which was owned by Anton Schindler and later sold to the collector Carl Meinert, an industrialist from Dessau who owned a remarkable assortment of musical curios and manuscripts.
2. Whereas Tchaikovsky reported his awe at this "*holy* musical *object!*" in a letter to Nadezhda von Meck on May 31, 1886, we learn of Rossini's encounter less directly through anecdotal lore shared by members of the Viardot family in 1856 and again in 1922. Regardless of whether he truly uttered the words "I am going to genuflect in front of this holy relic," or whether Louis Viardot exaggerated the encounter to spin a hagiography around the family heirloom, the incident affirms how the Viardot family behaved as gatekeepers of a sacred experience. See Everist, *Mozart's Ghosts*, 157–90; on Tchaikovsky's response, see 169–70; on Rossini's, see 171–73. See also Louis Viardot, "Manuscrit autographe du *Don Giovanni* de Mozart," *L'illustration* 27 (1856): 10–11.
3. Teresa Barnett calls this process the "waning of the relic," in Barnett, *Sacred Relics*, 163–96.
4. A three-volume collection traces the intellectual history of art-religion: A. Meier, Costazza, and Laudin, *Kunstreligion*. See also E. Kramer, "Idea of *Kunstreligion*."
5. Williamson, *Longing for Myth in Germany*. While some call this a "secular religion," or claim that religion superficially gilded the secular, Williamson asserts that Christianity was essential to

ostensibly secular domains such as Bildung and the longing for new mythology; for a counterargument, see Brachmann, *Kunst, Religion, Krise*, 85.

6. Grewe, "Aesthetic Religion, Religious Aesthetics, and the Romantic Quest for Epiphany."
7. Leistra-Jones, "Hans von Bülow and the Confessionalization of *Kunstreligion*."
8. Brodbeck, *Defining* Deutschtum.
9. Herdt, *Forming Humanity*.
10. Hobsbawm, "Who's Who or the Uncertainties of the Bourgeoisie," in Hobsbawm, *Age of Empire*, 165–91. See also Sumner Lott, *Social Worlds of Nineteenth-Century Chamber Music*, 9–13.
11. Pocket-sized biographies were published in abundance even before the travel literature necessitated by mass transit; examples include Oulibischeff, *Mozarts Leben und Werke* and Nohl, *Beethovens Brevier*.
12. William Weber has been a pioneer in this area with his monograph *The Great Transformation of Musical Taste*. Recent studies include Pietschmann and Wald-Fuhrmann, *Der Kanon der Musik*; Shadle, *Orchestrating the Nation*; and Gibbons, *Building the Operatic Museum*.
13. Horlacher, *Educated Subject and the German Concept of Bildung*.
14. Rehding, *Music and Monumentality*; Gottdang, "Porträts, Denkmäler, Sammelbildchen."
15. Dović and Helgason, *National Poets, Cultural Saints*, 71–96; on "postulators," see 81. For a seminal study of mythological and saintly tropes in artists' biographies, see Kris and Kurz, *Legend, Myth, and Magic in the Image of the Artist*.
16. Hobsbawm, "Mass-Producing Tradition: Europe, 1870–1914," in Hobsbawm and Ranger, *Invention of Tradition*, 263–307.
17. Nora, "Between Memory and History: *Les lieux de lémoire*," 12. This article was the first English-language translation (by Marc Roudebush) of Nora's introduction to the large-scale "lieux de mémoire" project of 1984. The book was later presented in English as *Realms of Memory: Rethinking the French Past*, edited by Lawrence D. Kritzman and translated by Arthur Goldhammer. Nora's project participated in a turn to historical consciousness at the height of postmodernism, which likewise compelled Paul Ricœur to weigh a long philosophical tradition that parsed history from memory, and to develop a new ethics of forgetting in an age of systemic violence; see Ricœur, *Memory, History, Forgetting*.
18. Derrida, *Archive Fever*, 19.
19. Nora, "Between Memory and History," 12; Derrida, *Archive Fever*, 11. Derrida refers here to portraiture, but I would extend his claim to a wider range of popular products that mask death in beauty, including the laurel-fringed death masks I will discuss in chapter 4.
20. MacLeod, "Sweetmeats for the Eye"; see also Henke, Kord, and Richter, *Unwrapping Goethe's Weimar*.
21. Full-fledged operas like Pfitzner's *Palestrina* of 1917 were the dressy counterpart to plays like Richard Plattensteiner's *Beethoven: The Great Musician in God's Honor* from 1916, which prefigures the 1924 Bruckner pasticcio called *The Musician of God*, cowritten by Viktor Léon.
22. Bräunig, "Der Mythos Mozart: Kitsch und Vermarktung."
23. Fritz Zerritsch, "Beethoven vor dem Löwenkäfig in Schönbrunn," in *Damenspende zum Concordia-Ball*, n.p.; also at BHB, Nq 5 / 1927 Dame.
24. Bartsch, *"Schwammerl,"* 199.
25. Postcards qualify as ephemera not least because they remain the domain of private collectors and are best accessed through eBay; over the years I have encountered several morbid

Chopin scenes designed by Polish and Czech painters, such as Józef Feliks Męcina-Krzesz, 1905; František Klimeš, 1919; Tadeusz Korpal, n.d.; and Adam Setkowicz, n.d.

26. Lutz, *Relics of Death*, 4.

27. For a comprehensive study of German-language music biographies and anecdotes in cultural memory, see Unseld, *Biographie und Musikgeschichte*.

28. McManus, *Brahms in the Priesthood of Art*.

29. Fine, "Assimilating to Art-Religion."

30. "Die Beethovengedächtnisfeier im Schwarzspanierhause," *Ostdeutsche Rundschau*, November 16, 1903, 4.

31. Rost, *Musik-Stammbücher*; Goldberg, "Chopin's Album Leaves."

32. Bonds, *Beethoven Syndrome*. On texted arrangements, see chapter 4, below.

33. Mathew, "Review of *The Beethoven Syndrome*."

34. Musicology has had no single material or corporeal turn; rather, its richly varied turns include the study of paratexts, histories of the book, biological anatomies, instrumental technologies, commodification and the marketplace, and the ontology of sound as matter in the history of science and philosophical materialism.

35. Brown, "Thing Theory." For an example of object biography, see Martin, *Beethoven's Hair*.

36. Latour, *Reassembling the Social*; J. Bennett, *Vibrant Matter*, e.g., viii–xi, 1–19. For examples of these approaches in music studies, see Bates, "Social Life of Musical Instruments"; and the articles in Mathew and Smart, "Quirk Historicism."

37. McCormick, "Agency of Dead Musicians." While my work is motivated by some of the same questions explored in McCormick's study, my methodology is historical, archival, and geographically bounded, whereas her project offers a sociological synthesis of composers, geographies, and time periods.

38. J. Bennett, *Vibrant Matter*, xvi.

39. Mathew, *Haydn Economy*, 124.

40. R. Morris, *Returns of Fetishism*, 309–19.

41. Charles de Brosses, "On the Worship of Fetish Gods: Or, a Parallel of the Ancient Religion of Egypt with the Present Religion of Nigritia" (1760), in R. Morris, *Returns of Fetishism*, 48.

42. T. Bennett, *Museums, Power, Knowledge*, esp. 39–43, 103–34. Relics, talismans, and cults of the dead have continued in contemporary celebrity culture; see Rojek, *Celebrity*, esp. 58–63.

43. Treitel, *Science for the Soul*; see also Charet, *Spiritualism and the Foundations of C. G. Jung's Philosophy*. In 1930, Freud's pupil Otto Rank posited a fourth dimension that might prove the autonomy of the immortal soul; see Rank, *Psychology and the Soul*.

44. Hau, *Cult of Health and Beauty in Germany*.

45. Leistra-Jones, "(Re-)Enchanting Performance."

46. F. Max Anton, Generalmusikdirektor der Stadt Bonn, to unknown recipient ("sehr geehrter Meister"), June 4, 1927, BHB Akten, sig. 116.

47. Minor, *Choral Fantasies*, 73. On civic ritual as religion, see Jefferies, *Imperial Culture in Germany*; and Unowsky, *Pomp and Politics of Patriotism*. On Denkmalwut, see Dović and Helgason, *National Poets, Cultural Saints*, 51–58.

48. Festival cantatas could be recycled for different occasions, and they fall into three stylistic categories: military-style fanfares that operate like fireworks (such as Joseph Gabriel Rheinberger's *Festchor*, JWV 49, 1855, which bugles "rufet Heil" in double-dotted arpeggios);

Lutheran-style chorales for pensive or erudite events (such as Felix Mendelssohn's *Lobgesang* to celebrate the invention of printing in Leipzig, 1840); and ambitious, multimovement oratorios (such as Liszt's two cantatas for Beethoven, the *Cantate zur Inauguration des Beethoven-Monuments* of 1845 and *Beethoven-Cantate* of 1870). Examples of piano arrangements include Eduard Lassen, "Beethoven-Ouverture: Für grosses Orchester," arr. piano solo and piano four-hands (Leipzig: Brenau und Hainauer, ca. 1871); Peter Cornelius, "Beethoven-Lied," op. 10 (E. W. Fritzsch, 1871); and Henri Gobbi, "Vorspiel der Festcantate zu Franz Liszt's 50 jährigem Künstler-Jubiläum in Pest (November 1873)" (Leipzig: C. F. Kahnt, n.d.).

49. The story of the failed musical monument in the Karlskirche is fascinating, if outside the scope of this book. The project was spearheaded by the WGM from 1815 to 1860, and its design began with Mozart alone; soon afterward, Gluck and Haydn were added; starting in the 1850s, Gluck got the boot, and Beethoven entered the fold, followed by Schubert, who supplanted Haydn; and after continued disagreement for decades, the project was disbanded in 1860. A chronicle of the failed project can be found in the archive of the WGM, Records of the Gluck-Haydn-Mozart-Beethoven-Schubert Monument Fond. In 1887, when Haydn's monument was unveiled in front of the Mariahilferkirche in Vienna, its semisacred location owed a debt to the Karlskirche project. A few years later, when the Mozart monument committee convened to discuss a location, they floated the idea of erecting this in front of the Karlskirche; Telesko, *Kulturraum Österreich*, 177.

50. Minor, *Choral Fantasies*; Rehding, *Music and Monumentality*.

51. Rojek, *Presumed Intimacy*; Baym, *Playing to the Crowd*.

52. Roach, *It*, 55.

53. Mole, *Byron's Romantic Celebrity*, 16.

54. Lynch, *Loving Literature*, 39. This quotation, which runs as a red thread throughout Lynch's book, hails from Coleridge's *Biographia Literaria*, 1:15.

55. North, *Domestication of Genius*, 1.

56. Westover, *Necromanticism*; Watson, *Author's Effects*, 228; Watson, *Literary Tourist*, 186. It makes historical sense to compare English literary tourism with German and Austrian practices because readers and listeners were often one and the same, rounding out their education in liberal *Bildung*. Watson justifies her focus on literature with the argument that celebrity intimacy was more urgent for arts that were mechanically reproduced, which excludes music; I would argue that music read at the parlor piano afforded the same claims to the composer's spirit as the home library did for Goethe or Byron. Watson, *Author's Effects*, 235n18.

57. Richards, *Temple of Fame and Friendship*; Fine, "Towards a History of the Eccentric Artist."

58. Applegate, "Mendelssohn on the Road."

59. Applegate and Potter, *Music and German National Identity*; Sadie and Sadie, *Calling on the Composer*, 71–88.

60. Cultural critics and psychoanalysts have posited explanations for the collector's obsession far beyond the parasocial: as a material outlet for spiritual longing, as Freudian misanthropy, as a quest for Benjaminian aura, and as a preoccupation with authenticity, to name a few. See Pascoe, *Hummingbird Cabinet*.

61. Anton Halm's wife allegedly sought one of Beethoven's locks via Karl Holz, who pranked her with goat hair as if to teach a lesson about soliciting a badge of distinction without a personal acquaintance; see Kerst, *Die Erinnerungen an Beethoven*, 2:162–63.

62. BHB, Nachlassverzeichnis Ludwig van Beethovens, NE 103, III, 11. Several of Beethoven's effects, including his washbasin, ended up in the vast musical collections of the Frankfurt winemaker Friedrich Nicolas Manskopf; see BHB, Akten (Tresor), sig. 276.

63. Schindler expressed gratitude that he was out of town for these "outrageous proceedings"; see Schindler, *Beethoven as I Knew Him*, 506. Schindler kept most effects for himself but offered the Maelzel metronome, two medicine spoons, and salt and pepper pots to the pianist Marie Pachler-Koschak; see Pachler, *Beethoven und Marie Pachler-Koschak*, 27.

64. Letter from Hermann Breit to "Annchen," August 19, 1910, BHB, Akten (Tresor), sig. 254. In a similar gesture of mistrust, W. Schmitz refused to donate a Beethoven lock as he did not think this was the "rightful place" for relics; see letter to Gerhard Schmidt in Bonn, September 23, 1926, BHB, Akten (Tresor), sig. 279.

65. As relayed in a clipping of the *Hamburger Nachrichten*, February 4, 1923, BHB, Z118m/29.

66. Haydon, *Diary*, 3:452. See also Semmel, "Reading the Tangible Past."

67. See Higgins, "Art, Genius, and Racial Theory in the Early Nineteenth Century."

68. Novello, *Mozart Pilgrimage*, 73–74 and 83.

69. Watson, *Author's Effects*, 94–116. On Mozart forgeries for profit, see Plath, "Gefälschte Mozart-Autographen." For facsimiles of Mozart's autographs cut apart and offered as gifts, see Konrad, *Wolfgang Amadeus Mozart*, esp. Fr 1781a on p. 59.

70. In 1932, a Bonn resident who grew up in the Beethoven-Haus reported that foreign visitors destroyed the cradle and floor in search of "a priceless relic," and the hotelier was known to sell them for a tip. Heinrich Baum, "Das Bett Beethovens," *General-Anzeiger* (Bonn), July 2, 1932, 341.

71. Transnational examples include the secular pantheons of London's Westminster Abbey, Ludwig I's Walhalla, St. Leonard's crypt in the Wawel Cathedral of Kraków, and the nationalistic repatriation of organs, such as the hearts of Shelley, Grétry, and Chopin. See McCormick, "Agency of Dead Musicians," 330–31.

72. Richard Wagner, "A Pilgrimage to Beethoven" (1840), in R. Wagner, *Richard Wagner's Prose Works*, 7:21–45. Wagner's novella was published first in French in the *Revue et gazette musicale* in November and December 1840, then appeared in 1841 in German in the *Abend-Zeitung* (Dresden). The anti-English sentiment in this novella might have appealed to its French readership and implied a caricature of Jewish greed. Vazsonyi, *Richard Wagner*, 37.

73. The British Library contains a wide array of English-language texted arrangements of Beethoven's instrumental works in binder's volumes. To name a few: "Rosalie!" arr. of "Adelaïde" (London: B. Williams, n.d.); *Songs of the Seasons: Adapted to Melodies by Beethoven by J. Pittman* (London, [1854]); and "Heaven!" arr. of Beethoven, Piano Sonata in F Minor, op. 23 (1886), British Library, H.2430.c.(17).

74. Nußbaumer, *Musikstadt Wien*.

75. The mistrust of the gullible middle class prefigured midcentury Anglophone polemics against the middlebrow. Macdonald, *Masscult and Midcult*; Guthrie, *Art of Appreciation*.

76. Karl Lueger's populist party was vocally skeptical of Bildung as the calling card of an out-of-touch liberal class who dreamed of old Vienna, the city of Schubert and Beethoven; see, for instance, Salten, *Das österreichische Antlitz*.

77. Irving, *Sketch Book of Geoffrey Crayon*, 251. On this American's playful encounters with British heritage, see Westover, *Necromanticism*, 106–41.

78. Hanslick, *Musikalisches Skizzenbuch*, 313–16.

79. Hanslick, *On the Musically Beautiful*, 65–66; and Hanslick, "Wagner-Kultus," 338–49.

80. Karnes, *Music, Criticism, and the Challenge of History*.

81. Walton, "Quirk Shame."

82. LaPorte, *Victorian Cult of Shakespeare*, 152. LaPorte has shown how "bardology" extended to the earliest literary criticism, which absorbed Shakespeare's plays into Victorian sermons and biblical philology. It is no large leap to extend this observation to musicology, whose institutions likewise took shape as islands of rigor surrounded by seas of hyperbole, or what LaPorte calls the "strange cousins or crazy uncles of today's scholarly practices" (8).

Chapter One

1. Self-identified as "David from the U.K.," June 20, 2018. My thanks to the Haydn House curators for allowing me to look through this book, which was on display under glass.
2. Frank Ernest Williams, August 11, 1890, BHB, Museum Visitors' Books.
3. Margry, *Shrines and Pilgrimage in the Modern World*.
4. Katzenstein, *Disjoined Partners*, 97–131.
5. Must, "Origin of the German Word *Ehre*, 'Honor.'" My thanks to Heghine Hakobyan for her assistance with etymology.
6. According to the corpus database of the *Digitales Wörterbuch der deutschen Sprache*, a project of the Union der deutschen Akademien der Wissenschaften, the usage of "Ehrenpflicht" increased sharply in the mid-nineteenth century, which corresponds with the establishment of state-sponsored historic preservation bureaus such as Germany's Historischer Verein für Niedersachsen (1835) and Austria's Bundesdenkmalamt (1853). While words like *Pietät* and *Ehrenpflicht* are unique to German and Austrian heritage projects, the climate of competition (central vs. regional, private vs. state sponsored) was similar in France and England; for a comparative approach, see Swenson, *Rise of Heritage*.
7. Unidentified newspaper clipping, 1889, BHB Akten, VBH 136.
8. *Bonner Zeitung* 60, March 1, 1889, front page.
9. Breuning, *Memories of Beethoven*, 98.
10. "Schwarzspanierhaus und Beethoven-Zimmer," *Neues Wiener Tagblatt (Tages-Ausgabe)*, December 31, 1903, 5.
11. Landau, *Erstes poetisches Beethoven-Album*.
12. Landau, xi.
13. Landau, xi, xii, and xiii, respectively.
14. Karl Kösting, "Germanias Weihnachtzeit," in Landau, *Erstes poetisches Beethoven-Album*, 435. See also Munich's festival prologue on pp. 361–62.
15. Ruth Solie, "Beethoven as Secular Humanist," in Solie, *Music in Other Words*. For examples of Beethoven as ethical teacher and Messiah, see the prologues for Prague, Stuttgart, and Vienna: Landau, *Erstes poetisches Beethoven-Album*, 384, 374, and 400.
16. Karl Simrock, who famously translated the *Niebelungenlied* into modern German, authored Bonn's prologue to consecrate the Beethovenhalle, while critic Franz Gehring wrote the review of that event. Vienna's prologue was by "Dr. Beck," likely Karl Isidor Beck. Landau, *Erstes poetisches Beethoven-Album*, 303–9 and 388–89, respectively.
17. Landau, *Erstes poetisches Beethoven-Album*, 389.
18. Messing, "Vienna Beethoven Centennial Festival of 1870."
19. Martina Nußbaumer's account of Vienna's monuments does not emphasize the city's poor reputation in this climate of competition and antagonism; see Nußbaumer, *Musikstadt Wien*.
20. Aichner, "Ludwig August Frankl."

21. Ludwig August Frankl, "Heiligenstadt und Bonn," in Landau, *Erstes poetisches Beethoven-Album*, 227.

22. "Die Enthüllung des Beethoven-Monumentes," *Fremden-Blatt*, May 1, 1880 (Abendblatt), 1.

23. Hanslick, "Beethoven in Wien," 3; Ed. H. [Eduard Hanslick], "Zur Enthüllung des Beethoven-Denkmales," *Neue Freie Presse*, May 1, 1880, front page.

24. Herbert Oakeley, "A Monument to Beethoven," *Musical Times and Singing Class Circular*, June 1, 1880, 282.

25. It was not uncommon for such projects to be framed explicitly as a response to German pressures. A new concert hall in Vienna, built between 1911 and 1913, was likewise described in terms of competition with German venues; see Nußbaumer, *Musikstadt Wien*, 84 and 92–153.

26. Hanslick, *Musikalisches Skizzenbuch*, 313–16.

27. Hermann Jakob Doetsch, cited in Kämpken and Ladenburger, *Bewegte und bewegende Geschichte*. See also the exhibition's companion website: https://internet.beethoven.de/de/ausstellung/125-jahre-beethoven-haus (accessed November 13, 2024).

28. "Priest of music," appears throughout McManus, *Brahms in the Priesthood of Art*; on purity, see 19–56.

29. Leistra-Jones, "Staging Authenticity," 398; see also Leistra-Jones, "Improvisational Idyll." On the New German School, see Vazsonyi, "Beethoven Instrumentalized." This explains why Joachim was dismayed in 1869 that Liszt and Wagner would spearhead the festival in Vienna, and his description of Beethoven's memory prefigures his ambitions a decade later: "the mention of those two famous men destroys my vision of Beethoven as a sublime yet simple spirit, whose unassuming majesty has gradually conquered the world." See Bickley, *Letters to and from Joseph Joachim*, 386–87, cited in Messing, "Vienna Beethoven Centennial Festival of 1870," 60.

30. Opening lines of the VBH's 1889 announcement and request for donations. See BHB Akten, VBH 179.

31. Graf von Bülow, *Deutsche Reichzeitung* 286 (June 4, 1905), BHB Akten, sig. 136, newspaper clipping collection.

32. Sent in a letter of thanks from Sanitätsrat Dr. Herschel to the Geheimrat P. A. Schmidt, submitted alongside a patriotic song of his own composition, April, 27, 1922, BHB Akten (Tresor), 292.

33. Handwritten note by Wilhelm Kuppe, BHB Akten, VBH 116.

34. *Godesburger Zeitung*, October 26, 1889, BHB Akten, VBH 136.

35. Vazsonyi, *Richard Wagner*, 16–27.

36. Bourdieu, *Rules of Art*, 148; see also Vazsonyi, 26.

37. Ernst Rudorff, "Über das Verhältnis des modernen Lebens zur Natur," *Preussische Jahrbücher* 45 (1880), 261–76, at 276. Rudorff, "Ave Maria am Rhein," op. 38, for soprano solo, women's choir, and orchestra, unpublished manuscript, Staatsbibliothek zu Berlin, Nachlass Ernst Rudorff.

38. Rudorff's ideas about nature's moral impact on Bildung were drawn from Friedrich Schiller. Rudorff, "Über das Verhältnis," 268; Lekan, "'Noble Prospect,'" 842–45.

39. Lekan, "'Noble Prospect,'" 856.

40. Rudorff, "Über das Verhältnis," 272.

41. *Bonner Zeitung* (1889), undated newspaper clipping, collection of Kuppe, BHB Akten, VBH 136. A similar complaint about the market cries was expressed in the prologue for Bonn's chamber festival in 1890; *Beethoven-Feier*, 2.

42. *Beethoven-Haus Bonn*, 91.

43. The program book for the festival of 1913 described this event in these stirring words and referred to a recited prologue by Ernst von Wildenbruch as a "Weiheakt": *Beethoven-Feier*

veranstaltet vom Verein Beethoven-Haus in Bonn: XI. Kammermusik-Fest Bonn; 27 April–1 May, 1913 (Bonn: Beethoven-Haus, 1913), 24 and 35, BHB no. 241 / 1913. The entire event of 1903 was described as a "consecration of the house" in its respective program book: *Beethoven-Feier veranstaltet vom Verein Beethoven-Haus in Bonn: Zweites Kammermusikfest zur Weihe des Hauses; Mai 1893*, BHB no. 241 / 1903. For the twenty-fifth-anniversary celebration of the Verein, the Gürzenich quartet reenacted the event with another rendition of the Cavatina on those same instruments.

44. BHB Akten, VBH 292.

45. Leistra-Jones, "(Re-)Enchanting Performance."

46. As quoted in *Beethoven-Haus Bonn*, 45–46.

47. VBH, "Aufruf" reprinted in the 1890 chamber festival program book, BHB, HCB P / 1878 Klav.

48. Like many of Beethoven's belongings, these instruments ended up in Berlin after Anton Schindler collected and sold them. Their relocation was spurred by a letter from the Prussian minister for Wissenschaft, Kunst und Volksbildung, Berlin, December 7, 1923. This document and the flurry of responses cited here are all held in the BHB Akten, VBH 54.

49. Clipping from the *Kölner Zeitung*, July 15, 1896, BHB Akten, VBH 136.

50. VBH letter from December 22, 1923, in BHB Akten, VBH 54.

51. VBH letter from December 22, 1923, in BHB Akten, VBH 54.

52. Letter from Minister Boelitz to the Ministerium für Wissenschaft, Kunst und Volksbildung, Berlin, February 15, 1924. Shortly thereafter, Franz Ries, the grandson of Ferdinand Ries and donor of Beethoven's viola, thanked the Beethoven-Haus for preserving the instrument; see BHB Akten (Tresor), 278. The instruments were played on special occasions, such as during the music festival in Bonn in 1927; see letter from secretary Friedrich Knickenberg, December 11, 1926, BHB Akten, VBH 173.

53. "Prologue," in program book, *Beethoven-Feier veranstaltet vom Verein Beethoven-Haus in Bonn: 1. Mai 1890; Kammermusik-Fest; Beethoven-Ausstellung*. Bonn: Peter Neusser, 1890, 2, BHB no. 141 / 1890.

54. *Musical Times and Singing Class Circular*, October 1, 1890, 591–92.

55. A Belgian visitor on September 7, 1902, marvels that "such a great man was born in such a tiny room!" (Un si Grand Homme / Parti d'une si petite chambre!). On September 19, 1910: "Behold this modest and awe-inspiring little room and tell me that Great Men come out of Great Spaces!" (Regardez cette modeste & impressionnante petite chambre & dites moi si les Grands Hommes sortent des Grands Espaces!). On April 20, 1927, Richard Loewenthal wrote that "the greatest spirits are not always born in palaces" (Die größten Geister werden nicht immer in Schlössern geboren), BHB, Museum Visitors' Books.

56. "Aus jeder kleinen Hütte / Kann irdsche [sic] Gröss entstehen / Dies zeigt uns hier aufs neue / Die Stätte wo wir stehen." Entry signed "Fred Morian and Natalie Krabbe" from March 14, 1926, BHB, Museum Visitors' Books.

57. Visitors' book entry by J.K. from Nuremberg, July 10, 1921: "Ewig Beethoven hoch—im *niederen* Raum geboren—empor gerungen zu *höchster* Höhe menschlichen Könnens durch Genien &—Fleiß!—"

58. Carl Schmidt, a professed portrait painter, wrote: "Keine Kirche hat mir innen so tiefen Eindruck gemacht wie dies Geburtszimmer." Entry from October 30, 1926, BHB, Museum Visitors' Books.

59. Hubert Binhold from Sauerland wrote: "Wie von Jahrtausend zu Jahrtausend immer wieder die Sonne ihre Strahlen sendet hinein in das kl. Dunkel deiner Geburtenstätte, so durchklingen und durchjauchzen deine Töne der Welten All, Götter verwandter König der ewigen Generationen." Entry from May 28, 1928, BHB, Museum Visitors' Books.

60. One such reimagining can be found in the *Bonner Tagebuch*, November 1, 1889, n.p., clipping in the collection of Wilhelm Kuppe, BHB Akten, VBH 136.

61. Johann Peter Lyser, "Denkblatt zur Beethovenfestes zur Bonn, 1845," BHB, B 534; see also Lange, *Musikgeschichtliches*, 14–15; Jordan, *Geselschap*, 40.

62. Poems that feature the dual crowns include Friedrich Mosengeil, "Poetische Erläuterung der Musik von Ludwig van Beethoven zu Goethe's Egmont"; H. Meier, "Zur Beethoven-Feier"; and (Josef Weil Ritter von) Weilen, "Prolog" to Vienna's centenary: all in Landau, *Erstes poetisches Beethoven-Album*, 46, 413–16, and 402–3, respectively.

63. Letter from papermaker Alexander Flinsch accompanying the submission of Geselschap's two watercolor sketches of Beethoven's birth, Berlin, December 14, 1900, BHB Akten (Tresor), 307b.

64. "Durch Kunst zum Sieg! / Die Dornen müssen spornen / Die Rosen jubelnd zu erringen, / Wohl hast du, Meister, viel gelitten— / Und durch den Kranz Unsterblichkeit erstritten!" Wilfried Edler from Weimar on May 15, 1917, BHB, Museum Visitors' Books.

65. Stadler, "Ein musikalisches Märchen," 115.

66. Hermine Bovet, "Zu dem Bild: Beethovens Geburt (im Beethoven-Museum zu Bonn), dem Beethoven-Verein gewidmet," BHB, Z586817. Bovet hailed from Bad Honnef, a spa town on the Rhine.

67. These records were used to generate tourist statistics during fund-raising drives; see, for instance, the visitor report in BHB Akten (Tresor), VBH 115. While tourism to the museum was international, the majority of poems and messages were written by German-speaking visitors. Unless otherwise indicated, all dated entries cited in this section stem from the BHB, Museum Visitors' Books. Author names or initials are provided when legible.

68. Habermas, *Structural Transformation of the Public Sphere*, 43.

69. Knittel, "Pilgrimages to Beethoven."

70. "Lieber Meister: heute kam ich herauf, dich zu besuchen—da ritt mich der Teufel, daß ich zuerst hier dieses breitspurige Buch durchstöbern mußte.—Was du wohl zu diesem umfangreichen Bagagi von Namen- und Titelaffen sagen würdest? . . . Daß ich zuguterletzt dich selbst noch entdeckt und mit dir zwischen gehalten hab! Ewigen Dank dafür! Einstweilen ein herzlich Lebewohl. Ich muß jetzt noch ein Weilchen [durch] führen." Signature illegible, June 30, 1918.

71. "Guter Freund, wer du auch seist, an heiliger Stätte ziert auch dir Bescheidenheit! K. Sch." June 30, 1918.

72. Poem by Hubert Reuter of London translated below by Dr. Helserstock, August 3, 1922. See also the correction made by a visitor on September 20, 1910, in which the German-nationalist comment "O Deutschland, du bist reich an—Großen!" is corrected in pencil by a nameless visitor, who draws attention to Beethoven's Dutch ancestry: "Er stammte von Nederlanden durch seiner Vater und von Deutschland durch seiner Mutter."

73. In her travelogue, a woman named Elsbeth Pushee recalled paging through the book in search of a friend while feeling rushed by another visitor waiting in line; Elsbeth Pushee Travel Diary (1890), Allison-Shelley Manuscript Collection (3862), Rare Books and Manuscripts, Special Collections Library, Pennsylvania State University. In what looks like a playful conversation

between husband and wife, who share a surname, two authors wrote the first and second movements of the Fifth Symphony in the visitors' book on January 1, 1928.

74. An example of brief allusions to sacredness, which were common: Schroder Hansmann wrote, "diese geweihten Räume in ehrfürchtiger Verehrung am 14. Oktober 1911."

75. Entry from October 10, 1897, by Emil Büchner, a painter and craftsman who studied in Munich and made his career in Leipzig, not to be confused with Adolf Emil Büchner from the Wagner circle. "Beethovenhaus, die Stätte deutscher Größe, / Du Wiege der, der eine Welt erschuf. / Steh unversehrt noch in den fernsten Zeiten / Gleich ihm, den hier der Gottheit Kuß erwartet. / Nach zehntägiger Gastfreunschaft / der Abschiedsgruß von Deinem Hofmaler."

76. Entry from June 26, 1902: "Beethoven, du bist / Mein liebster Componist / Drum reim ich da dies Verslein auch / Was bei Verehren ist der Brauch / Doch muß ich aber wieder gehn / Da ich noch müsst' dein Haus besehn. / Herr Baumgarten aus Freiburg."

77. Message by one Schotta, March 29, 1922: "Wie kann man anders, als mit Andachtsschauern durch—bäh. Beethoven und *ich* dichten? Hut ab, du gehst durch heilige Räume."

78. Visiting on a Sunday after church, Regina Ulißen wrote: "Aus dem Gotteshause in das Beethovenhaus! Ich beuge mich in Andacht u. Ehrfurcht, ein Andenken an den Tonmeister dem der Ruhm der Unsterblichkeit für alle Zeiten sicher ist!" November 7, 1911. M. Mittelschull of Cologne was moved to prayer: "Voll scheuen Zagens bin ich eingetreten / Und stehe nur, der scheuen Ehrfurcht voll, / Es tragt der Blick, wohin er wenden soll, / Und meine Seele rüstet still zum Beten." June 6, 1927.

79. "Man sagte mir, / ich hätte Ähnlichkeit mit dir / Meister der Töne / Mein Gesicht ist *nicht* das deiner, / doch mein *Gehör*. / Ich wollte immer zum einmal / deine IX hören, / auch ich bin taub wie du. / Das Unterschied: *Du* konntest sie, wenn auch nicht hören, / So doch schaffen, / Und das ist eine ganze Menge mehr." Signature illegible, September 5, 1926.

80. "Als Schwerhöriger stand ich in tiefsten Ergriffenheit auf dem heiligen Boden des Geburtshauses meines größten Leidensgefährten!" Rudolf Mänz from Saxony, May 22, 1927.

81. "Geisterhaft, bezaubernd . . . macht die Anwesenheit in diesen Räumen tiefen Ehrfurcht." Signature illegible, September 24, 1924.

82. "Die Seele des Beethovens fült [sic] das ganze Haus." Signature illegible, March 31, 1907. Similar sentiments were expressed in odes from July 22, 1925, and August 28, 1927; a poem by Erasmus from Erlangen from August 5, 1910, describes being deeply moved by the spirit of beauty: "In fröhlicher Fahrt / nach Sucherart / bin ich auch *deinem* Hause genaht, / gegrüßt von dem Geist, / der die Schönheit preist, / fühl ich mich wohl auf den Ewigen Pfad."

83. "In diesem Hause scheint die Zeit keine Macht zu haben. Hier leben ewige Schöpfungen in einem ganz eigenen Sinne. Nicht ist es die Baulichkeit, noch irgend ein Symbol, die sie zum Ausdruck bringen, nein, man fühlt in diesen Räumen das Ewige so nah und selbstverständlich wie die Luft, die man atmet. Beethoven ist eine der tiefsten Offenbarungen des Weltwesens; dessen wird man mit Nachdruck da inne, wo sein Genius sich erstmalig den Gesetzen der Erscheinungen dieser Welt gab." Wilhelm Kircher from Frankfurt am Main, September 14, 1924.

84. Hans Maria Saget, "Träumerei in Beethoven's Geburtshaus" (Reverie in Beethoven's birth house), BHB, B 1501/40. While this poem remains undated, its folder location suggests that it was written between 1930 and 1950.

85. For example, Beethoven's "Spring" Sonata, No. 5, op. 24 for violin and piano appears on September 20, 1896.

86. What little is known about Carl Berg (1870–1923) has been assembled by Kristina Krämer in the database *Musik und Musiker am Mittelrhein* (MMM2). After his studies in Frankfurt,

Berg took a position as chapel master in Offenbach and later directed the Kurorchester of Bad Neuenahr—a position indicated in his visitors' book entry but not yet recorded in MMM2. He composed a variety of choral works, songs, military music, and piano works with delightful titles such as "Töff! Töff! Automobil-Galopp" (1904). He attempted to take his own life in 1915 but survived, and he visited the Beethoven-Haus both before and after this tragic event. The work he inscribed into the visitors' books has been lost and does not appear in MMM2. http://mmm2.mugemir.de/doku.php?id=bergc&s[]=carl%20berg (accessed November 13, 2024).

87. Youens, "Hugo Wolf and the Operatic Grail."
88. Rojek, *Presumed Intimacy*.
89. Brümmer, "Koelman, Margarete," 62; Wild, *Ein Liebesschicksal in Liedern*.
90. "Weih, eh du eintrittst, Herz u. Hand— / Hier ist in Wahrheit heilges Land. / Zieh deines Alltags Schuhe aus." Excerpt from Margarete Koelman, pseud. Irene Wild, "O Mensch, in diesem Heiligtum," BHB, Museum Visitors' Books, May 23, 1903.
91. "Und dein Antlitz bleibt ernst u. traurig, / Schwarzgefurcht von der Menschheit Wehe, / Nieder auf mein armseliges Tun. / Sieh, heut stell ich / Dich unter Blüten, dem Winter entschlossen, / Weihe ein Blumenopfer dir. / Liebend trag ich dich dann zurücke, / Daß deine Nähe den Schlaf mir hüte, / Schwiege mein Haupt an dein hehres Antlitz: / Mein großer Toter, verlass mich nicht!" Irene Wild, "Beethoven's Büste," BHB, Museum Visitors' Books, May 23, 1903.
92. "Und in ihm sang und tönte— / Ein irdisch Ohr vernahm es nie— / Die er so heiß ersehete, / Die *zehnte* Symphonie. / Mit angehaltnem Atem / Lauscht die Natur dem Sterbelied. / Ein letzter Blitz—ein Donner: / Das helden Geist entflieht." Excerpt from Irene Wild, "Beethovens Tod," handwritten poem laid on the floor of the birth room, March 25, 1912, BHB, Z 5868, 2.
93. "Laßt seine Melodien / In Euch erklingen, / Wenn hohe Tage Euch bewegen— / Zu seinem Antlitz schaut hinauf, / Wenn dunkle Stunden / das Herz bedrungen. / Schöpft Kraft und Trost / Erhöhstes Leben / Aus seines Geistes erhabnem Flug!" Irene Wild, "Hier lebt vor meiner Seele," BHB, Museum Visitors' Books, May 21, 1906.
94. Excerpt from Margarete Koelman, pseud. Irene Wild, "Jüngst hielt ich deinen Neunte Sinfonie," BHB, Museum Visitors' Books, July 21, 1911.
95. On concert guides and musical literacy, see Thorau, "Werk, Wissen, und touristisches Hören"; and Botstein, "Listening through Reading."
96. Wild, "Dschang und Dschau." See also Fine, "Assimilating to Art-Religion," 24.
97. Margarete Koelman, pseud. Irene Wild, "Beethoven," BHB, Museum Visitors' Books, July 15, 1911.
98. J. Baruda, R.A.S.C., Manchester, UK, BHB, Museum Visitors' Books, August 1, 2025.
99. While this Old Testament word is in Hebrew, its usage in the period was secular and not associated with Jewish practice. Select examples of the many poems by this name include B. G. Ambler, "Mizpah," *Lloyd's Magazine* 6, no. 6 (August 1879): 369–70; and Jean H. Macnair, "Mizpah," *Chambers's Journal* 3, no. 123 (April 7, 1900): 304. Denver's "Mizpah arch" was unveiled at Union Station in 1906 and dismantled in 1931 to reshape the city's infrastructure for automobiles.
100. Brooks, Stephens, and Thormahlen, *Sound Heritage*; see especially therein Linda Young, "All about House Museums," 59–72.
101. Koshar, *German Travel Cultures*; Tümmers, *Rheinromantik*.
102. Lekan, *Imagining the Nation in Nature*, 24–30.
103. On the population surge and subsequent developments, see Parsons, *Vienna: A Cultural History*, 219.
104. Nußbaumer, *Musikstadt Wien*; Phillips, "Exhumations, Honorary Graves," 306.

105. Nora, "Between Memory and History," e.g., 7–8.

106. "Die Beethovengedächtnisfeier im Schwarzspanierhause," *Ostdeutsche Rundschau*, November 16, 1903, 4.

107. Zweig, *World of Yesterday*, 18.

108. As reported in "Beethoven Feier," *Neues Wiener Tagblatt*, November 16, 1903, 5.

109. On the effort to preserve pieces of the building from Beethoven's time, see the letter from October 13, 1903, Wien Museum Archive, Akt. 704/03 and the protocols from December 11, 1903, Wien Museum Archive, A Z. 77/1904. Photographs of every inch of the house, which include the furnishings of its inhabitants, were assembled into an album housed in the Wien Museum Archive. The phrase "sarcophagal fragments" (*särgliche Fragmente*) stems from a skeptical article that claimed the project was an empty promise meant to appease the protesters: "Schwarzspanierhaus und Beethoven-Zimmer," *Neues Wiener Tagblatt (Tages-Ausgabe)*, December 31, 1903, 5. See also Pötschner, *Das Schwarzspanierhaus*, 63–77.

110. Several critics expressed this view, including Anton August Raaf, "Die Beethoven-Feier im Schwarzspanierhause zu Wien," *Die Lyra: Allgemeine deutsche Kunstzeitschrift für Musik und Dichtung*, December 1, 1903, 3–4.

111. "Ein Genius rang in diesen Mauern aus, / Um diese Wände flog sein letztes Schauen, / So wie des Firnenwand'rers Blick im Graus / Der Wolke sucht den Weg aus Licht hinaus, / Zu ruhen nicht, auf neue Bahn zu bauen!" and the final lines, "Hier sinkt auf dieses Haus die Ewigkeit, / Die Mauer weicht, ein Geist will heimwärts schreiten." Hermann Hango, "Beethoven (anläßlich der Demolierung seines Sterbehauses)" (Vienna: Druck von Paul Gerin, 1903), BHB, Z 6382.

112. Gregor Pöck's speech was reproduced in "Beethoven-Feier," *Reichspost* (Vienna), November 17, 1903, 5.

113. Raaf, "Die Beethoven-Feier," 3.

114. "Beethoven-Gedächtnisfeier," *Die Zeit* (Vienna), November 16, 1903, 407–8.

115. Eduard Pötzl, "Beethoven und das Schwarzspanierhaus," *Neues Wiener Tagblatt*, September 27, 1903, 1–2.

116. "Das Ende des 'medizinischen Kaffeehauses,'" *Deutsches Volksblatt*, September 1, 1903, newspaper clipping in Wien Museum Archive, Akt 705/1903.

117. Max Dvořák, "Monumenta deperdita," *Kunstgeschichtliches Jahrbuch der Zentral-Kommission für Erforschung und Erhaltung der kunst- und historischen Denkmale* (Vienna), Beiblatt to vol. 4, 1910, 176–78, at 176–77.

118. Zweig, *World of Yesterday*, 16–17.

119. The Bösendorfer stone is housed in the archives of the WGM, in a box labeled "Erinnerung an den Bösendorfer Saal, 1872–1913" (In memory of the Bösendorfer Hall, 1872–1913), WGM ER Bösendorfer-Saal 1. The visage stone belonged to Gustav Schütz, who took two stones. On one, he mounted a bronze mask of Beethoven, which he later gifted to Pablo Casals, and on the other, he had an artist carve the mask with light rays; these are archived alongside his recollections in WGM, B, 3301. Two further stones from the Schwarzspanierhaus were taken as relics by unknown collectors: WGM, ER Beethoven 5 and ER Beethoven 15. The former is a piece of wallpapered plaster from the death room, mounted onto a piece of wood with a photograph and the words "Aus Beethovens Sterbezimmer." The latter is more interesting: a stone from the courtyard of the house, painted metallic copper with the first ten measures of the first violin part of the String Quartet, op. 135, movement 1; holes and a cord are installed to hang the relic on the wall.

120. Heinrich Penn, "Beethoven. (Zur Demolierung seines Sterbehauses.)," unidentified newspaper clipping, BHB, Z 2931, 14.

121. Lest this seem like an interpretative stretch: in Landau's album, the poem "Beethoven's Grab" by M. G. Saphir depicts an angel who lays flowers on Beethoven's grave while the nightingales sing "Adelaïde"; see Landau, *Erstes poetisches Beethoven-Album*, 199–205.

Chapter Two

1. Koch, "Chronik des Denkmals," 16; see also Hanslick, "Das Monument."
2. *Musical Times and Singing Class Circular*, June 1, 1896, 390–91.
3. Hummel, "Das Mozart-Album der Internationalen Mozartstiftung."
4. "Euch hat die Mitwelt froh den Kranz gespendet, / Gethan, daß man im Tod auch noch bewundert, / Doch liegt ein andrer Meister hier begraben, / An dem sich müde zeugte ein Jahrhundert, / Den nur der Herr nach tausend Jahren sendet! / Wo ist die Ruhestatt, die sie ihm gaben, / Das Monument? Sie haben, / Fürwahr, sie haben keines ihm geschichtet! / Gut machen müsset ihr, was Eure Ahnen / Verbrochen an des Künstlers heil'gen Manen, / Ihr ehret euch, wenn ihr es ihm errichtet; / Doch bald, dass nicht des Meisters Geist erscheine / Und furchtbar mahne, wie der Gast aus Steine." Frankl, excerpt from "Allerseelentag," 1834, handwritten entry in the Mozart Album from February 4, 1883, MGA, no accession number.
5. "Ihr ruft, noch war Er nicht erkaltet, / 'Ein Denkmal Ihm!' in lautem Chor. . . . Wer giebt dem Todten das Geleite? / Niemand! Kein Kranz, kein Fackelschein! / Kein Lied ertönt und keine Saite, / Es knarrt das Fuhrwerk nur allein. . . . Wisst Ihr es, wo sie ihn verscharrten? / Habt Ihr gesühnt der Väter Thun? / Umsonst! Wir harren und wir harrten— / Wie mögt Ihr, Dankvergessen ruhn? . . . Du Volk von Wien, besinne Dich!" Ludwig August Frankl, "Merk's Wien," 1883, handwritten entry in the Mozart Album from February 4, 1883.
6. The majority of archival material cited in this chapter hails from two boxes at the MGA: A IV 3 and A IV 4, which were assembled for an exhibition about the *Magic Flute* cottage in 1950. Accession numbers are provided where available; some material remains unnumbered or belongs to less formally catalogued collections of the MGA.
7. Exchange between the Mozart-Stiftung and the Viennese architect Rudolph Payer in July and August 1873, MGA, box A IV 4.
8. "Das Musikfest in Salzburg," *Innsbrucker Tagblatt*, July 23, 1877, 3. It was reported here that Emperor Franz Joseph I had graced the cause with his contribution to the Mozart Album. See also Schurich, *Das Zauberflötenhäuschen*. For a roster of contributors to the Mozart Album, see Hummel, "Das Mozart-Album."
9. Examples include Schumann's study in Zwickau, Scriabin's in Moscow, Tchaikovsky's in St. Petersburg, or Britten's in Aldeburgh.
10. Painter, "Mozart at Work."
11. Translated by Painter, "Mozart at Work," 224.
12. Großpietsch, "Pilgerreisen zu Mozart und ein Salzburger Bürgerhaus."
13. Novello, *Mozart Pilgrimage*.
14. Letter from Franz Schubert to his brother in 1825, in Deutsch, *Franz Schubert's Letters and Other Writings*, 105.
15. Morgenstern, "Constanze Nissen in Salzburg"; K. Wagner, *Das Mozarteum*.
16. Mielichhofer, *Das Mozart-Denkmal zu Salzburg*, 16–17.
17. The festival program was reprinted in the *Niederrheinische Musik-Zeitung* 4, no. 34 (August 23, 1856): 274.

18. Engl, *Genesis der Internationaler Mozart Stiftung*; Angermüller, "Die Bedeutung der ISM Salzburg."

19. A competitive tone was already evident in Freisauff, *Das erste Salzburger Musikfest*, 23. Festival cofounder Friedrich Gehmacher called Salzburg the "Austrian Bayreuth" in 1913: "Antrag an das Kuratoreum der Internationalen Stiftung Mozarteum," reprinted in Holl, "Dokumente zur Entstehung der Salzburger Festspiele," 152. In 1935, festival director Erwin Kerber claimed that Wagner considered Salzburg before he settled on Bayreuth; see Kerber, *Ewiges Theater*, 37. On the comparison of Salzburg with Bayreuth, see also Hoffmann, "Vom Mozartdenkmal zur Festspielgründung," 413–19.

20. These notes from the committee are reprinted in the exhibit catalog, *Die Salzburger Festspiele*, 14. Wagner made this remark in a letter to Franz Liszt on January 30, 1852; R. Wagner, *Sämtliche Briefe*, 4:270. See Kreuzer, *Curtain, Gong, Steam*, 17. The design for the new hall was featured on the cover of the *Neue illustrirte Zeitung* 18, no. 2 (September 7, 1890): 1.

21. A photographic postcard from the festival shows Mozart's bust from behind as it gazes over the rooftops, and the festschrift for the event, which devoted a lengthy chapter to the cottage, proclaimed that "the genius of the great master will hover over this site and will look down upon Salzburg with blessings, the city that he loved so little during his lifetime." Souvenir postcard, "Zur Erinnerung an das Erste Salzburger Musikfest," 1877, A. Czurda, Salzburg, MGA; see also Freisauff, *Das erste Salzburger Musikfest*, 52–53.

22. Single-page flyer, *Cook's Personally Conducted Tour / to Belgium, the Rhine Nürnberg, Munich, / Salzburg, for the Grand Mozart Festival, / Lake Constance, the Bernese Oberland, / The Lake of Geneva, and Paris / Leaving London, Monday, July 7th, 1879* (London, 1879), MGA, no accession number.

23. Writing in the cottage visitors' book in September of 1882, Emma Marleni from Philadelphia substituted one form of piety for another: "After being refused to enter the Kapuziner-Kloster, I enjoyed the reccolection [sic] and memorie of dear, beloved Mozart, thinking of many happy hours his great spirit has given me." Regrettably, all the visitors' books except one (1880–83) were destroyed in World War II. On pilgrimage sites and relics in Salzburg and its environs, see Gugitz, *Österreichs Gnadenstätten in Kult und Brauch*, 5:189–98.

24. Louisa Lergetporer, "Erinnerungen an der Mozartfest 1877," handwritten poem, MGA. Louisa was likely a relative of Alois Lergetporer, mayor of Salzburg from 1831 to 1847; the exact relation is unknown, as she is not listed among his children or daughters-in-law. Poetic tributes written after a festival were common; see, for instance, the bound volume gifted by Johannes Goebels to the Beethoven-Haus shortly after the 1927 death centennial, BHB Z 2931.

25. D., "Salzburger Mozartiana," *Salzburger Volksblatt*, September 17, 1885, 2–4.

26. "Dort eben steht es nun, so stolz und frei"; "Hail [sic] ihm, der dieß so edel ausgedacht, / Dem Freiherrn *Sterneck*, sei ein Hoch gebracht." Lergetporer, "Erinnerungen."

27. Horner, *Die internationale Wallfahrt zum Mozart-Häuschen*.

28. Religious rhetoric is still used today to encourage philanthropy; see Habisch, "Spiritual Capital."

29. Childhood memoir by Frau Alfred Heidl, sent to the IMS president on October 16, 1909, MGA. The author's first name is not specified, and she is absent from biographical indices.

30. Letter from Sterneck on behalf of the IMS to the City of Salzburg, requesting to display the cottage in the Mirabellgarten, January 19, 1874, MGA, box A IV 3, no. 17.079, pp. 1–2.

31. Alfred Walter, "Die Geburtsstätte der Zauberflöte," *Die Gartenlaube* 4 (1874): 70–71.

32. H. M. Schuster, "Die Internationale Stiftung 'Mozarteum' in Salzburg," *Allgemeine Kunst-Chronik* 6 (November 4, 1882): 608–10, at 608.

33. "Inventar über die Mozarthäuschen am Kapuzinerberge zu Salzburg befindliche Gegenstände der Internationalen Mozart-Stiftung," MGA, box A IV 4.

34. Letter from Starhemberg to the IMS from August 4, 1890, MGA, A IV 3.

35. Bauer, *Josef Hoffmann*.

36. A handwritten note on the back of Hoffmann's watercolor sketch recommends Egyptian-looking plants that would thrive in an Alpine climate; MGA. The need to protect the cottage was likely amplified by the valuable objects on display inside it, which could be susceptible to damp conditions.

37. This statement was affixed to a draft of a formal letter that the IMS considered sending to the Silver Kings. See MGA, A IV 3, no. 138.

38. IMS memo, July 13, 1878. MGA, A IV 3, no. 137.

39. IMS notice sent to the local newspapers, January 8, 1880, MGA, A IV 3, no. i.6 and no. 145/1880. Sociologists have shown that donors see giving as a duty when they are active in a religious congregation, even as they prefer to earmark funds for select causes. Ostrower, *Why the Wealthy Give*, 16.

40. Prominent Viennese musicians spread the word: Albert Weltner offered to solicit donations, and thereafter his friends appeared on the 1881 roster (see his letter to Sterneck, Vienna, September 29, 1879), while the composer and pianist Louise Adolpha Le Beau held a charity concert, contributed her proceeds, and requested a spot on the Salzburg program in return (letter to IMS from Munich, November 20, 1880), MGA, A IV 3, no. 343.

41. Ostrower, *Why the Wealthy Give*, 69–85.

42. "Gutachten des 'Technischen Clubs' über eingereichte Pläne für eine Umhüllung des Mozart-Häuschens, verfaßt von dem hierzu bestellten 'Comite', bestehend aus den Herren Prof. V. Berger, Arch. C. Demel und Baumeister J. Christof, gezeichnet vom Städt. Oberbaurat Dauscher und von Ing. Hans Müller," 1883, MGA, A IV 3, no. 390.

43. "Gutachten des 'Technischen Clubs.'"

44. "Lokalnotiz von 1980," *Der Lugenschippel*, January 31, 1880, newspaper clipping, MGA, A IV 3. This humor periodical, whose name means "the big liar" in Austrian dialect, must have had fairly limited circulation, as it is not found in the collections of most Austrian libraries.

45. Nora, "Between Memory and History."

46. A list of visitors and dates was collated from press clippings by Josef Hummel, former archivist of the Internationale Stiftung Mozarteum, MGA, A IV 3.

47. Otto Elben dismissed Metternich's remarks as a misunderstanding of these choirs' true purpose, which was "a sacred duty to become a beacon of refinement and Bildung." Elben, *Der volkstümliche deutsche Männergesang*, 468; on men's choirs in Austria, see 115–22.

48. A survey of choral concerts in the Lower Rhine region between 1828 and 1878 shows that, surprisingly, only five percent of songs were patriotic in nature. See Blommen, *Anfänge und Entwicklung des Männerchorwesens am Niederrhein*, 200.

49. Challier, *Grosser Männergesang-Katalog*.

50. Wagner composed several works for the Dresden men's choir: *Liebesmahl der Apostel* (1843), *Der Tag Erscheint* (1843), and commissions for festive occasions.

51. Brusniak, "Involvement of Freemasons in the 'Erstes Deutsches Sängerfest.'"

52. As Ryan Minor puts it, these works not only were of their moment but "[captured] that moment's self-awareness *as* a moment." Minor, *Choral Fantasies*, 119. A well-known example is Franz Liszt's cantata for the unveiling of Beethoven's monument in 1845, which opens with a self-congratulatory tone: "What brings together the masses here? What has called you to gather?

One might believe from this crowd that today must be a festive day." "Was versammelt hier die Menge? / Welch Geschäft rief euch herbei? / Glaubt man doch an dem Gedränge, / daß ein Festtag heute sei." O. L. B. Wolff (text) and Franz Liszt (music), *Cantata for the Inauguration of the Bonn Beethoven Monument* (1845). See Rehding, *Music and Monumentality*, esp. 71.

53. Max von Weinzierl and Dr. Märzroth [Moritz Barach], "Des Künstlers Genius / Gedicht von Dr. Märzroth / zur Eröffnung des Mozarthäuschens am Kaupziner- / berge zu Salzburg / am 19. Juli 1877," Landesbibliothek Coburg, Ms Mus 944.

54. Text by Märzroth as reprinted in Freisauff, *Das erste Salzburger Musikfest*, 53.

55. "*Das* nur hat den *rechten* Klang, / Was dem *Volk* zu *Herzen* drang!" Contribution dated July 18, 1877, MGA, Mozart Album.

56. Painter, "W. A. Mozart's Beethovenian Afterlife," 125–32.

57. Engl, announcement to the press on behalf of IMS, May 28, 1877, MGA, A IV 3. On Goethe's cottage, see Springer, *Die klassische Stätten von Jena und Ilmenau*, 27–32.

58. Freisauff, *Das erste Salzburger Musikfest*, 47.

59. Blurb by Alfred Walter of the IMS board affixed to a souvenir photograph, April 20, 1877, MGA.

60. To protest the garden location of Mozart's monument in 1896, Max von Kalbeck wrote: "[Mozart's] place was not out in open nature, nor in an enclosed garden. . . . The dramatist belonged in the midst of the hustle and bustle of the city, on the public, teeming streets, where the colorful play of life made a meaningful impression on him." Max Kalbeck, "Mozart in Wien," *Neues Wiener Tagblatt*, April 21, 1896, 1.

61. ". . . Natur vereinte das Freyte und Heitere, / So entsteht die Schönheit und die Rose einst / ihr Symbol: zu finden / auf dem 'Kapuzinerberg.'" Entry by U. U. Berbaut, July 3, 1881, MGA, Zauberflötenhäuschen Visitors' Book.

62. Lergetporer, "Erinnerungen."

63. Lergetporer's poem bears the year 1877, and the painting was displayed starting in October 1877; it is difficult to say whether she was familiar with the image before drafting her ode. An IMS inventory notes that prints of the painting, which has aroused great interest from the public, could be purchased for forty Kronen and serve as valuable *Erinnerungsblatt* (souvenir leaf) for visitors; "Inventar über die Mozarthäuschen am Kapuzinerberge zu Salzburg befindliche Gegenstände der Internationalen Mozart-Stiftung," MGA Box A IV 4. When the painting was unveiled, Romako left a message in the Mozart Album describing his deep childhood affection for the *Magic Flute* and his amazement that Mozart, despite being so poor (as reflected by the humble cottage), was nonetheless rich in ingenuity. Anton Romako, entry in the Mozart Album, October 26, 1877. It is notable that a similar design appeared two decades later as a runner-up for Mozart's monument in Vienna; Mozart sits at a spinet in the rejected monument design by Edmund Hellmer; see Telesko, *Kulturraum Österreich*, 178.

64. Brodbeck, *Defining* Deutschtum.

65. Hiebl, "German, Austrian, or 'Salzburger'?" For Hoffmann's allusion to Bavaria, see Gallup, *History of the Salzburg Festival*, 11.

66. Alexander Rehding, "Collective Historia," in Rehding, *Music and Monumentality*, 141–68. See also Gur, "Music and 'Weltanschauung.'" Given that the chain ends with Liszt and Wagner, with Brahms absent, one might surmise that the artist sympathized with the New German School. One could counterargue that Brahms was still focused on chamber music and smaller-scale piano and vocal works when this image was created.

67. Louis Schneider and Wilhelm Taubert wrote the libretto, expanded Mozart's original numbers, and appended orchestrations of Mozart's Lieder; K. Wagner, *Das Mozarteum*, 24. Mark Everist has studied the afterlife of this arrangement in France in *Mozart's Ghosts*, 54–74.

68. R. Wagner, "Das Publikum in Zeit und Raum." On Wagner's elevation of *Die Zauberflöte* to position himself as heir, see Daverio, "Mozart in the Nineteenth Century," 182.

69. Nohl, *Der Geist der Tonkunst*, 197; and Nohl, "Die Entstehung der Zauberflöte."

70. Silverman, *Becoming Austrians*.

71. Steinberg, *Meaning of the Salzburg Festival*, 37.

72. Schorske, *Fin-de-Siècle Vienna*. See also Steinberg, *Meaning of the Salzburg Festival*, x–xii, 26–36, and 42.

73. Hanslick was charmed by the 1862 Volksfest on the Mönchsberg; see his review, "Das deutsche Künstlerfest in Salzburg, 10.9.1862," in *Hanslick: Sämtliche Schriften*, 1:132–37.

74. Speech recited at the cottage on August 15, 1879, printed in "Das Zauberflötenhäuschen," in *Erster Jahresbericht der Internationalen Stiftung Mozarteum in Salzburg*, 74.

75. Hugo Hofmannsthal, "Die Mozart-Zentenarfeier in Salzburg," 1891, in Hofmannsthal, *Gesammelte Werke*, 1:42–46.

76. Halse, "Literary Idyll in Germany, England, and Scandinavia 1770–1848."

77. Quoted in Gallup, *History of the Salzburg Festival*, 13.

78. Hugo Hofmannsthal, "Der erste Aufruf zum Salzburger Festspielplan," 1919, in Hofmannsthal, *Festspiele in Salzburg*, 33–38.

79. Kerber, *Ewiges Theater*, 20.

80. Bahr, "Die Hauptstadt von Europa." Later writings described Mozart's music, and Salzburg, as the perfect balance of German and Italian: Bahr, "Die Mozartstadt"; Bahr, *Salzburg*.

81. *Salzburger Volksblatt*, June 16, 1902; cited in Sayler, *Max Reinhardt and His Salzburg*, 200.

82. Kerber, *Ewiges Theater*, 40.

83. See Botstein, "Aesthetics and Ideology."

84. W. H. Auden, "Metalogue to the *Magic Flute*," in Auden, *Collected Poems*, 576–78.

85. Mahler composed in two other cottages at Steinbach am Attersee and Maiernegg. In 1981, the International Gustav Mahler Society petitioned for Austria's historic preservation registry to extend formal protections to these sites.

86. This local history has not yet been published outside of Toblach and lives primarily through oral history and materials associated with the Mahler Weeks. Josef Lanz, "Musikwochen in memoriam Gustav Mahler," essay published on event website, "The Gustav Mahler Musik Weeks," https://www.kulturzentrum-toblach.eu/smartedit/documents/_mediacenter/musikwochen-josef-lanz_de.pdf (accessed November 13, 2024).

87. Eventually, the festival committee persuaded the RAI to televise a scaled-down arrangement of *Das Lied von der Erde* performed in the small museum that occupied Mahler's former apartments.

88. In 2011, percussionist and composer Nebojša Jovan Živković posted a YouTube tour of the premises that complains bitterly about the mistreatment of this "shrine," and YouTube comments reflect not only the perceived piety of the site, but the shame associated with *Ehrenschuld*. See https://www.youtube.com/watch?v=Ar-KkCKuy4w (accessed December 25, 2016).

89. Saskia Santer has led this project; my thanks to Saskia and Herbert Santer for offering a tour of the premises during my visit to Toblach.

90. I include this perspective with explicit permission from my interlocutor, who was informed of the purposes of the conversation in advance, and who read and approved this passage in German translation. Conversation on August 13, 2022.

Chapter Three

1. Alex Ross, "Chopin's Heart," *New Yorker*, February 5, 2014, https://www.newyorker.com/culture/culture-desk/chopins-heart.

2. Reiter, "Causes of Beethoven's Death and His Locks of Hair"; Senior, "Did Beethoven Die of Lead Poisoning?"; Stadlbauer et al., "History of Individuals of the 18th/19th Centuries." In a recent twist, the original assertions of lead poisoning from two decades ago turned out to be correct once Beethoven's authenticated locks were tested. A high concentration of lead was often added as a sweetener to inexpensive wines, which Beethoven consumed in large quantities, thinking it medicinal. Rifai et al., "Letter to the Editor."

3. Witt et al., "A Closer Look at Frédéric Chopin's Cause of Death"; Perciaccante et al., "Did Frédéric Chopin Die from Heart Failure?"; Charlier et al., "Heart of Frédéric Chopin (1810–1849)"; Duclos-Vallée et al., "Is Frédéric Chopin's Death Elucidated?"

4. Davies, *Romantic Anatomies of Performance*, 41–65.

5. Goldberger, Whiting, and Howell, "Heartfelt Music of Ludwig van Beethoven," 286.

6. François Martin Mai values Beethoven's conversation books as a rare document of doctor-patient interactions; Mai, *Diagnosing Genius*.

7. Niemack, "Herzschlag und Rhythmus"; Schweisheimer, *Beethovens Leiden*, 120.

8. Adolphe, "'Where Thought Touches the Blood.'"

9. Yang, *Planet Beethoven*, 39–57.

10. Ludendorff, *Der ungesühnte Frevel an Luther, Lessing, Mozart, Schiller*; and Ludendorff, *Mozarts Leben und gewaltsamer Tod*. Dieter Kerner wrote over a dozen articles on Mozart's poisoning, two coauthored books for the Ludendorffs' press, and two large volumes on composers' ailments, including Kerner, *Krankheiten großer Musiker*. For Otto Erich Deutsch's take on Kerner, see Stafford, *Mozart Myths*, 38. Kerner was nonetheless deemed "sophisticated medically and musicologically": Saffle and Saffle, "Medical Histories of Prominent Composers," 81. Mozart has attracted a disproportionate number of both pathographies and conspiracy theories because his prodigy implied that his talent was inborn, and the unknown whereabouts of his body allowed free speculative rein.

11. A valuable history of the term "pseudoscience" has been outlined by Frietsch, "Boundaries of Science/Pseudoscience." While "pseudoscience" appears first in the scientific revolutions of the early modern period, it was not formally defined until Gall's skeptics did so. In the seventeenth century, the prefix "pseudo-" participated in a Janus-faced polemics: theologians denounced scientific heresy as "pseudo-philosophy" while natural scientists denounced magic as "pseudo-science." In his 1824 rebuke of phrenology, François Magendie clarified that these methods were grounded in belief, not observation, and in a desire to entertain; this text would later, in translation, become a staple of American medical school education, leading American anatomists to reject phrenology more consistently than their continental colleagues; see Magendie, *Précis élémentaire de physiologie*, 1:202. In the 1930s, Karl Popper refined the definition of pseudoscience to describe methods that reject the vital stage of falsification, or the capacity to be disproven. See Popper, *Conjectures and Refutations*, 33.

12. McMahon, *Divine Fury*.

13. Straus, *Broken Beauty*, 12–13. The archetype of the "Saintly Sage" was first named as such in a study of historical cinema: Norden, *Cinema of Isolation*, 131–33.

14. Admittedly, the line between hoarding and collecting is not straightforward. Rebecca R. Falkoff argues that hoarding begins with control over the material world and culminates in submission, as possession of things gives way to a helpless possession *by* things. Falkoff, *Possessed*, 14–18.

15. Adorno, *Introduction to the Sociology of Music*, 88.

16. On men of the professions as amateur musicians, see Sumner Lott, *Social Worlds of Nineteenth-Century Chamber Music*, 79–106. A prominent collector of rare musical autographs was the anatomist and violinist Guido Richard Wagener.

17. Hau, "Holistic Gaze," 499.

18. Penelope Gouk, "Sister Disciplines? Music and Medicine in Historical Perspective," in Gouk, *Musical Healing in Cultural Contexts*, 171–96, at 181.

19. Billroth also hosted chamber music evenings with physiologist Theodor Wilhelm Engelmann and Bernard Naunyn, whose wife was a professional singer and whose medical pupil Otto Loewi was a devoted Wagnerian. Similarly, the Leipzig physiologist Carl Friedrich Wilhelm Ludwig hosted chamber music gatherings, and his medical colleague Friedrich Trendelenburg authored a comprehensive anatomy of bow hold and string timbre.

20. Julius Tandler, "Über den Schädel Haydns," in *Mittheilungen der anthropologischen Gesellschaft in Wien*, vol. 39 (1909), 272. The following year, Tandler's lecture made a splash in the international papers, and it remained influential a decade later, as in E. Jentsch, "Die lokalisation der musikalischen Anlage im Schädel," *Zeitschrift für die gesamte Neurologie und Psychiatrie* 48, no. 1 (1919): 263–93. For a secondhand account of the network of anatomists invested in composers' remains, see Stockert-Meynert, *Theodor Meynert und seine Zeit*.

21. A curated selection includes the physician Johann Heinrich Feuerstein, who took up the project of writing Mozart's biography when it was left unfinished by Georg Nikolaus von Nissen; the Swedish orthopedist Franz Berwald, who doubled as a composer; Alfred von Bary, who sang leading roles at Bayreuth to supplement his income as a psychiatrist; the Kolisch family, into which Arnold Schoenberg married, and whose members were equal parts medical and musical; and Walter Bransen, a physician and composer who fled the Nazis with his wife, Dorothee Manski, an opera singer who became a professor of voice at Indiana University while her husband worked as a film composer under various pseudonyms. Archival records of medical schools yield a wealth of more obscure figures, such as the Viennese surgeon Georg Lotheissen, who wrote songs on medical topics for his quartet society; see the Lotheissen's diaries in the Josephinium Archive, Vienna, 2699, with songs such as "Des jungen Chirurgen Reisegesang," nos. 28 and 33, along with pages (such as no. 21, p. 5) that record his active concert attendance.

22. Vincent, "Doctors Look at Music," 243. See also Garrison, "Medical Men Who Have Loved Music"; and Gillespie, "Doctors and Music."

23. Csiszar, *Scientific Journal*. On specialization in German institutions, see Ringer, *Decline of the German Mandarins*, 102–13.

24. Early modern music-medical discourse theorized the sympathetic vibrations of the nerves, resonances that stir the humors and the passions, and dissonances supercharged with pain: see Varwig, *Music in the Flesh*; Head, "C. P. E. Bach 'in Tormentis.'" On anecdotes and scientific collecting, see Carroll, *Science and Eccentricity*.

25. Letter quoted in Hanslick's introduction to Billroth, *Wer ist musikalisch?*, 5.

26. Sacks, *Musicophilia*. An early predecessor of the genre hails from German physician Ernst Anton Nicolai; see Nicolai, *Die Verbindung der Musik mit der Arztneybelahrheit*.

27. Billroth, *Wer ist musikalisch?* See also Kahler, *Theodor Billroth als Musikkritiker*.
28. Billroth, *Wer ist musikalisch?*, 162.
29. Feis, *Studien über die Genealogie und Psychologie der Musiker*. The Dr. Hoch in question was not a medical doctor but Joseph Hoch, a lawyer and amateur musician who founded the conservatory with a large inheritance.
30. Feis, 58 and 52, respectively.
31. Lombroso, *Genie und Irrsinn*; on the Bach family, see 72; on Schumann, 87–88; on mental illness among Jews, 68–72. Berlioz is absent in this book but appears in a section on degeneracy and genius in Lombroso, *Man of Genius*, 27. On Lombroso's speculative approach, see Hiller, "Lombroso and the Science of Literature and Opera." On Lombroso's influence, see Person, *Der pathographische Blick*.
32. Grey, "Wagner the Degenerate"; Vetter, "Wagner in the History of Psychology"; Steger and Thiery, "Theodor Puschmann"; Dreyfus, *Wagner and the Erotic Impulse*, 117–74; James Kennaway, "Modern Music and Nervous Modernity," in Kennaway, *Bad Vibrations*, 63–98.
33. Waldvogel, *Auf der Fährte des Genius*, 7.
34. Jeffrey Kallberg, "Small Fairy Voices: Sex, History, and Meaning in Chopin," in Kallberg, *Chopin at the Boundaries*, 62–88. Pathographers debated whether Chopin's music reflected his feverish creativity and alleged sexual deviance, such as rival pathographies by Bordes, *La maladie et l'oeuvre de Chopin*, and Ganche, *Souffrances de Frédéric Chopin*.
35. Waldvogel, *Auf der Fährte des Genius*, e.g., 61–67. See the discussion of Waldvogel in Köhne, *Geniekult in Geisteswissenschaften und Literaturen*, 74–88.
36. As one pathographer put it, "illness, whether psychic or physical, releases soulful emotions in the creative artist that might not have been known to him otherwise, and that become powerful forces in his creative exertions." Gustav Ernest, "Friedrich Chopin: Genie und Krankheit," *Die medizinische Welt* 20 (May 17, 1930): 723. The idea that suffering begets art was amplified by the reception of Beethoven's confessions in his Heiligenstadt Testament; see Bonds, *Beethoven Syndrome*, 128–34.
37. Lange-Eichbaum, *Problem of Genius*, xvii–xviii. See also his *Genie, Irrsinn und Ruhm*.
38. Lange-Eichbaum, *Genie, Irrsinn und Ruhm*, 146–50. On the "nimbus," see 22 and 153.
39. Lange-Eichbaum, 151–52. Anticipating later discourses in persona studies, Lange-Eichbaum argues that art does not represent a harmonious self, but a divided array of theatrical masks; see 244–54. Lange-Eichbaum's relationship with National Socialism is elusive. His second edition in 1956 acknowledges how racialism has tarnished pathography, but, to support select claims, he cites "oral testimonials" by Walther Rauschenberger, the racialist psychologist who analyzed composers' faces; see entry no. 1358, p. 501.
40. "Inländische Korrespondenz. I. Aus Wien. Wien den 1. May 1798," *Der neue teutsche Merkur* 69 (1798): 180–84, at 180.
41. Finger and Eling, *Franz Joseph Gall*, 177. Skull theft was easier in Vienna, where a late eighteenth-century ordinance left funerals unsupervised; Hagner, *Geniale Gehirne*, 62. Anton Schindler reported that Beethoven's gravedigger was bribed unsuccessfully; see letter no. 477 in Albrecht, *Letters to Beethoven and Other Correspondence*, 3:214–17. Ironically, Gall's own skull capacity was far below the declared average for men of genius, which his admirers pointedly ignored; see Gould, *Mismeasure of Man*.
42. Finger, *Origins of Neuroscience*, 32–50.
43. Finger and Eling, *Franz Joseph Gall*, 197–219.
44. Poskett, *Materials of the Mind*.

45. Poskett, 39.

46. Hagner, *Geniale Gehirne*, 64. On Soemmerring's critical stance toward the theories of Gall, see Oehler-Klein, *Die Schädellehre Franz Joseph Galls*, 60–64. On anatomists collecting each other, see Hagner, "Skulls, Brains, and Memorial Culture."

47. "Inländische Korrespondenz. I. Aus Wien. Wien den 1. May 1798," 180.

48. In the late nineteenth century, the word *Reliquienhascherei* referred to celebrity souvenir hunting as well as relic theft from churches. Select examples include the travelogue by Richard Andree, *Vom Tweed zur Pentlandföhrde: Reisen in Schottland*, 87; and Stein, "Eine heilige Stätte," 295.

49. Lavater, *Essays on Physiognomy*, 2:212, 284–85. See also Gray, *About Face*, 10. It is compelling that the mid-eighteenth century scientist Emanuel Swedenborg, who postulated cortical localization well over a century before the concept was widely adopted, abandoned neuroscience for theology after he experienced a series of mystic visions.

50. Quoted in McMahon, *Divine Fury*, 77n18.

51. The idea of "eyes of light" can be found in ancient Zoroastrian and Jewish sources such as the Dead Sea Scrolls, in the Gospel of John, and later in Gnostic and Anabaptist texts. In the Book of Enoch, the Messiah is said to have eyes of light, and a similar metaphor introduces the Gospel of John. Hans Denck, a prominent Anabaptist, espoused a theology of "inner light" and "inner voice." See Estes, "Dualism or Paradox?"; Denck and Bauman, *Spiritual Legacy of Hans Denck*, 36. My thanks to Anne Kreps for these insights. Lavater frequently refers to the "fiery gaze"; in one passage, it "penetrates with the irresistibility of lightning; as it irradiates, dazzles, and, with the rapidity of thought, assumes at pleasure a robe of light, or shrouds itself in darkness." Lavater, *Essays on Physiognomy*, 1:99.

52. Gray, *About Face*, 8; on Lavater's methods of dissemination, see xxxii–xxxiv.

53. Accounts by Karl Bursy in 1816, Daniel Amadeus Atterbohm in 1819, and Antonie Brentano in 1811; see Kopitz and Cadenbach, *Beethoven aus der Sicht seiner Zeitgenossen*, 1:172, 1:39, and 1:99, respectively, quoted in Fine, "Towards a History of the Eccentric Artist," 579. For comparable examples regarding Haydn, see Davison, "Face of a Musical Genius"; on Gall's influence on the portraiture of artists, see Davison, "Franz Liszt and the Physiognomic Ideal in the Nineteenth Century"; and Davison, "Musician in Iconography."

54. On Rellstab, see Leitzmann, *Ludwig van Beethoven*, esp. starting on p. 335; on Schubert, see Clark, *Analyzing Schubert*, 39–52. On physiognomy as a measure of Jewishness, see Kimber, "Never Perfectly Beautiful"; and Knittel, *Seeing Mahler*.

55. In a curious line of reasoning, he also declared elephants, camels, and bears to be the most musical animals. Gr. (Georg August von Griesinger), "Ueber ein physiologisches Kennzeichen des musikalischen Talents. Nach Hrn D. Galls Entdeckungen," *Allgemeine musikalische Zeitung*, October 28, 1801, 66–69, at 68.

56. Frigau Manning, "Phrenologizing Opera Singers."

57. Finger and Eling, *Franz Joseph Gall*, 21–22.

58. Finger and Eling, 198–200.

59. Paudler, "Die Rasse Beethovens"; Rauschenberger, "Rassenmerkmale Beethovens und seiner nächsten Verwandten"; Rauschenberger, "Über die Rassichen Grundlagen der Deutschen Tonkunst"; and Rauschenberger, *Erb- und Rassenpsychologie schöpferischer Persönlichkeiten*. Rauschenberger wrote an unnerving letter to Max Auer to inquire about Bruckner's hair and eye coloring and to request a portrait of the composer's father, admitting his inkling there might be traces of Black features there. Letter from Walther Rauschenberger to Max Auer, Frankfurt, May 13, 1945, ÖNB, Estate of Max Auer, 31 Auer, 455.

60. Lavater, *Essays on Physiognomy*, 1:193.

61. Gray, *About Face*; see the chapter "Goethe as Found(l)ing Father of Modern German Physiognomics," 137–76.

62. Hagner, *Geniale Gehirne*, 70–72.

63. J. A. W. v. Goethe, "Rede bei Niederlegung von Schillers Schädel," 1472–73.

64. J. W. v. Goethe, "'Bei Betrachtung von Schillers Schädel,'" 155–57.

65. Dickey, *Cranioklepty*, 1–11.

66. Ludwig August Frankl, "Mozart's Schädel ist gefunden," *Neue Freie Presse* (Morgenblatt), January 8, 1892, 1–3.

67. Nohl, *Beethovens Leben*, 3:963n338.

68. Horace, *Odes* 4.8.28.

69. Breuning, "Skulls of Beethoven and Schubert," 60; orig. Gerhard von Breuning, "Die Schädel Beethoven und Schubert," *Neue Freie Presse* (Feuilleton), September 17, 1886. Mozart's skull was advertised by the museum as a major attraction: MGA, box A IV 4. On Hyrtl's donation to Philadelphia's Mütter Museum, see his letter to Thomas Hewson Bache, October 29, 1873, Mütter Museum archive, Philadelphia. See also Tichy, "Zur Anthropologie des Genies."

70. Karhausen, "Mozarteum's Skull."

71. Ottokar Janetschek concluded his 1924 biofictional Mozart novella with relief that Mozart largely escaped the relic snatching of mad doctors because his body was sucked directly into heaven. Janetschek, *Mozart*, 351.

72. Ashbrook, *Donizetti and His Operas*, 671n26. While the museum records confirm that the skull cap was discovered and reinterred, the point about loose change is entirely anecdotal; in a competing anecdote, the cap stored blotting sand. Slonimsky, *Slonimsky's Book of Musical Anecdotes*, 98.

73. WGM, ER Beethoven 11.

74. Moritz von Schwind, "Schubertiade," Pencil, 1868, Wien Museum Archive.

75. Breuning, *Memories of Beethoven*, 117.

76. Meredith, Special issue; see especially Meredith, "History of Beethoven's Skull Fragments: Part One."

77. Meredith, "History of Beethoven's Skull Fragments: Part One," 29. The professor of surgery was Victor Goodhill, who had been concertmaster of the USC Symphony Orchestra and studied at the New England Conservatory; he hosted the documentary *Beethoven: Triumph over Silence*, directed by Hamid Naficy (Santa Monica, CA: Pyramid Film and Video, 1985), film reel, 37 min.

78. See Meredith, "History of Beethoven's Skull Fragments: Part One," esp. 12.

79. Bankl and Jesserer, *Die Krankheiten Ludwig van Beethovens*. See also Meredith, "History of Beethoven's Skull Fragments: Part Two."

80. Reiter, "On the Authenticity of Beethoven's Skull Fragments."

81. Begg et al., "Genomic Analyses of Hair from Ludwig van Beethoven." This international collaboration among thirty-two researchers was funded in part by the American Beethoven Society.

82. Martin, *Beethoven's Hair*; Rifai et al., "Letter to the Editor."

83. Reece, *Forgery in Musical Composition*.

84. Lianne Kolirin and Ashley Strickland, "Fragments of Skull Believed to Be Beethoven's Returned to Vienna from US for Scientific Analysis," *CNN*, July 21, 2023, https://www.cnn.com/style/article/beethoven-skull-fragments-vienna-scli-intl-scn/index.html; "Seligmann Fragments Back

in Vienna after Decades," website of the Medical University of Vienna, July 20, 2023, https://www.meduniwien.ac.at/web/en/about-us/news/2023/news-in-july-2023/seligmann-fragments-back-in-vienna-after-decades/.

85. Correspondence between the Ministerium der geistlichen, Unterrichts- und Medizinalangelegenheiten, the WGM, and the University of Vienna spanning from May 5 to June 16, 1904, WGM.

86. Laqueur, *Work of the Dead*, 80–106.

87. Perhaps Schaaffhausen earned Clara's support by pledging to update the autopsy report of Franz Richarz, which attributed Robert's mental illness to a litany of physiological abnormalities. Franz Richarz and Eberhard Peters, "Sektionsbefund der Leiche von Robert Schumann," Robert-Schumann-Haus, Zwickau, 6133-A3; see also Richarz, "Robert Schumann's Krankheit," *Kölnische Zeitung* 40 (August 30, 1873). Schumann's lock of hair bears the inscription: "Hopfhaare Rob. Schumanns a. d. Besitze Geh. Rats Schaaffhausen." Robert-Schumann-Haus, Zwickau, 3576-B3. Despite that Schaaffhausen replaced the skull after his measurements, as specified in protocols from July 10, 1879, held by the Robert-Schumann-Haus in Zwickau, a stubborn rumor spread that the skull vanished and remains lost to this day: Worthen, *Robert Schumann*, 385; and Peter F. Ostwald, quoted in Tibbetts, *Schumann*, 98–99.

88. Phillips, "Exhumations, Honorary Graves.'"

89. Deathridge, "Elements of Disorder," 321; Yearsley, *Bach and the Meanings of Counterpoint*, 209–38.

90. Poskett, *Materials of the Mind*; Hagner, *Geniale Gehirne*, 100–101; Hagner, "Skulls, Brains, and Memorial Culture," 209–10.

91. Treitel, *Science for the Soul*, 29–55.

92. Klages, *Die Grundlagen der Charakterkunde*. Klages's *Prinzipien der Charakterologie* was first published in 1910 and subsequently included in several revised and expanded editions, the fifth of which is cited here.

93. Hau, "Holistic Gaze," 512–13; Böhme, "Physiognomie als Begriff der Ästhetik." One of Klages's methods was the pseudoscience of "graphology," the analysis of character through handwriting, which motivated autograph collectors. Stefan Zweig, a dedicated collector of musical manuscripts, described these as "earthly shadows of genial personalities"; Zweig, *World of Yesterday*, 161–62.

94. Kreissle von Hellborn, *Franz Schubert*, 466n1.

95. Breuning, *Memories of Beethoven*, 115. On qualms about Schubert's left temple, see Otto Erich Deutsch, "Schubert vergiftet! Eine Entgegnung," Wienbibliothek im Rathaus, H.I.N. 243209, 4.

96. Mundy, "Evolutionary Categories and Musical Style." See also Mugglestone, "Guido Adler's 'The Scope, Method and Aim of Musicology.'"

97. Adler writes that the musicologist should "distinguish the individual style traits of the independent artist from the conventions of a school." Adler, "Style Criticism," 175; see also Adler, "Schubert and the Viennese Classic School."

98. Theodor von Frimmel, "Beethoven's Schädel (Mit Abbildungen)," *Neue illustrirte Zeitung*, no. 13 (December 19, 1880): 199.

99. "Die Cranien dreier musikalischer Koryphäen."

100. "Die Cranien dreier musikalischer Koryphäen," 34.

101. In 1885, Schaaffhausen's presentation of famous heads to the Deutsche Gesellschaft für Anthropologie was distilled in his "Einige Reliquien berühmter Männer."

102. See also A. Weisbach, C. Toldt, Th. Meynert, "Bericht über die an den Gebeinen Ludwig van Beethoven's gelegentlich der Uebertragung derselben aus dem Währinger Orts-Friedhofe auf den Central-Friedhof der Stadt Wien am 21. Juni 1888 vorgenommene Untersuchung," *Mittheilungen der anthropologischen Gesellschaft in Wien* 18 (1888): 73–76.

103. Weisbach, Toldt, and Meynert; Theodor von Frimmel, "Beethovens Bildnisse," *Die Presse* (Vienna), October 20, 1884, 4; Frimmel, *Neue Beethoveniana*, 314–18.

104. His, *Johann Sebastian Bach*.

105. His, *Johann Sebastian Bach*, 4–6; Yearsley, *Bach and the Meanings of Counterpoint*, 212.

106. Yearsley, *Bach and the Meanings of Counterpoint*, 209–38.

107. His, "Anatomische Forschungen." His's arguments about Bach's Germanness anticipate remarks by psychiatrist Ernst Kretschmer in his *Geniale Menschen*, 94.

108. His, "Anatomische Forschungen," 390 and 392n2.

109. Later generations of anatomists did indeed mine the data. In 1905, the racialist anatomist Richard Weinberg compared the gyrification of Bach and Beethoven, noting uncommonly deep folds and massive development around sensory centers and the auditory cortex. These traits he defined as the "organic layout of genius" that reflects the "musical soul," in contrast with the mechanical folds of diligent practicing. Richard Weinberg, "Gehirnform und Geistesentwicklung," *Politisch-anthropologische Revue* 3 (1905): 686–98. Weinberg reprised claims about Beethoven's brain by the influential neuroanatomist Paul Flechsig; see Flechsig, *Gehirn und Seele*, 102n53 and 689.

110. D.V. [Ernst Klotz], *Das fragwürdige Todtenbein von Leipzig*.

111. Noble, *That Jealous Demon, My Wretched Health*; Hutcheon and Hutcheon, *Four Last Songs*.

112. Schaffer, "How Disciplines Look."

113. Raffa, *Dante's Bones*, 199–208.

114. Dyer, *White*, 208 and 211, respectively.

115. Verdery, *Political Lives of Dead Bodies*, 29.

116. Peritz, "Castrato Remains"; Stanyek and Piekut, "Deadness: Technologies of the Intermundane."

Chapter Four

1. According to Theodor von Frimmel, the confusion was already widespread by 1827; Frimmel, *Neue Beethoveniana*, 273.

2. Product catalog from Micheli Brothers advertising the sale of Beethoven's masks, ca. 1900, n.p., BHB (no accession number). On the appeal of masks in the decorative arts, see Ulmer, "Zwischen Geniekult und Existenzmaskerade."

3. As Carl put it, "because the barber's apprentice said that he could never use the razor again after he had shaved a dead man with it, I bought it from him." See the facsimile and the commentary by Otto Biba in Danhauser, *Nach Beethovens Tod*, n.p.

4. Josef Danhauser, *Franz Liszt am Flügel phantasierend*, oil on wood (Berlin: Nationalgalerie, 1840). In a sense, this painting is a masterpiece of advertising: Danhauser managed to satisfy a commission for piano maker Conrad Graf, whose gleaming instrument catches the eye, while also promoting his superior sculpting skills.

5. See the advertisement flyer for Gebrüder Micheli and the company's letter to Fritz Knickenberg, September 22, 1911, BHB Akten, VBH 81–82.

6. Saliot, *Drowned Muse*.

7. Waddell, *Moonlighting*; Corbineau-Hoffmann, *Testament und Totenmaske*; Leppert, "Musician of the Imagination"; Comini, "Visual Beethoven" and Comini, *Changing Image of Beethoven*.

8. Bonds, *Beethoven Syndrome*, 140.

9. Fine, "Beethoven's Mask."

10. McMullan, *Shakespeare and the Idea of Late Writing*; McMullan and Smiles, *Late Style and Its Discontents*; Notley, *Lateness and Brahms*; Hutcheon and Hutcheon, *Four Last Songs*; Bonds, "Irony and Incomprehensibility"; M. Spitzer, *Music as Philosophy*.

11. Gibbs, "Performances of Grief."

12. On the Crucifixion: "The earth shook, the rocks split and the tombs broke open." Matthew 27:51-52 (NRSV). On the apotheosis of Elijah: "As they continued walking and talking, a chariot of fire and horses of fire separated the two of them, and Elijah ascended in a whirlwind into heaven." 2 Kings 2:11 (NRSV). The apotheosis of Romulus comes from Ovid, *Metamorphoses* 14:805-28 and *Fasti* 2:481-512. Some saints could summon inclement weather, while others protected devotees from storms or had feast days that align with seasonal cold snaps; examples include St. Scholastica, St. Swithun, St. Medard, St. Godelieve, and the so-called Ice Saints of German and Austrian folklore.

13. Rau, *Beethoven: Historischer Roman*, 4:375-76.

14. Elbertzhagen, *Die Neunte*, 42-50 and 92.

15. Dammert, *Das Wunderkind*, 190.

16. Lux, *Franz Schuberts Lebenslied*, 280.

17. Pointon, "Casts, Imprints, and the Deathliness of Things," 176; on the withdrawal of the gaze, Pointon quotes Grootenboer, *Treasuring the Gaze*, 121.

18. Ariès, *Hour of Our Death*, 474. While Ariès's study is several decades old, sprawling in its ambitions, and heavily Francophone, it remains a respected study whose arguments resonate with the accounts of composers' deathbeds cited here, which suggests that paradigms of death could be transnational and transconfessional.

19. On eighteenth-century chorale traditions and the *ars moriendi*, see Yearsley, *Bach and the Meanings of Counterpoint*, 1-41.

20. Sketchbook of Josef Teltscher, including drawings of Beethoven on his deathbed, 1827, British Library, Zweig MS 207:1827.

21. First published in the *Wiener Zeitung* 30 (1842); English translation in Landon, *Beethoven*, 391. See also the comparison of these accounts by E. Kramer, "Idea of *Kunstreligion*," 203-7.

22. Anselm Hüttenbrenner, memoir published in *Tagespost* (Graz), October 23, 1868; first written as a letter to Alexander Wheelock Thayer, Hallerschloß zu Graz, August 20, 1860. English translation in Landon, *Beethoven*, 392. Biographers have speculated that "Frau van Beethoven" was Beethoven's maid Sali, mistaken by Hüttenbrenner for sister-in-law Johanna.

23. Ariès, *Hour of Our Death*, 471-72.

24. Ariès, 473.

25. English translation in Landon, *Beethoven*, 393.

26. Huneker, *Chopin*, 77-78.

27. Felix Joseph Barrias, *The Death of Chopin*, oil on canvas (Kraków: National Museum, 1893). Collections that feature a print of this work include Dole, *Famous Composers*; Scobey and Horne, *Stories of Great Musicians*; Kobbé, *Loves of Great Composers*. Teofil Kwiatkowski, like Teltscher, made several sketches of Chopin's face shortly before and after his death as models

for his painting *The Last Moments of Frédéric Chopin*, oil on canvas (Warsaw: Frédéric Chopin Museum, 1849–50).

28. Liszt, *Life of Chopin*, 197.

29. Lutz, *Relics of Death*. With a focus on English literature, Lutz offers a local explanation for death relics as secularized practices that were displaced from saint veneration during the iconoclasm of the English Reformation. Their later import into Germany might have reflected a similar climate in the Lutheran North.

30. Pointon, "Casts, Imprints, and the Deathliness of Things," 187.

31. Harriet Goodhue Hosmer, *Clasped Hands of Robert and Elizabeth Barrett Browning*, bronze, 1853, Metropolitan Museum of Art, accession no. 1986.52; see Pointon, "Casts, Imprints, and the Deathliness of Things," 188–91.

32. Hallam and Hockey, *Death, Memory, and Material Culture*.

33. For an overview of the English-language literature on Mozart's death and requiem, see Keefe, *Mozart's Requiem*, 11–43; the German-language literature has not yet been thoroughly explored. On myths surrounding Mozart's deathbed, see Stafford, *Mozart's Death*, 31–55.

34. "Oh, dies Requiem, Ihr meine Lieben, / Soll meines Erdenwallens Schlußstein sein! / Das war der *Todesbote*! Seht Ihr ihn? / Mein Grabgesang sei dieses Requiem; / Und wenn ich sterbe, soll's an meinem Sarge / Ertönne. Er ist wahrlich mir sehr lieb." [Alfons] Wermonty, "Mozart," Trauerspiel in 5 Aufzügen, GSA, sig. 59/155, 109. To my knowledge, this handwritten play has never yet been studied; its ahistorical twists can be understood in the context of Salieri "gossip" explored by Franseen, "'Everything You've *Heard* Is True.'"

35. Eisen, "Mozart's Leap in the Dark."

36. Davison, "Painting for a Requiem."

37. See, for instance, Knittel, "'Late,' Last, and Least."

38. E. Kramer, "Idea of Transfiguration."

39. On these terms, among others, to describe unfinished works, see R. Kramer, *Unfinished Music*, esp. 311–13.

40. Jeffrey Kallberg, "Chopin's Last Style." On "Last Thoughts" miniatures, see 303–4.

41. Before its attribution to Carl Reißiger, Weber's alleged last waltz appeared as a parlor song: Carl Collmick, arr. "Letzte Idee von C. M. v. Weber" (Offenbach a.M.: Johann André, 1832); the best-known version was published three years later and attributed, oddly, to both Reißiger and Weber: "Letzte Idee von C. M. v. Weber" (Offenbach a/m: André, 1835). Beethoven's WoO 62, an unfinished sketch for a string quintet, was arranged for piano by Anton Diabelli, "Letzte musikalische Gedanke" (Vienna, 1838). Beethoven's Bagatelle in B-flat Major was published as "Dernière pensée musicale" and alternately as "Une pensée de Louis van Beethoven: Air du tremolo," by Schlesinger in 1840, as well as in Prague by J. Hoffmann, ca. 1840 and reprinted by the subsequent owner of Hoffmann's printing business (Leipzig: Friedrich Hofmeister, 1850). In 1842, Schlesinger published his *Dernière pensée et une pensée de i. Charles Marie de Weber, ii. Louis van Beethoven, iii. Vincenzo Bellini, etc.*, British Library, Music H.3691.g.(2).

42. Examples include Tzschirner, *Tzschirner's letzte Worte an heiliger Stätte gesprochen*; and Schulz, *Edle Charakterzüge, schöne und große Handlungen*. Franz Liszt composed an ode based on Goethe's last words, "Licht, mehr Licht!" for Goethe's birthday celebration in Weimar in 1849; *Fest-Album zur Säcular-Feier von Goethe's Geburtstag*.

43. Numerous examples of this vocabulary can be found in by Landau, *Erstes poetisches Beethoven-Album*; see, for instance, pp. xii, 83, 109, 154, 302, 348, and 383, as well as Hugo Klein, "Allerlei von Beethoven," *Neue Musik-Zeitung* 25, no. 5 (December 17, 1904): 104–6.

44. Bruck, *Beethoven der Deutsche*, 16. On the term "Erlösungskraft" in German Romanticism, see McMahon, *Divine Fury*, 192–93; McMahon here quotes and discusses Türck, *Der geniale Mensch*, 271, as among the first extended meditations on the shared qualities of Napoleon and Christ.

45. Sullivan, *Beethoven: His Spiritual Development*; see also Korsyn, "J. W. N. Sullivan and the *Heiliger Dankgesang*," 140–41.

46. For a fuller account of this discourse, see Fine, "Beethoven's Mask."

47. Theodor W. Adorno's essay "Late Style in Beethoven" (1937) concludes with the pronouncement that "in the history of art late works are the catastrophes." Adorno, *Essays on Music*, 564–68, at 567. This essay, along with Adorno's "Alienated Masterpiece: The Missa Solemnis" (1959)—found in Adorno, *Essays on Music*, 569–83—has become a classic text in late-style criticism. Its claims were revived by Edward Said in his own last work, *On Late Style: Music and Literature against the Grain*.

48. These dichotomies, among others, were laid out by McMullan, *Shakespeare and the Idea of Late Writing*, 46–47.

49. Franz Grillparzer's funeral oration is reprinted in Kopitz and Cadenbach, *Beethoven aus der Sicht seiner Zeitgenossen*, 1:389–92, at 389.

50. Adorno, "Late Style in Beethoven," 566. See also McMullan, *Shakespeare and the Idea of Late Writing*, 564–65.

51. Leistra-Jones, "Hans von Bülow and the Confessionalization of *Kunstreligion*."

52. Hermann Kretzschmar had stern words for Nohl in "Beethoven als Märtyrer." Nohl authored biographies of Mozart, Haydn, Beethoven, Liszt, Wagner, and Spohr; and ideologies of the New German School guide volumes like *Der Geist der Tonkunst*; *Die Beethoven-Feier und die Kunst der Gegenwart*; and *Beethoven, Liszt, Wagner*.

53. Knittel, "Wagner, Deafness, and the Reception of Beethoven's Late Style." For Nohl's description of Beethoven as an anchorite and "Musikgottessohn" (musical son of God), see Nohl, *Die Beethoven-Feier*, 12–13; and Nohl, *Der Geist der Tonkunst*, 225–29.

54. Ordinarily, the word *Grabgesang* referred to functional funeral music that emulated the Requiem Mass, with homophonic, minor-mode dirges and texts that consoled the dying or invited the community to mourn. Nohl's art song is closer to the practice of texting instrumental music—which dates to Beethoven's own lifetime, as Helmut Loos has shown—than to this genre of collective mourning. The closest precedent for Nohl's arrangement is the Adagio of the String Quartet, op. 127 into a song for soprano entitled *Beethovens Heimgang* (Beethoven's homecoming) with a text by Friedrich Schmidt in which Beethoven's spirit, finally at peace, rises to rejoin God. See H. Loos, "Die Texierung Beethovenscher Instrumentalwerke."

55. "In conclusion, we would like to share the tune of this adagio, and we have tried to give it words that resound with its inner meaning. Perhaps in living song, the living essence of Beethoven's nature came forth, which cannot be entirely represented by word and image alone—his sorrow-borne joy in the happiness and existence of our race, his *'unrequited love' of humanity!*" Nohl, *Eine Stille Liebe zu Beethoven*, 263–66.

56. Nohl was not the only one to package Beethoven's late adagios as freestanding miniatures for piano or voice; around the same time, a transcription of the *Lento assai* from the String Quartet, op. 135 was published by Mortier de Fontaine (Munich: Jos. Aibl, ca. 1870).

57. Hamburger, *Visual and the Visionary*, 325–36; Kuryluk, *Veronica and Her Cloth*.

58. Kessler, "Configuring the Invisible by Copying the Holy Face"; Hahn, "Absent No Longer."

59. Kristeva, *Severed Head*, 41–42.

60. The Feast of the Holy Face of Jesus was established in the nineteenth century; see Corbin, "Les Offices de la Sainte Face."

61. Bynum, *Christian Materiality*, 98.

62. Theodor Haas, "Gedanken bei Betrachtung der Beethoven-Maske," *Neue Musik Zeitung* 43 (1922): 391; Alessandra Comini arrives independently at the same notion in "The Visual Beethoven."

63. Wade, *Domenico Brucciani*; Kammel, "Der Gipsabguss."

64. Micheli Brothers advertisement, "Beethoven-, Wagner-, Liszt-Masken" *Neue Freie Presse* (Vienna), January 28, 1913, and the company's catalogue, *Preis-Verzeichnis plastischer Bildwerke*.

65. The fashion for ethnographic sculpture was spurred by Charles-Henri Cordier. See Dechant and Goldscheider, *Goldscheider: Firmengeschichte und Werkverzeichnis*. Beethoven wall masks include nos. 2202 and 2285 (1900); the head on a pedestal is no. 2318 (1902).

66. Dechant and Goldscheider, nos. 1171 and 1172 (1898).

67. The private collection owned by the Carrino family contains dozens of images with floating Beethoven masks, especially ex libris plates; several examples are shown in the BHB museum catalogue: Carrino and Carrino, *Eine Beethoven-Wunderkammer in Italien*.

68. Silke Bettermann, "Die Beschäftigung mit Beethovens Lebendmaske," in Bettermann, *Franz von Stuck und Beethoven*, 35–78.

69. Rilke, *Die Aufzeichnungen des Malte Laurids Brigge*, 69; Saliot, *Drowned Muse*, 130. Rilke's pairing is reiterated in other literary sources, such as Louis Aragon's *Aurélien*: "There are copies of this mask [the Inconnue] all over the place, in all the shops where they sell plaster masks and statuettes. It always turns up alongside the 'Boy with the Thorn' and Beethoven's death mask" (267). The Boy with the Thorn was a replica of the famous Greco-Roman bronze in the Uffizi Gallery of Florence. Comini notes this passage in her study of Beethoven's mask but makes no mention of the Inconnue; Comini, "Visual Beethoven," 288.

70. Saliot, *Drowned Muse*.

71. Compare Willy Zielke's 1929 photograph of the Inconnue (Saliot, *Drowned Muse*, 13) with Joseph Schneider's undated photograph of Beethoven's life mask (BHB B 437). Further funerary interpretations of Beethoven's mask are discussed in Bettermann, *Franz von Stuck und Beethoven*, 35–78.

72. Ferdinand Eckardt, "Beethoven Sterbehaus in der Schwarzspanierstraße," WGM Archive, B 3306.

73. Industrialist and arts patron Gustav Schütz cribbed two large slabs of stone from the site of demolition. On one, he mounted a bronze cast of Beethoven's life mask, and on the other he commissioned this same mask carved into the stone; Gustav Schütz, 1906, WGM Archive, B 3301.

74. "Meine Lieder werden leben / Wenn ich längst entschwand: / Mancher wird vor ihnen beben, / Der gleich mir empfand. / Ob ein Andrer sie gegeben, / Oder meine Hand: / Sieh, die Lieder durften leben, / Aber ich entschwand!" Annette Droste-Hülshoff, "Am fünften Sonntag in der Fasten," in Droste-Hülshoff, *Das geistliche Jahr*.

75. Quoted in Brayshaw, "Oskar Schlemmer's Kitsch," 14. On the memento mori as kitsch, see Goodwin, *Kitsch and Culture*.

76. My assessment of these visual tropes stems from more than one hundred images culled from three main sources: the collections of the Beethoven-Haus in Bonn, including the Carrino collection; and Bettermann, *Beethoven im Bild*.

77. Correspondence between the Gebrüder Micheli company and Fritz Knickenberg, secretary of the Beethoven-Haus, between 1908 and 1914, BHB Akten, VBH 81–82. The plaster

company sent a letter from the Goethe Museum (November 13, 1908) to persuade the BHB. For a photograph of the final product, see BHB, P 61.

78. Letter from the Gebrüder Micheli company to the BHB, September 24, 1908, BHB Akten, VBH 81-82.

79. Alfred Polgar, "Beethoven Masks" (1921) in Segel, *Vienna Coffeehouse Wits*, 262-63. Polgar jests that the mask adds a spiritual quality when hung like a crucifix behind the bed or the kitchen cupboard.

80. Lewis, *Tarr*, 4; see Waddell, *Moonlighting*, 147.

81. Foreign tourists were said to leave Paris with a miniature Eiffel Tower, an Arc de Triomphe paperweight, prints of the jewels of the Louvre, and the Inconnue; see Saliot, *Drowned Muse*, 5.

82. Huyssen, "Mass Culture as Woman."

83. Huelsenbeck, *Dada Almanach*.

84. See Simmons, "Chaplin Smiles on the Wall," 7-9.

85. On the erotic muse broadly, see Downes, *Muse as Eros*.

86. Fidus's best-known sketch for a Beethoven temple shows a female nude touching his oversized face (BHB B296); the less often discussed sketch shows a messianic, wide-eyed face of Beethoven streaming with light rays, illuminating a naked man, woman, and child: BHB B 296 and B2570. On Fidus's influence in German life reform, and later National Socialism, see Hau, *Cult of Health and Beauty in Germany*, 198.

87. Bettermann, *Beethoven im Bild*, 62-67 and 75-79. Prechtl's satire was used in 1974 an advertisement for the music publisher Schirmer.

88. Saliot, *Drowned Muse*, 28.

89. Walter Benjamin writes about dust in several passages of his unfinished Arcades Project: as a coating of dust on Victorian interiors that transforms kitsch into melancholy, and as the enemy of the Victorians, who sought to freeze time and hide evidence of its passage. Benjamin, "Boredom: Eternal Return"; and Benjamin, "Dream Kitsch."

90. The composer looks almost demonic in Hermann Torggler, *Beethoven-Phantasieportrait* (1902), in Bettermann, *Beethoven im Bild*, 29. Stuck portrayed Beethoven's mask with open eyes in a series of plaster reliefs (Bettermann, 296-97).

91. Max Klinger, *Pietà: Maria und Johannes trauernd am Leichnam Christi* (Dresden: Staatliche Kunstsammlungen Dresden, 1890). The original painting was destroyed in World War II.

92. It is frequently held that Klinger's large-scale painting of the same year, *The Crucifixion of Christ*, also depicts Beethoven as St. John. While the face may look Beethovenian from afar or in miniature, when one stands before the canvas in the Museum of Fine Arts in Leipzig, the comparison is less convincing. Max Klinger, *Kreuzigung Christi* (Leipzig: Museum der bildenden Künste, 1890).

93. Comini, *Changing Image of Beethoven*: on "colossal kitsch," 9; on Klinger's *Pietà*, 410.

94. Karnes, *Kingdom Not of This World*, 37-65; Celenza, "Darwinian Visions"; Comini, *Changing Image of Beethoven*, 388-415.

95. Morton, *Max Klinger and Wilhelmine Culture*, 38-50.

96. Max Klinger, *Dead Mother*, engraving from series *On Death, Part 2, Opus XIII* (Cambridge: Harvard Art Museum, 1889, Singer 239); *Death as Savior*, likewise from *On Death, Part 2* (Melbourne: National Gallery of Victoria, 1889, inv. no. 188/1); *Death*, engraving from the series *A Love* (New York: Metropolitan Museum of Art, 1887-1903, 52.586.1[10]).

97. Frisch, *German Modernism*, 93-106.

98. Maier-Graefe, *Der Fall Böcklin*.

99. H.H. "Die Brillenprobe: Beethovens Totenmaske als Modell," *Volkszeitung* (Vienna), August 2, 1926, BHB, Z338 / 40.

100. Georg Mühlen-Schulte, "Es brennt bei Beethoven," *Lustige Kölner Zeitung*, May 10, 1933, BHB, Z170ak / 1.

101. Bon Lehnan, "Die Totenmaske," *Erstes Beiblatt*, no. 151 (June 25, 1932), BHB, Z38a / 10.

102. Massimo Bontempelli, "Beethovens Totenmaske," *Dichtung und Welt: Beilage zur Prager Presse*, August 15, 1926, BHB, Z256 / 14.

103. "London: Die zehnte Beethoven'sche Sinfonie ist wahr . . . ," *Der Klavier-Lehrer* 18:10 (May 15, 1895, Berlin): 124–25.

104. Desider Kosztolányi, "Beethoven," *Der Brenner* 8 (February 1, 1914): 390–97.

105. Chua, "Beethoven's Other Humanism." Adorno's comment comes from Adorno, *Beethoven*, 164. Benjamin's remarks come from his "Little History of Photography" (1931), cited in Hansen, "Benjamin's Aura," 343.

106. Walter Benjamin, "Work of Art in the Age of Its Technological Reproducibility," 27; Barthes, *Camera Lucida*, 14.

107. Chua, "Beethoven's Other Humanism," 627.

108. Benkard, *Das ewige Antlitz*.

109. In a clipping collection from the BHB one finds over a dozen press features on death masks published after Benkard's book, such as Oskar Maurus Fontana, "Das letzte Antlitz," *Der Tag* (Vienna), December 12, 1926; Michael Grusemann, "Beethovens inneres und äußeres Gesicht," *Berliner Börsenzeitung*, March 26, 1927; and "Totenmasken," *Das interessante Blatt*, October 28, 1928.

110. Saliot, *Drowned Muse*, 61.

111. These ruminations appear in Döblin's provocative introduction to August Sander's blunt photographic portraits of nameless Germans. Just as death levels the faces of luminaries in Benkard's book, he argues, social class levels the individual to create nameless "types"; Döblin, "Faces, Images, and Their Truth" (originally published as "Von Gesichtern, Bildern und ihrer Wahrheit," in August Sander, *Antlitz der Zeit* [Munich: Transmare Verlag, 1929], 5–15). It should be noted that caricatured figurines of anonymous people were called "types," such as Goldscheider's famous "Wiener Typen," or statuettes of Hungarian shepherds, Viennese urbanites, Romani street musicians, and so on, which mirrored the diversity of the Austrian Emprie. Numerous examples of this inventory are shown in Dechant and Goldscheider, *Goldscheider: Firmengeschichte und Werkverzeichnis*.

112. Döblin, "Faces, Images, and Their Truth," 8–9; Pointon, "Casts, Imprints, and the Deathliness of Things," 175.

Chapter Five

1. "Musik," *Kikiriki: Humoristisches Wochenblatt für Stadt und Land* (Petrovgrad) 2, no. 37 (January 21, 1934): 5. This rustic satirical paper centered in Petrovgrad in modern-day Serbia can be found in the collections of the ÖNB, 668723-C, and should not be confused with the better-known *Kikeriki: Humoristisches Volksblatt* of Vienna.

2. For an overview of Adorno's essays on mass culture, popular music, and kitsch, see Richard Leppert's commentary in Adorno, *Essays on Music*, 327–72. In 1939, Dwight Macdonald, though uncredited, heavily edited Clement Greenberg's famous essay "Avant-Garde and Kitsch";

during the war, he admired Adorno's writings on the topic; and after the war, he authored essays such as "A Theory of Mass Culture," published in *Diogenes* in 1953, and "Masscult and Midcult," the crux of his 1962 collection of essays: Macdonald, *Masscult and Midcult*. See esp. the introduction to this volume by Louis Menand, vii–xxii.

3. Adorno, "Theorie der Halbbildung (1959)," in Adorno, *Gesammelte Schriften*, 8:93–121. For an English translation by Deborah Cook, see Adorno, "Theory of Pseudo-culture (1959)."

4. A. Loos, "Ornament and Crime"; Hermann Broch, "Einige Bemerkungen zum Problem Kitsches," in Broch, *Gesammelte Werke*, 6:295–309. Journalist and music critic Adolf Weissmann decried the mechanization of jazz, radio, film, and listeners a decade before Adorno did; Weissmann, *Die Entgötterung der Musik*.

5. Rehding, *Music and Monumentality*, 77–78 and 105.

6. Schorske, *Fin-de-Siècle Vienna*; Beller, *Rethinking Vienna 1900*.

7. Oskar Schlemmer, "True Kitsch Is Beautiful" (unpublished manuscript, 1922), in Brayshaw, "Oskar Schlemmer's Kitsch," 14.

8. An admittedly anachronistic take: there is something of 1960s pop art here, like the cut edges of the icons on the cover of *Sgt. Pepper's Lonely Hearts Club Band*. The cover for this 1967 album by the Beatles was codesigned by Jann Haworth and Peter Blake.

9. On art criticism's anachronistic and overlapping sense of historical time, see Didi-Huberman, "Before the Image, before Time."

10. On the etymology of "kitsch," see Calinescu, *Five Faces of Modernity*, 234.

11. Huyssen, "Mass Culture as Woman."

12. Hermann Broch, "Notes on the Problem of Kitsch," in Dorfles, *Kitsch*, 49–76, at 49.

13. Greenberg, "Avant-Garde and Kitsch."

14. See Leslie, "Souvenirs and Forgetting."

15. Benjamin, "Work of Art in the Age of Its Technological Reproducibility"; and Benjamin, "Dream Kitsch."

16. Dorfles, *Kitsch*.

17. Pazaurek, *Guter und schlechter Geschmack*, 354. Pazaurek's guide to bad taste was published three years after the 1909 museum exhibit and reached its third edition by 1919. See also Doherty, "What Is There to Be Learned from Kitsch?"

18. Gillo Dorfles, "Myth and Kitsch," "Monuments," and "Religious Trappings" in Dorfles, *Kitsch*, 37–48, 79–83, and 141–43; Karl Pawek, "Christian Kitsch," in Dorfles, *Kitsch*, 143–50.

19. Ngai, *Our Aesthetic Categories*. The postmodern vein of kitsch studies focuses on affect, notably sweet sentiments and the aloof detachment that signals postmodern irony; see Kundera, *Unbearable Lightness of Being*, 251–52; Eco, "Postmodernism, Irony, the Enjoyable"; and M. Morris, *Persistence of Sentiment*. On gift shop wares as postmodern objects of high culture verkitscht, see Fliedl, *Wa(h)Re Kunst*, published after the exhibition *Wa(h)re Kunst*, Offenes Kulturhaus, Linz, December 1996 to January 1997.

20. Binkley, "Kitsch as a Repetitive System"; Binkley's argument builds on the theory of modernity's "disembeddedness" by Giddens, *Modernity and Self-Identity*, e.g., 18. The influential concept of "embeddedness" was first posited in 1944 by Karl Polanyi in *The Great Transformation*.

21. Pazaurek, *Guter und schlechter Geschmack*, 354.

22. Schubert cards often bear the signature yellow pants and blue jacket of Goethe's Werther, a nod to his melancholy temperament that elicits a dual nostalgia for the Weimar classics market, or what Burkhard Henke calls "Goethe®": see Burkhard Henke, "Goethe®: Advertising, Marketing, and Merchandising the Classical," in Henke, Kord, and Richter, *Unwrapping Goethe's*

Weimar, 15–35. Schubert appears in Werther's signature outfit in card no. 126 for Eckstein-Halpaus cigarettes and card no. 38 from the "Deutsches Denken und Schaffen" series of Wagner margarine.

23. Adorno, "Commodity Music Analyzed."

24. Benjamin argues that, in a new age of proximity, when objects draw us near, the aesthetic of surrealism probes the dreams of a new kind of human; Benjamin, "Dream Kitsch."

25. One of the aphorisms is titled "Kitsch That Writes," in Bloch, *Heritage of Our Times*, 17. It should be noted that Benjamin was critical of Bloch's book, which he held to be sanctimonious and out of step with the moment: see Rabinbach, "Unclaimed Heritage."

26. Olalquiaga, *Artificial Kingdom*.

27. Ender, "'Sie müssen doch sehen, wie Alban lebt.'" The BHB replicated and sold the bust of Brutus kept on Beethoven's desk: letter from Gebrüder Micheli to Fritz Knickenberg, October 17, 1911, BHB Akten, sig. 81. A funereal landscape fashioned from Beethoven's hair was in the possession of Anton Schindler, who sold it to the industrialist Carl Meinert (BHB, R4); the object was still on display in the Beethoven-Haus exhibits in 1911 as an exemplar of "the taste of the time" but had disappeared from the exhibit catalogues by 1926, which may mark the moment when Victorian hair art began to look verkitscht. See the exhibit catalogues edited by Ferdinand August Schmidt and Fritz Knickenberg, *Das Beethoven-Haus in Bonn*, BHB 261/1911 and *Kleiner Führer durch das Beethoven-Haus in Bonn*, BHB 261/1926. The pearled bouquet of Liszt's hair hails from the estate of Eduard Liszt; see Eckhardt, *Liszt Ferenc Gedenkmuseum*, 62.

28. Lutz, "Dead Still among Us."

29. D. Spitzer, *Verliebte Wagnerianer*.

30. D. Spitzer, 35.

31. This influential concept hails from Thomas à Kempis's fifteenth-century devotional text *De Imitatione Christi*. The term is not to be confused with the *imitatio* of classical rhetoric, which is closer in some ways to homage or style study.

32. Hanslick, "Wagner-Kultus," 339.

33. Tappert, *Ein Wagner-Lexicon*.

34. Tappert's compendium includes two mentions of Bayreuth as "Götzentempel" (*Neues Berliner Tageblatt*, 1875; rpt. in Tappert, *Ein Wagner-Lexicon*, 15); and "Tempel zu Bayreuth" (*Neue Berliner Musikzeitung*, 1875; rpt. in Tappert, *Ein Wagner-Lexicon*, 17); and one reference to "Wagner Nono," a play on Pope Pio Nono (*Tribüne* [Berlin], 1873; rpt. in Tappert, *Ein Wagner-Lexicon*, 42).

35. Nordau, *Entartung*. See also Dreyfus, *Wagner and the Erotic Impulse*, 117–74.

36. Vazsonyi, *Richard Wagner*.

37. Bourdieu, *Rules of Art*, 81–84; Vazsonyi, *Richard Wagner*, 21–27.

38. Richard Wagner, "Eine Pilgerfahrt zu Beethoven," in R. Wagner, *Gesammelte Schriften und Dichtungen*, 1:115–41.

39. Green, *Dedicating Music, 1785–1850*.

40. On Wagner's production at Dresden, see Vazsonyi, "Beethoven Instrumentalized." On Liszt's approach to the visible transcriber, see Kregor, *Liszt as Transcriber*, and esp. 257, on his all-Beethoven program.

41. Rehding, *Music and Monumentality*, 67; Minor, *Choral Fantasies*.

42. On Liszt's self-perception as a Christ figure, see Pesce, *Liszt's Final Decade*.

43. Franz Liszt, *Am Grabe Richard Wagners* (1883). Liszt later transcribed the piece twice, once for string quartet and harp and again for organ. See Hamilton, "Wagner and Liszt."

44. McManus, *Brahms in the Priesthood of Art*. See also Minor, *Choral Fantasies*, 68.

45. GSA, 59/173, 5 and 59/140, 1, respectively. The ode in question was by Patrick Parry (not to be confused with John Orlando Parry), who heard Liszt play this sonata at a salon hosted by Duke Carl Alexander in Weimar. Examples of decorated odes, or "toasts," abound in Liszt's effects in the GSA's folder 59/174, the most interesting being a certificate of thanks for Liszt's charity recitals after the Great Fire of Hamburg in 1842, which included a medallion made from smelted cathedral bells (59/174, 3).

46. GSA 59/176, 3. Folders no. 175 and 176 contain a wealth of further examples of handwritten poems mailed to Liszt by his admirers.

47. *Fest-Gesang, nach der Weise des Schiller-Fest-Liedes*, text by Franz von Dingelstedt, music by Franz Liszt, performed in Weimar on October 22, 1860. A stanza from the ode by Louisabeth Röckel reads: "Wir weihen Dir's wieder—Ein Kranz.—Ein Denkmal, das Selbst Du errichtest, ein Sternbild, mit ewigem Glanz!"

48. The event was spearheaded by the Archbishop Lajos Haynald, and its festival committee boasted Hungary's leading figures in the arts. In addition to the two public events outlined here, the festival concluded with a semiprivate grand banquet for Austro-Hungarian high society, with formal speeches, albums and gifts, and telegrams from around the world read aloud.

49. Hugo Wittmann, "Das Liszt-Jubiläum: I" and "Das Liszt-Jubiläum: II," *Neue Freie Presse* (Feuilleton), November 11 and 13, 1873, 1 and 2.

50. GSA 59/265, 1–8; these remarks were made in the *Ungarischer Lloyd* and *Neues Pester Journal*, respectively. To support Lina Ramann's biographical project, Liszt maintained and hand-corrected a large selection of newspaper clippings that detail the events held in his honor.

51. Henri Gobbi, "Vorspiel der Festcantate zu Franz Liszt's 50 jährigem Künstler-Jubiläum in Pest (November 1873)," arr. piano four hands (Leipzig: C. F. Kahnt, n.d.), Anna Amalya Bibliothek, Weimar, Magazin L 1697. The original cantata was written for choir, soloists, harp, and two pianos.

52. "Sehet in des Meisters Seele / Himmelsträume sanft erstehen / Lieder, süß, wie Vesperläuten, / Tiefe Andacht zu verbreiten, / Nahen aus den Himmelshöhen, Wo Sein Geist sich hinverlor. / Dort nur lebt Er, dort nur webt Er, / Lauscht der Seraphinen Chor." The full text appears in an undated newspaper clipping, Emil Ábrányi, "Jubel-Cantate zur Lißtfeier (Musik von Heinrich Gobbi)," *Sonntags-Blatt*, in GSA 59/174, 145.

53. Wittmann, "Das Liszt-Jubiläum: II," 2.

54. Wittmann, "Das Liszt-Jubiläum: I," 1.

55. Wittmann, 1.

56. Franz Liszt's letter to Carolyne Sayn-Wittgenstein (Budapest, March 18, 1879) can be found alongside all other emendations to his last will and testament in Schnapp, *Liszts Testament*; and GSA, L 2007.

57. Franz Liszt, letter to Carolyne Sayn-Wittgenstein (Rome, November 27, 1869), in Schnapp, *Liszts Testament*, n.p.

58. Ryan Minor argues that Liszt's participation in German or Hungarian national events masked his cosmopolitanism and necessitated Selbstinszenierung; see Minor, "Prophet and Populace in Liszt's 'Beethoven' Cantatas." A cousin to these exhumation petitions was the ritual consecration of Liszt's home in Weimar in 1887, two years before the founding of the Beethoven-Haus in Bonn.

59. Caricature, *Bolond Istók* 6 (March 6, 1887): 5; and GSA 59/372, 1. On the debate that spurred this caricature, see Walker, *Franz Liszt*, 3:524n4. The figure crushed under Liszt's tomb appears to be the historian Kálmán von Thaly, who argued against prioritizing Liszt's

exhumation; the figure crushed by Rákóczy is either Edmund Steinacker, who sided with the Liszt camp, or possibly Prime Minister Tisza himself.

60. Mathew, *Political Beethoven*, 55.

61. Roach, *It*, 30.

62. Hust, *August Bungert*. Friedrich Nietzsche met Bungert in 1883 in Genoa and struggled to determine whether Bungert was a genuine talent or a fraud; Hust, 144–54.

63. Cited in Hust, 351–53.

64. Cited in Hust, 344–47.

65. Review of *Odysseus' Heimkehr* in the *Berliner Börsenzeitung* 123 (April 1, 1898), cited in Hust, *August Bungert*, 355.

66. Hust, 347.

67. "Tagesgeschichte: Musikerbriefe. Dresden," *Das musikalische Wochenblatt* 1, no. 30 (1896): 4, cited in Hust, *August Bungert*, 342.

68. Maier-Graefe, *Der Fall Böcklin*.

69. August Bungert, "Moderne Musik," *Der Bund: Ein Monatschrift für den Bungert-Bund und seine Freunde* 3, nos. 4–5 (February and March 1914): 90. The journal's final issue before the outbreak of war ends with Chop railing against the use of foreign words in German. After the war, he found a new outlet for those sentiments as editor for the *Signale für die musikalische Welt*. Max Chop, "Deutsch sein!," *Der Bund: Ein Monatschrift für den Bungert-Bund und seine Freunde* 4, no. 8 (June 1915): 121–28.

70. Willibald Nagel, "?," *Rheinische Musik- und Theater- Zeitung* (Cologne) 13, no. 16 (April 27, 1912): 245–46.

71. Baudrillard, *Simulacra and Simulation*.

72. Maack, *Wie Steht's mit dem Spiritismus*. See also Treitel, *Science for the Soul*.

73. Georg Simmel reviewed Gustave Le Bon's *Psychologie des foules* (Paris: F. Alcan, 1895) in Vienna's *Die Zeit* before authoring his own essays on mass culture. See Jonsson, *Crowds and Democracy*, 12 and 54–61.

74. Zilsel, "Mozart und die Zeit"; Heuss, "Ein Tag in Sankt Ludwig."

75. Timms, *Karl Kraus, Apocalyptic Satirist*, 1:123–24. On the richness of musical life in the pages of Vienna's feuilletons, see McColl, *Music Criticism in Vienna*.

76. Overy, "'Whole Bad Taste of Our Period.'"

77. Hermann Broch, "Hofmannsthal and His Time," in Broch, *Hugo von Hofmannsthal and His Time*, 33–81, at 81. See also Hargraves, *Music in the Works of Broch, Mann, and Kafka*, 55.

78. "After the toils and troubles of the day we go to Beethoven or to Tristan. This my shoemaker cannot do. I must not deprive him of his joy, since I have nothing else to put in its place. But anyone who goes to the *Ninth Symphony* and then sits down and designs a wallpaper pattern is either a con man or a degenerate." A. Loos, "Ornament and Crime," 175.

79. Watkins, "Schoenberg's Interior Designs."

80. Karl Kraus, "Der Fortschritt," *Die Fackel*, March 1909; see Timms, *Karl Kraus, Apocalyptic Satirist*, 1:142. Kraus's periodical *Die Fackel* ran from 1899 to 1936.

81. Edward Timms, "The Crisis of Musical Culture," in Timms, *Karl Kraus, Apocalyptic Satirist*, 2:412–32.

82. See especially Edward Timms, "Dreams and Nightmares: The Visionary Satirist," in Timms, *Karl Kraus, Apocalyptic Satirist*, 1:209–25.

83. Zilsel, "Mozart und die Zeit." On Kraus's influence on *Der Brenner* and other periodicals beyond Vienna, see Timms, *Karl Kraus, Apocalyptic Satirist*, 1:200. Zilsel's best-known work is

his *Die sozialen Ursprünge der neuzeitlichen Wissenschaft*, which was left unfinished upon his death and published posthumously.

84. Zilsel, "Mozart und die Zeit," 268.

85. Zilsel, 271.

86. Horkheimer and Adorno, *Dialectic of Enlightenment*.

87. Zilsel, *Die Geniereligion*.

88. Zilsel, 3–4.

89. Zilsel, *Die Geniereligion*, 104; Zilsel, "Mozart und die Zeit," 271. See also Fine, "Assimilating to Art-Religion."

90. Jonsson, *Crowds and Democracy*, 88.

91. Elias Canetti, who authored a major treatise on crowds after he participated in the riots of 1927 Vienna (*Masse und Macht*, 1960), called Bildung the "quarantine zone for the individual against the mass in his own soul." Canetti, *Auto-da-Fé*, 38.

92. On mass culture and decadence, see Brantlinger, *Bread and Circuses*.

93. Alfred Polgar, "Orchester von Oben," 1921, in Segel, *Vienna Coffeehouse Wits*, 263–65, at 264.

94. Wedekind, *Kitsch*, 2:212.

95. The character Nachtigall is the descendent of an Australian Aboriginal tenor and a Viennese Jewish socialite. Throughout the play he is referred to as a "Niggerjud," abbreviated as "NJ," both as Wedekind's own shorthand and in dialogue when he leaves the room. This jarring label points ahead to the degenerate Jazz Jew of the infamous poster for the "Degenerate Music" exhibition in Düsseldorf in 1938.

96. Wedekind, *Kitsch*, 2:214 and 2:220, respectively.

97. Broch, *Massenwahntheorie*; see also Sterling, *Hermann Broch and Mass Hysteria*; and Klebes, "Glauben und Wissen."

98. Hermann Broch, "Mythos und Altersstil" (1947), in Broch, *Hermann Broch: Kommentierte Werkausgabe*, 9/2:227; see also Hargraves, *Music in the Works of Broch, Mann, and Kafka*, 14–15.

99. Broch, "Notes on the Problem of Kitsch." The sociologist Norbert Elias argued that kitsch emerged from a bourgeois appropriation of the aesthetic of the courts; Romanticism's aspirations are a big risk that results in formlessness, or distorted extremes of scale. Elias, "Kitsch Style and the Age of Kitsch."

100. Broch, "Notes on the Problem of Kitsch," 52. On Beethoven and Bach as paragons of formal cohesion, in contrast with *Verkitschung*, see Broch's "Mythos und Altersstil" (1947).

101. Broch, *Hugo von Hofmannsthal and His Time*, 55–59.

102. Hargraves, *Music in the Works of Broch, Mann, and Kafka*, 1–5.

103. Attfield, *Challenging the Modern*, 70–105.

104. Discussed in Attfield, 99–105, transcribed and translated on 212–15, at 212 and 213. Attfield translates "Kitsch und Geschmier" as "a crowdpleasing scrawl" whereas I here furnish "kitschy scrawlings."

105. Haiger's temple concealed its performers to create a transcendent listening experience, adapting an idea from an essay by Paul Marsop that proposed an acousmatic screen to hide performers behind foliage; see Kane, *Sound Unseen*, 104.

106. Forsyth, *Buildings for Music*, 140; McVeigh, "Concert Series," 297.

107. François Garas, "Temple of Thought, Dedicated to Beethoven, under Construction," pen and ink, watercolor, 1897–1914, Paris, Musée d'Orsay, ARO 2002 41. Nude figures symbolized

collective humanity for the *art nouveau*. In a comparable etching by Arthur Paunzen, named for Beethoven's Ninth Symphony, a towering cliff is ringed with naked figures whose arms stretch to the heavens; Arthur Paunzen, "Phantasien über Beethoven-Symphonien," Vienna, 1918. Verlag der Buchhandlung Richard Lanvi, WGM Archive, I.N. 22068.

108. Hutschenruyter, *Das Beethovenhaus*.

109. On Anton Schindler's forgeries, see Lockwood, *Beethoven's Lives*, 18–30.

110. Heuss, "Ein Tag in Sankt Ludwig," 135.

111. Verdery, *Political Lives of Dead Bodies*; see also Rév, "Parallel Autopsies"; and Gal, "Bartók's Funeral."

112. The ossuary was in place centuries before Bruckner's interment; its arrangement in rows hails from early nineteenth-century anthropometrists who ordered the remains by size.

113. See Bruckner's last will and testament in Göllerich, *Anton Bruckner*, 2:322.

114. Coreth, *Pietas Austriaca*.

115. Grasberger, Partsch, and Harten, *Bruckner—skizziert*, 158–61, at 161.

116. Anton Bruckner, letter to Felix Weinwurm, January 16, 1868, in Nowak, *Anton Bruckner, Briefe* 1, no. 78. On this episode as it relates to Bruckner's feelings about the sublime and immeasurable, see Kohrs, *Anton Bruckner*, 103–5. There was some precedent for European royalty to be embalmed and displayed under glass. In the royal crypt of Weimar, one can find Carl Friedrich von Sachsen-Weimar-Eisenach (1783–1853) and Karl Bernhard von Sachsen-Weimar-Eisenach (1792–1862) displayed like Maximilian.

117. Kohrs, *Anton Bruckner*, 95–123.

118. The arrangement had been worked out decades prior, but not without some quiet discord among the monks; Buchmayr, "Prälatengang Nr. 5."

119. St. Florian, Stiftsarchiv, Bruckner-Archiv, XIII/13, Beisetzung des Dr. Anton Bruckner in der Gruft der Stiftskirche, 15.10.1896.

120. Lindner, "Die Posthume Reisen des Anton B.," 131. Lindner's argument was reiterated in an email from Elisabeth Maier, here quoted and translated with permission: "Here one should add that Bruckner *had* to be embalmed: this was necessary for the transportation of a body, and he wanted to be buried in his *Heimat*." Correspondence between Elisabeth Maier and Abigail Fine, May 1, 2022.

121. "S. Sarg schaffte er sich selbst bei Lebzeiten an in Wien, sehr weit, wie s. Kleider, d. in d. ganz. Länge d. Person ein Glas eingelassen, so daß man ihn sehen kannte. (Auer sah ihn, sehr gut erhalten, weil sehr gut einbalsamiert.)" Notebooks of August Göllerich, ÖNB, F28 Göllerich 386/13, n.p.

122. There are casual reports of ticketed bus excursions to Bruckner's crypt for festivals in surrounding cities; see the excursion announcement for the 1933 Fest-Tagung in Salzburg in *Bruckner-Blätter* 5, no. 1 (1933): 3. Visiting conductors had special treatment at St. Florian, including accommodations in Bruckner's former apartments; see Buchmayr, "Prälatengang Nr. 5."

123. "Zur Bruckner-Feier in St. Florian," *Volksfreund* 34 (August 23, 1924): 4. For a more complete account of events, see "Am Grabe Bruckners: Die Sängerfahrt des Wiener Männer-Gesangvereines," *Linzer Volksblatt* 148 (June 29, 1924): 7–8.

124. "Bruckner Wallfahrt," *Tages-Post* (Linz) 257 (November 10, 1921): 6.

125. Bruckner's *Nachruf*, WAB 81a, was composed in 1877 in memory of Joseph Seiberl.

126. ÖNB, Estate of Franz Gräflinger, "Festzauber im Brucknerland," F30 Graeflinger 540. See also "Bruckner-Gedenkfeier in St. Florian. Anläßlich des 25. Todestages des Meisters," F30 Graeflinger 563.

127. ÖNB, Estate of Franz Gräflinger, "Bruckner und St. Florian," F30 Graeflinger 564.

128. In the early twentieth century, the catacombs of the Stephansdom were an appealing topic for lovers of the Gothic. The most extensive book on the subject was reissued in 1924, shortly after Tutankhamen was unearthed: Senfelder, *Die Katakomben bei St. Stefan in Sage und Geschichte.* See also the historical novel by Rudolf von Rosen, *Die Katakomben von Sankt Stephan,* published serially in various issues of the *Illustrierte Kronen-Zeitung* in 1916, along with articles such as "Die Wiener Katakomben," *Wiener Hausfrau,* November 6, 1911, 6; and Arthur Anders, "Die Wiener Katakomben," *Architekten- u. Baumeister-Zeitung,* November 19, 1911, 737–39.

129. Riggs, *Photographing Tutankhamun.*

130. Attfield, *Challenging the Modern,* 106–39.

131. Heuss, "Wie steht es heute um Bruckner?," 490. See Attfield, *Challenging the Modern,* 112.

132. Rudolph Steiner studied harmony and counterpoint with Bruckner in 1879–80 at the Technische Hochschule in Vienna. Steiner met regularly with Eckstein to discuss theosophy. While Göllerich clearly interacted with both Steiner and Eckstein, the full extent of his esoteric activities has not been charted. He was known to have met regularly at a vegetarian restaurant with various theosophists and life reformers; furthermore, in 1944, his widow, Gisela, purchased a great many volumes of spiritual, esoteric, and theosophical literature. See Maier, *Années de Pèlerinage,* 2:22–26. Steiner's drawing of Bruckner's seating position can be found in Göllerich's notebooks, ÖNB, Estate of August Göllerich, F28 Goellerich 386/10, n.p.

133. Painter, *Symphonic Aspirations,* 188–97 and 247.

134. Potter, *Art of Suppression.*

135. While this plan did not come to full fruition, it compelled Adolf Hitler and Joseph Goebbels to visit Bruckner's tomb and sites across Upper Austria in 1941, as I discuss shortly; see the citation of Goebbels's diary in Gilliam, "Annexation of Anton Bruckner," 587.

136. Letters from Adolf Wenusch to Abbot Vinzenz Hartl and Max Auer show a passionate campaign for preservation; writing to Auer on September 12, 1929, he notes that mold grows on the nose; and writing to Hartl on October 17, 1929, he calls himself a "veritabler Bruckner-Narr." See Maier and Grasberger, *Die Bruckner-Bestände,* vol. 2, group 14, 77–99, at 77 and 79.

137. Maier and Grasberger, vol. 2, group 14, 94.

138. Auer wrote in 1930 that "the whole matter revolves around the question of which is more pious: to allow the ephemeral earthly remains of the Master to succumb to natural decay or to subject the body to procedures that, even if they succeed, would only be effective for a limited time." Maier and Grasberger, vol. 2, group 14, 89. The abbot Vinzenz Hartl shared this view; a handwritten memo notes how rarely the lid is lifted; Maier and Grasberger, vol. 2, group 14, 96. Even so, two years later, another public descent to the tomb accompanied the consecration of the new Bruckner organ: "Massenbesuch bei Anton Bruckner," *Salzburger Volksblatt,* May 6, 1932, 7.

139. Letter from Norbert Wibiral to Fritz Rauch, Linz, January 13, 1961, in Maier and Grasberger, *Die Bruckner-Bestände,* vol. 2, group 14, 100–101.

140. Report by Bruno Kaufmann from the Anthropologisches Forschungsinstitut in Basel, 1996. Maier and Grasberger, 2:298–300.

141. Rehding, *Music and Monumentality,* 169–96.

142. Gilliam, "Annexation of Anton Bruckner."

143. "Dr. Goebbels als Musikhistoriker: Ein Nachklang zur Bruckner-Rede," *Sturm über Österreich,* June 20, 1937. Given that Auer kept a clipping of this article, he may have been

sympathetic to its sentiments. The German theologian Adolf Köberle argued in 1936 that Bruckner was "a child of medieval piousness" whose music was inextricable from his Catholic faith, which was not joyless and rule-bound but pastoral and life affirming; Köberle, *Bach, Beethoven, Bruckner als Symbolgestalten des Glaubens*, 30–39, at 31.

144. Max Auer, "Was sie aus St. Florian machen wollten," *Oberösterreichischer Nachrichten*, March 7, 1946, 2. For an extended history of the so-called Bruckner Bayreuth, see Kreczi, *Das Bruckner-Stift St. Florian*.

145. On Auer's cautious boundary and his embrace of the promise of support for the Internationale Bruckner Gesellschaft, see Gilliam, "Annexation of Anton Bruckner," 589.

146. ÖNB, Estate of Max Auer, Radio address given on Tuesday, October 6, 1936, at 2:30 p.m., F31 Auer 96, p. 14.

147. Kam, "Between Musicology and Mythology." On the comparable politics of the Vienna State Opera's reopening in 1955, see Stachel, "'Das Krönungsjuwel der österreichischen Freiheit'"; on the politics of the Zero Hour as apology and alibi, see Pollock, *Opera after the Zero Hour*.

148. Josef Häupl, "Bruckners Toten-Ehrung," *Linzer Volksblatt*, October 12, 1946, 2.

149. ÖNB, Estate of Max Auer, F31 Auer 105, p. 7.

150. Ludwig K. Mayer, "Festlicher Sonntagnachmittag in St. Florian," *Oberösterreichischer Nachrichten*, September 18, 1956, clipping in ÖNB F31 Auer 454-80. Auer referred to this as the "tönendes Grabmal" in his various calls to restore the organ in 1930; see for instance the *Bruckner-Blätter* 3:2/3 (1931): 31. On Goebbels's cinematic use of music for his ceremonies, see Rehding, *Music and Monumentality*, 169–96.

151. My critical reading of these rituals in historical context should not be mistaken for disrespect toward the Bruckner lovers who continue to make this pilgrimage to St. Florian, or any other listeners who journey in the name of a composer.

152. See, for instance, Peter Jan Margry, "Secular Pilgrimage: A Contradiction in Terms?," in Margry, *Shrines and Pilgrimage in the Modern World*, 13–48.

153. There is some disagreement among historians regarding the National Socialists' relationship with occult movements. Revisionist histories such as Treitel, *Science for the Soul*, suggest that the Nazis were largely interested in Ariosophy's racialism and banned the majority of occult societies, which represent a varied and decentralized subculture. Eric Kurlander insists that no other political party in history has been so strongly influenced by border sciences and fringe doctrines, and that the purge was piecemeal; see Kurlander, *Hitler's Monsters*.

Coda

1. The epigraphs are from Arendt, *Men in Dark Times*, 205–6; and Georges Perec, "Approaches to What?" (1973), in Perec, *Species of Spaces, and Other Pieces*, 205–7.

2. The masking function of piety may be a historical predecessor to the aesthetic moralism that justifies harm in present-day music culture; see Cheng, *Loving Music till It Hurts*.

3. Perec, "Approaches to What?," 207.

4. Nettl, *Heartland Excursions*, 11–42.

5. Sontag, "Notes on Camp," 282–83.

6. Morales, *Pilgrimage to Dollywood*.

7. Ngai, *Our Aesthetic Categories*.

Bibliography

Historical Periodicals

Abend-Zeitung (Dresden)
Abhandlungen der mathematisch-physischen Classe der Königlich Sächsischen Gesellschaft der Wissenschaften
Allgemeine Kunst-Chronik
Allgemeine musikalische Zeitung
Architekten- und Baumeister-Zeitung
Beilage zur Neuen Musik-Zeitung
Berliner Börsenzeitung
Bolond Istók
Bonner Tagebuch
Bonner Zeitung
Bruckner-Blätter
Chambers's Journal
Correspondenz-Blatt der deutschen Gesellschaft für Anthropologie
Das interessante Blatt
Das musikalische Wochenblatt
Der Auftakt
Der Brenner
Der Bund: Ein Monatschrift für den Bungert-Bund und seine Freunde
Der Klavier-Lehrer
Der Lugenschippel
Der neue teutsche Merkur
Der Tag (Vienna)
Deutsche Reichzeitung
Deutsches Volksblatt
Dichtung und Welt: Beilage zur Prager Presse
Die Fackel
Die Gartenlaube
Die Lyra: Allgemeine deutsche Kunstzeitschrift für Musik und Dichtung
Die medizinische Welt

Die Musik
Die Presse (Vienna)
Die Zeit (Vienna)
Erstes Beiblatt
Etude Magazine
Fremden-Blatt
General-Anzeiger (Bonn)
Godesburger Zeitung
Illustrierte Kronen-Zeitung
Illustrierte Welt
Illustrirte Zeitung (Leipzig)
Hamburger Nachrichten
Innsbrucker Tagblatt
Kikiriki: Humoristisches Wochenblatt für Stadt und Land (Petrovgrad)
Kölnische Zeitung (formerly the *Kölner Zeitung*)
Kunstgeschichtliches Jahrbuch der Zentral-Kommission für Erforschung und Erhaltung der kunst- und historischen Denkmale
L'illustration
Linzer Volksblatt
Lloyd's Magazine
Lustige Kölner Zeitung
Mittheilungen der anthropologischen Gesellschaft in Wien
Musical Times and Singing Class Circular
Neue Berliner Musikzeitung
Neue Freie Presse (Vienna)
Neue illustrirte Zeitung
Neue Musik-Zeitung
Neues Berliner Tageblatt
Neues Pester Journal
Neues Wiener Tagblatt
Niederrheinische Musik-Zeitung
Oberösterreichischer Nachrichten
Ostdeutsche Rundschau
Politisch-anthropologische Revue
Preussische Jahrbücher
Reichspost (Vienna)
Revue et gazette musicale
Rheinische Musik- und Theater-Zeitung (Cologne)
Salzburger Volksblatt
Salzburger Zeitung
Sturm über Österreich
Tagespost (Graz)
Tages-Post (Linz)
Tribüne (Berlin)
Ungarischer Lloyd
Volksfreund

Volkszeitung (Vienna)
Volk und Rasse
Vorwärts
Wiener Hausfrau
Wiener Zeitung
Zeitschrift für die gesamte Neurologie und Psychiatrie
Zeitschrift für Rassenkunde

Works Cited

Adler, Guido. "Schubert and the Viennese Classic School." Translated by Theodore Baker. *Musical Quarterly* 14, no. 4 (1928): 473–85.

———. "Style Criticism." Translated by Oliver Strunk. *Musical Quarterly* 20, no. 2 (1934): 172–76.

Adolphe, Bruce. "'Where Thought Touches the Blood': Rhythmic Disturbance as Physical Realism in Beethoven's Creative Process." In *The New Beethoven: Evolution, Analysis, Interpretation*, edited by Jeremy Yudkin, 78–88. Rochester, NY: Boydell and Brewer, 2020.

Adorno, Theodor W. *Beethoven: The Philosophy of Music*. Edited by Rolf Tiedemann. Translated by Edmund Jephcott. Cambridge: Polity, 1998.

———. "Commodity Music Analyzed." In *Quasi una fantasia: Essays on Modern Music*, translated by Rodney Livingstone, 37–52. London: Verso, 1998.

———. *Essays on Music*. Edited by Richard Leppert. Translated by Susan H. Gillespie. Berkeley: University of California Press, 2002.

———. *Gesammelte Schriften*. Edited by Rolf Tiedemann. 20 vols. Frankfurt: Suhrkamp, 1970.

———. *Introduction to the Sociology of Music*. Translated by E. B. Ashton. New York: Seabury, 1962.

———. "Theory of Pseudo-culture (1959)." Translated by Deborah Cook. *Telos* 95 (1993): 15–38.

Aichner, Herlinde. "Ludwig August Frankl—Politiker der Erinnerung." In *Ludwig August Frankl (1810–1894): Eine jüdische Biographie zwischen Okzident und Orient*, edited by Louise Hecht, 275–89. Cologne: Böhlau Verlag, 2016.

Albrecht, Theodore, ed. and trans. *Letters to Beethoven and Other Correspondence*. 3 vols. Lincoln: University of Nebraska Press, 1996.

Andree, Richard. *Vom Tweed zur Pentlandföhrde: Reisen in Schottland*. Jena: Hermann Costenoble, 1866.

Angermüller, Rudolph. "Die Bedeutung der ISM Salzburg für das Salzburger Kulturleben bis zum Ersten Weltkrieg." In *Bürgerliche Musikkultur im 19. Jahrhundert in Salzburg: Ein Symposion aus Anlaß des hundertjährigen Gründungstages der internationalen Stiftung Mozarteum Salzburg; 20. September 1980; Mozarteum Wiener Saal*, edited by Rudolph Angermüller, 58–92. Salzburg: Internationale Stiftung Mozarteum, 1981.

Applegate, Celia. "Mendelssohn on the Road: Music, Travel, and the Anglo-German Symbiosis." In *The Oxford Handbook of the New Cultural History of Music*, edited by Jane F. Fulcher, 228–44. Oxford: Oxford University Press, 2011.

Applegate, Celia, and Pamela M. Potter, eds. *Music and German National Identity*. Chicago: University of Chicago Press, 2002.

Aragon, Louis. *Aurélien*. Translated by Eithne Wilkins. New York: Duell, Sloane, and Pearce, 1947.

Arendt, Hannah. *Men in Dark Times*. San Diego: Harcourt Brace, 1968.
Ariès, Philippe. *The Hour of Our Death*. 1977. Translated by Helen Weaver. New York: Knopf, 1981.
Ashbrook, William. *Donizetti and His Operas*. Cambridge: Cambridge University Press, 1982.
Attfield, Nicholas. *Challenging the Modern: Conservative Revolution in German Music, 1918–33*. Oxford: Oxford University Press; British Academy, 2017.
Auden, W. H. *The Collected Poetry of W. H. Auden*. New York: Random House, 1945.
Bahr, Hermann. "Die Hauptstadt von Europa: Eine Phantasie in Salzburg." In *Essays*, 235–41. Leipzig: Insel-Verlag, 1912.
———. "Die Mozartstadt." *Musica Divina* 2, no. 8/9 (1914): 303–6.
———. *Salzburg*. Berlin: J. Bard, 1914.
Bankl, Hans, and Hans Jesserer. *Die Krankheiten Ludwig van Beethovens: Pathographie seines Lebens und Pathologie seiner Leiden*. Vienna: W. Maudrich, 1987.
Barnett, Teresa. *Sacred Relics: Pieces of the Past in Nineteenth-Century America*. Chicago: University of Chicago Press, 2013.
Barthes, Roland. *Camera Lucida: Reflections on Photography*. Translated by Richard Howard. New York: Hill and Wang, 1981.
Bartsch, Rudolf Hans. *"Schwammerl": Ein Schubert-Roman*. Leipzig: Verlag L. Staackmann, 1912.
Bates, Eliot. "The Social Life of Musical Instruments." *Ethnomusicology* 56, no. 3 (2012): 363–95.
Baudrillard, Jean. *Simulacra and Simulation*. Translated by Sheila Faria Glaser. Ann Arbor: University of Michigan Press, 1994.
Bauer, Oswald Georg. *Josef Hoffmann: Der Bühnenbildner der ersten Bayreuther Festspiele*. Munich: Deutscher Kunstverlag, 2008.
Baym, Nancy K. *Playing to the Crowd: Musicians, Audiences, and the Intimate Work of Connection*. New York: New York University Press, 2018.
Beethoven-Feier, Kammermusik-Fest. Bonn: Beethoven-Haus, 1890.
Begg, Tristan James Alexander, Axel Schmidt, Arthur Kocher, Maarten H. D. Larmuseau, Göran Runfeldt, Paul Andrew Maier, John D. Wilson, Rodrigo Barquera, Carlo Maj, András Szolek, Michael Sager, Stephen Clayton, Alexander Peltzer, Ruoyun Hui, Julia Ronge, Ella Reiter, Cäcilia Freund, Marta Burri, Franziska Aron, Anthi Tiliakou, Joanna Osborn, Doron M. Behar, Malte Boecker, Guido Brandt, Isabelle Cleynen, Christian Strassburg, Kay Prüfer, Denise Kühnert, William Meredith, Markus M. Nöthen, Robert David Attenborough, Toomas Kivisild, and Johannes Krause. "Genomic Analyses of Hair from Ludwig van Beethoven." *Current Biology* 33, no. 8 (2023): 1431–47.
Beller, Steven, ed. *Rethinking Vienna 1900*. New York: Berghahn Books, 2001.
Benjamin, Walter. "Boredom: Eternal Return." In *The Arcades Project*, edited by Rolf Tiedemann, translated by Howard Eiland and Kevin McLaughlin, 101–19. Cambridge, MA: Belknap Press of Harvard University Press, 1999.
———. "Dream Kitsch: Gloss on Surrealism." In *Selected Writings*, edited by Howard Eiland, Michael W. Jennings, and Gary Smith, translated by Rodney Livingstone, 2:3–5. Cambridge, MA: Belknap Press of Harvard University Press, 2002.
———. "The Work of Art in the Age of Its Technological Reproducibility." 2nd version. In *The Work of Art in the Age of Its Technological Reproducibility, and Other Writings on Media*, edited by Michael W. Jennings, Brigid Doherty, and Thomas Y. Levin, translated by Edmund Jephcott, Rodney Livingstone, Howard Eiland, et al., 21–55. Cambridge, MA: Belknap Press of Harvard University Press, 2008.

Benkard, Ernst. *Das ewige Antlitz: Eine Sammlung von Totenmasken.* Berlin: Frankfurter Verlags-Anstalt, 1926.
Bennett, Jane. *Vibrant Matter: A Political Ecology of Things.* Durham, NC: Duke University Press, 2010.
Bennett, Tony. *Museums, Power, Knowledge: Selected Essays.* London: Routledge, 2017.
Bettermann, Silke. *Beethoven im Bild: Die Darstellung des Komponisten in der bildenden Kunst vom 18. bis zum 21. Jahrhundert.* Bonn: Beethoven-Haus, 2012.
———. *Franz von Stuck und Beethoven: Musik in der Kunst des Münchner Jugendstils.* Bonn: Beethoven-Haus, 2013.
Bickley, Nora, trans. *Letters to and from Joseph Joachim. 1914.* New York: Vienna House, 1972.
Billroth, Theodor. *Wer ist musikalisch?* Edited by Eduard Hanslick. 4th ed. Berlin: Gebrüder Paetel, 1912.
Binkley, Sam. "Kitsch as a Repetitive System: A Problem for the Theory of Taste Hierarchy." *Journal of Material Culture* 5, no. 2 (2000): 131–52.
Blommen, Heinz. *Anfänge und Entwicklung des Männerchorwesens am Niederrhein.* Cologne: Arno Volk, 1960.
Bloch, Ernst. *Heritage of Our Times.* Translated by Neville Plaice and Stephen Plaice. Berkeley: University of California Press, 1991.
Böhme, Gernot. "Physiognomie als Begriff der Ästhetik." In *Perspektiven der Lebensphilosophie: Zum 125. Geburtstag von Ludwig Klages,* edited by Michael Großheim, 45–65. Bonn: Bouvier Verlag, 1999.
Bonds, Mark Evan. *The Beethoven Syndrome: Hearing Music as Autobiography.* Oxford: Oxford University Press, 2019.
———. "Irony and Incomprehensibility: Beethoven's 'Serioso' String Quartet in F Minor, op. 95, and the Path to the Late Style." *Journal of the American Musicological Society* 70, no. 2 (2017): 285–356.
Bordes, Marc. *La maladie et l'oeuvre de Chopin.* Lyon: Bosc Frères, M. et L. Riou, 1932.
Botstein, Leon. "Aesthetics and Ideology in the *Fin-de-Siècle* Mozart Revival." *Current Musicology* 51 (1993): 5–25.
———. "Listening through Reading: Musical Literacy and the Concert Audience." *Nineteenth-Century Music* 16, no. 2 (1992), 129–45.
Bourdieu, Pierre. *The Rules of Art: Genesis and Structure of the Literary Field.* Translated by Susan Emanuel. Stanford, CA: Stanford University Press, 1995.
Brachmann, Jan. *Kunst, Religion, Krise: Der Fall Brahms.* Kassel: Bärenreiter, 2003.
Brantlinger, Patrick. *Bread and Circuses: Theories of Mass Culture as Social Decay.* Ithaca, NY: Cornell University Press, 2016.
Bräunig, Angela. "Der Mythos Mozart: Kitsch und Vermarktung." In *Theater um Mozart,* edited by Bärbel Pelker, 209–36. Heidelberg: Universitätsverlag Winter, 2006.
Brayshaw, Emily. "Oskar Schlemmer's Kitsch (1922): A Contextualization and Translation." *Journal of Aesthetics and Culture* 13, no. 1 (2021): 1–15.
Breuning, Gerhard von. *Memories of Beethoven: From the House of the Black-Robed Spaniards.* Edited by Maynard Solomon. Translated by Henry Mins and Maynard Solomon. Cambridge: Cambridge University Press, 1992.
———. "The Skulls of Beethoven and Schubert." Edited by William Meredith. Translated by Hannah Liebmann. *Beethoven Journal* 20, no. 1/2 (2005): 58–60.
Broch, Hermann. *Gesammelte Werke.* 10 vols. Zurich: Rhein, 1952.

———. *Hermann Broch: Kommentierte Werkausgabe*. Edited by Paul Michael Lützeler. 6 vols. Frankfurt: Suhrkamp, 1994.

———. *Hugo von Hofmannsthal and His Time: The European Imagination, 1860–1920*. Edited and translated by Michael P. Steinberg. Chicago: University of Chicago Press, 1984.

———. *Massenwahntheorie: Beiträge zu einer Psychologie der Politik*. Frankfurt: Suhrkamp, 1979.

Brodbeck, David Lee. *Defining* Deutschtum: *Political Ideology, German Identity, and Music-Critical Discourse in Liberal Vienna*. Oxford: Oxford University Press, 2014.

Brooks, Jeanice, Matthew Stephens, and Wiebke Thormahlen, eds. *Sound Heritage: Making Music Matter in Historic Houses*. London: Routledge, 2022.

Brown, Bill. "Thing Theory." *Critical Inquiry* 28, no. 1 (2001): 1–22.

Bruck, Arthur Moeller van den. *Beethoven der Deutsche*. Minden: J. C. C. Bruns' Verlag, 1917.

Brümmer, Franz. "Koelman, Margarete." In *Lexicon der deutschen Dichter und Prosaisten vom Beginn des 19. Jahrhunderts bis zur Gegenwart*, 6th ed., 4:62. Leipzig: Philipp Reclam, 1913.

Brusniak, Friedhelm. "The Involvement of Freemasons in the 'Erstes Deutsches Sängerfest' in Frankfurt-on-Main in 1838: Observations from a Choral-Sociological Perspective." In *Choral Singing: Histories and Practices*, edited by Ursula Geisler and Karin Johansson, 113–21. Cambridge: Cambridge Scholars, 2014.

Buchmayr, Friedrich. "Prälatengang Nr. 5—Anton Bruckner als Gast im Stift St. Florian." In *Bruckner Jahrbuch 2011–2014*, edited by Andreas Lindner and Klaus Petermayr, 7–44. Linz: Anton Bruckner Institut, 2015.

Bynum, Caroline Walker. *Christian Materiality: An Essay on Religion in Late Medieval Europe*. Cambridge, MA: MIT Press, 2011.

Calinescu, Matei. *Five Faces of Modernity: Modernism, Avant-Garde, Decadence, Kitsch, Postmodernism*. Durham, NC: Duke University Press, 1987.

Canetti, Elias. *Auto-da-Fé*. Translated by C. V. Wedgwood. New York: Farrar, Straus, Giroux, 1984.

Carrino, Sergio, and Giuliana Carrino. *Eine Beethoven-Wunderkammer in Italien: Die Sammlung Carrino; Katalog zur Sonderausstellung im Beethoven-Haus Bonn; 14. Dezember 2012–20. Mai 2013*. Bonn: Beethoven-Haus, 2012.

Carroll, Victoria. *Science and Eccentricity: Collecting, Writing, and Performing Science for Early Nineteenth-Century Audiences*. London: Pickering and Chatto, 2008.

Celenza, Anna Harwell. "Darwinian Visions: Beethoven Reception in Mahler's Vienna." *Musical Quarterly* 93 (2010): 514–59.

Challier, Ernst. *Grosser Männergesang-Katalog*. Giessen: self-published by author, 1900.

Charet, F. X. *Spiritualism and the Foundations of C. G. Jung's Philosophy*. Albany: State University of New York Press, 1993.

Charlier, Philippe, Antonio Perciaccante, Marc Herbin, Otto Appenzeller, and Raffaela Bianucci. "The Heart of Frédéric Chopin (1810–1849)." Letter to the editor. *American Journal of Medicine* 131, no. 4 (2018): 173–74.

Cheng, William. *Loving Music till It Hurts*. New York: Oxford University Press, 2019.

Chua, Daniel K. L. "Beethoven's Other Humanism." *Journal of the American Musicological Society* 62, no. 3 (2009): 571–645.

Clark, Suzannah. *Analyzing Schubert*. Cambridge: Cambridge University Press, 2011.

Coleridge, Samuel Taylor. *Biographia Literaria*. Vol. 1. London: Fenner, 1817.

Comini, Alessandra. *The Changing Image of Beethoven: A Study in Mythmaking*. 1987. Rev. ed. Santa Fe, NM: Sunstone, 2008.

———. "The Visual Beethoven: Whence, Why, and Whither the Scowl?" In *Beethoven and His World*, edited by Scott Burnham and Michael P. Steinberg, 287–312. Princeton, NJ: Princeton University Press, 2000.

Corbin, Solange. "Les Offices de la Sainte Face." *Bulletin des études portugaises et de l'Institut français au Portugal* 11 (1947): 22–25.

Corbineau-Hoffmann, Angelika. *Testament und Totenmaske: Der literarische Mythos des Ludwig van Beethoven*. Hildesheim: Weidmann, 2000.

Coreth, Anna. *Pietas Austriaca*. Translated by William D. Bowman and Anna Maria Leitgeb. West Lafayette, IN: Purdue University Press, 2004.

Csiszar, Alex. *The Scientific Journal: Authorship and the Politics of Knowledge in the Nineteenth Century*. Chicago: University of Chicago Press, 2018.

Damenspende zum Concordia-Ball, 14. Februar 1927. Vienna: Elbemühl Papierfabriken und Graphische Industrie, 1927.

Dammert, Rudolf. *Das Wunderkind: Ein Mozartbuch*. Berlin: Gustav Weise Verlag, 1937.

Danhauser, Carl. *Nach Beethovens Tod: Erinnerungen von Carl Danhauser: Kommentiertes Faksimile des Autographs im Archiv der Gesellschaft der Musikfreunde in Wien*. Edited and explicated by Otto Biba. Vienna: Gesellschaft der Musikfreunde, 2001.

Daverio, John. "Mozart in the Nineteenth Century." In *The Cambridge Companion to Mozart*, edited by Simon P. Keefe, 169–84. Cambridge: Cambridge University Press, 2003.

Davies, James Q. *Romantic Anatomies of Performance*. Berkeley: University of California Press, 2014.

Davison, Alan. "The Face of a Musical Genius: Thomas Hardy's Portrait of Joseph Haydn." *Eighteenth-Century Music* 6, no. 2 (2009): 209–27.

———. "Franz Liszt and the Physiognomic Ideal in the Nineteenth Century." *Music in Art* 1, no. 2 (2005): 133–44.

———. "The Musician in Iconography from the 1830s and 1840s: The Formation of New Visual Types." *Music in Art* 28, no. 1/2 (2003): 147–62.

———. "Painting for a Requiem: Mihály Munkácsy's *The Last Moments of Mozart* (1885)." *Early Music* 39, no. 1 (2011): 79–92.

Deathridge, John. "Elements of Disorder: Appealing Beethoven vs. Rossini." In *The Invention of Beethoven and Rossini: Historiography, Analysis, Criticism*, edited by Nicholas Mathew and Benjamin Walton, 305–32. Cambridge: Cambridge University Press, 2013.

Dechant, Robert E., and Filipp Goldscheider. *Goldscheider: Firmengeschichte und Werkverzeichnis*. Stuttgart: Arnoldsche, 2007.

Denck, Hans, and Clarence Bauman. *The Spiritual Legacy of Hans Denck*. Edited and translated by Clarence Bauman. Leiden: Brill, 1991.

Derrida, Jacques. *Archive Fever: A Freudian Impression*. Translated by Eric Prenowitz. Chicago: University of Chicago Press, 1998.

Deutsch, Otto Erich, ed. *Franz Schubert's Letters and Other Writings*, translated by Venetia Savile. London: Faber and Gwyer, 1928.

Dickey, Colin. *Cranioklepty: Grave Robbing and the Search for Genius*. Denver, CO: Unbridled Books, 2010.

Didi-Huberman, Georges. "Before the Image, before Time: The Sovereignty of Anachronism." In *Compelling Visuality: The Work of Art in and out of History*, edited by Claire Farago and Robert Zwijnenberg, 31–44. Minneapolis: University of Minnesota Press, 2003.

"Die Cranien dreier musikalischer Koryphäen." *Mittheilungen der anthropologischen Gesellschaft in Wien* 17, no. 4 (April 1887): 33–36.

Die Salzburger Festspiele: Ihre Vorgeschichte und Entwicklung (1842–1960); Ausstellung anläßlich der Eröffnung des neuen Festspielhauses in Salzburg in den Räumen der Residenz; Juli bis September 1960. Salzburg: Internationale Stiftung Mozarteum, 1960.

Doherty, Brigid. "What Is There to Be Learned from Kitsch? Walter Benjamin and the 'Furnished Man.'" *Cabinet* 39 (2010). https://www.cabinetmagazine.org/issues/39/doherty.php.

Döblin, Alfred. "Faces, Images, and Their Truth." In *Face of Our Time: Sixty Portraits of Twentieth-Century Germans*, by August Sander, translated by Michael Robertson, 7–16. Munich: Schirmer/Mosel Verlag, 1994.

Dole, Nathan Haskell. *Famous Composers*. New York: Cromwell, 1902.

Dorfles, Gillo, ed. *Kitsch: The World of Bad Taste*. New York: Bell, 1969.

Dović, Marijan, and Jón Karl Helgason. *National Poets, Cultural Saints: Canonization and Commemorative Cults of Writers in Europe*. Leiden: Brill, 2017.

Downes, Stephen. *The Muse as Eros: Music, Erotic Fantasy, and Male Creativity in the Romantic and Modern Imagination*. Aldershot: Ashgate, 2006.

Dreyfus, Laurence. *Wagner and the Erotic Impulse*. Cambridge, MA: Harvard University Press, 2010.

Droste-Hülshoff, Annette von. *Das geistliche Jahr*. Stuttgart: Cotta, 1851.

Duclos-Vallée, Jean-Charles, and Serge Erlinger. "Is Frédéric Chopin's Death Elucidated?" Letter to the editor. *American Journal of Medicine* 141, no. 4 (April 2018): e171.

Dyer, Richard. *White*. 20th anniversary ed. London: Routledge, 2017.

Eckhardt, Mária, ed. *Liszt Ferenc Gedenkmuseum: Katalog der ständigen Ausstellung*. Budapest: Liszt Ferenc Gedenkmuseum und Forschungszentrum, 2011.

D.V. [Ernst Klotz]. *Das fragwürdige Todtenbein von Leipzig: Satire auf die tieftraurige Historie vom Leben, Sterben, und Ausgraben der Gebeine J. S. Bach's*. Leipzig: Paul de Wit, 1906.

Eco, Umberto. "Postmodernism, Irony, the Enjoyable." In *Postscript to the Name of the Rose*, 65–72. San Diego: Harcourt Brace, 1984.

Eisen, Cliff. "Mozart's Leap in the Dark." In *Mozart Studies*, edited by Simon P. Keefe, 1–24. Cambridge: Cambridge University Press, 2006.

Elben, Otto. *Der volksthümliche deutsche Männergesang*. 2nd ed. Tübingen: H. Laupp, 1887.

Elbertzhagen, Theodor Walter. *Die Neunte: Eine Beethoven-Legende*. Berlin: Wilhelm Limpert Verlag, 1933.

Elias, Norbert. "The Kitsch Style and the Age of Kitsch." In *Early Writings*, translated by Edmund Jephcott, 85–96. Dublin: University College Dublin Press, 2006.

Ender, Daniel. "'Sie müssen doch sehen, wie Alban lebt . . .': Bergs Wohnräume und die Inszenierung des Authentischen." In *Helene Berg und das Erbe Alban Bergs: Erinnerung Stiften; Bericht zur internationalen Tagung am 16. und 17. März, 2017*, edited by Daniel Ender, Martin Eybl, and Melanie Unseld, 120–52. Vienna: Universal Edition, 2018.

Engl, Johann Evangelist. *Genesis der Internationaler Mozart Stiftung*. Salzburg: Internationale Stiftung Mozarteum, 1873.

Erster Jahresbericht der Internationalen Stiftung Mozarteum in Salzburg. Salzburg: Internationale Stiftung Mozarteum, 1881.

Estes, Douglas. "Dualism or Paradox? A New 'Light' on the Gospel of John." *Journal of Theological Studies* 71, no. 1 (2020): 90–118.

Everist, Mark. *Mozart's Ghosts: Haunting the Halls of Musical Culture*. Oxford: Oxford University Press, 2012.

Falkoff, Rebecca R. *Possessed: A Cultural History of Hoarding*. Ithaca, NY: Cornell University Press, 2021.

Feis, Oswald. *Studien über die Genealogie und Psychologie der Musiker*. Wiesbaden: J. F. Bergmann, 1910.

Fest-Album zur Säcular-Feier von Goethe's Geburtstag am 28en August 1849 in Weimar. Hamburg: Schuberth, 1849.

Fine, Abigail. "Assimilating to Art-Religion: Jewish Secularity and Edgar Zilsel's *Geniereligion* (1918)." *Yale Journal of Music and Religion* 6, no. 2 (2020): 10–32.

———. "Beethoven's Mask and the Physiognomy of Late Style." *Nineteenth-Century Music* 43, no. 3 (2020): 143–69.

———. "Towards a History of the Eccentric Artist: Beethoven's Bad Manners and the Lure of the Anecdote." *Music and Letters* 104, no. 4 (2023): 567–91.

Finger, Stanley. *Origins of Neuroscience: A History of Explorations into Brain Function*. New York: Oxford University Press, 1994.

Finger, Stanley, and Paul Eling. *Franz Joseph Gall: Naturalist of the Mind, Visionary of the Brain*. Oxford: Oxford University Press, 2019.

Flechsig, Paul. *Gehirn und Seele*. 2nd ed. Leipzig: Veit, 1896.

Fliedl, Gottfried, ed. *Wa(h)Re Kunst: Der Museumsshop als Wunderkammer; Theoretische Objekte, Fakes und Souvenirs*. Frankfurt: Anabas Verlag, 1997.

Forsyth, Michael. *Buildings for Music: The Architect, the Musician, the Listener from the Seventeenth Century to the Present Day*. Cambridge, MA: MIT Press, 1985.

Franseen, Kristin. "'Everything You've Heard Is True': Resonating Musicological Anecdotes in Crime Fiction about Antonio Salieri." *Journal of Historical Fictions* 4, no. 1 (2022): 41–60.

Freisauff, Rudolf von. *Das erste Salzburger Musikfest*. Salzburg: Internationale Mozart Stiftung, 1877.

Frietsch, Ute. "The Boundaries of Science/Pseudoscience." Europäische Geschichte Online. January 14, 2015. Leibniz Institute of European History, Mainz. http://www.ieg-ego.eu/frietschu-2015-de.

Frigau Manning, Céline. "Phrenologizing Opera Singers: The Scientific 'Proofs of Musical Genius.'" *Nineteenth-Century Music* 39, no. 2 (2015): 125–41.

Frimmel, Theodor von. *Neue Beethoveniana*. Vienna: Carl Gerold's Sohn, 1888.

Frisch, Walter. *German Modernism: Music and the Arts*. Berkeley: University of California Press, 2005.

Gal, Susan. "Bartók's Funeral: Representations of Europe in Hungarian Political Rhetoric." *American Ethnologist* 18, no. 3 (1991): 440–58.

Gallup, Stephen. *A History of the Salzburg Festival*. Topfield, MA: Salem House, 1987.

Ganche, Édouard. *Souffrances de Frédéric Chopin: Essai de médecine et de psychologie*. Paris: Mercure de France, 1935.

Garrison, Fielding H. "Medical Men Who Have Loved Music." *Musical Quarterly* 7, no. 4 (1921): 527–48.

Gibbon, Elaine Fitz. "Beethoven Returns to Bonn: Origins, Belonging, and Misuse in Mauricio Kagel's *Ludwig van* (1969)." *Current Musicology* 107 (2020): 29–61.

Gibbons, William. *Building the Operatic Museum: Eighteenth-Century Opera in Fin-de-Siècle Paris*. Rochester, NY: University of Rochester Press, 2013.

Gibbs, Christopher H. "Performances of Grief: Vienna's Response to the Death of Beethoven." In *Beethoven and His World*, edited by Scott Burnham and Michael P. Steinberg, 227–85. Princeton, NJ: Princeton University Press, 2000.

Giddens, Anthony. *Modernity and Self-Identity: Self and Society in the Late Modern Age*. Stanford, CA: Stanford University Press, 1991.

Gillespie, W. F. "Doctors and Music." *Canadian Medical Association Journal* 33, no. 6 (1935): 676–79.

Gilliam, Bryan. "The Annexation of Anton Bruckner: Nazi Revisionism and the Politics of Appropriation." *Musical Quarterly* 78, no. 3 (1994): 584–604.

Goethe, Johann Wolfgang von. "'Bei Betrachtung von Schillers Schädel,' Alt. Titled 'Schillers Reliquien.'" In *Goethe, the Lyrist: 100 Poems in New Translations Facing the Originals*, edited and translated by Edwin H. Zeydel, 155–57. Chapel Hill: University of North Carolina Press, 1955.

Goethe, Julius August Walther von. "Rede bei Niederlegung von Schillers Schädel auf der Großherzoglichen Bibliothek in Weimar." In Johann Wolfgang von Goethe, *Goethes poetische Werke*, vol. 8: Autobiographische Schriften Teil 1, 1471–73. Stuttgart: Cotta, 1982.

Goldberg, Halina. "Chopin's Album Leaves and the Aesthetics of Musical Album Inscription." *Journal of the American Musicological Society* 73, no. 3 (2020): 467–533.

Goldberger, Zachary D., Steven M. Whiting, and Joel D. Howell. "The Heartfelt Music of Ludwig van Beethoven." *Perspectives in Biology and Medicine* 57, no. 2 (2014): 285–94.

Göllerich, August. *Anton Bruckner: Ein Lebens- und Schaffens-Bild*. Edited by Max Auer. 4 vols. Regensburg: G. Bosse, 1974.

Goodwin, Sarah Webster. *Kitsch and Culture: The Dance of Death in Nineteenth-Century Literature and Graphic Arts*. New York: Garland, 1988.

Gottdang, Andrea. "Porträts, Denkmäler, Sammelbildchen: Ein visueller Kanon der Musik?" In *Der Kanon der Musik: Theorie und Geschichte*, edited by Klaus Pietschmann and Melanie Wald-Fuhrmann, 832–57. Munich: Richard Boorberg Verlag, 2013.

Gouk, Penelope, ed. *Musical Healing in Cultural Contexts*. Aldershot: Ashgate, 2000.

Gould, Stephen Jay. *The Mismeasure of Man*. 2nd ed. New York: Norton, 1996.

Grasberger, Renate, Erich Wolfgang Partsch, and Uwe Harten. *Bruckner—skizziert: Ein Porträt in ausgewählten Erinnerungen und Anekdoten*. Vienna: Musikwissenschaftlicher Verlag, 1991.

Gray, Richard T. *About Face: German Physiognomic Thought from Lavater to Auschwitz*. Detroit: Wayne State University Press, 2004.

Green, Emily H. *Dedicating Music, 1785–1850*. Rochester, NY: University of Rochester Press, 2019.

Greenberg, Clement. "Avant-Garde and Kitsch." 1939. In *Art and Culture: Critical Essays*, 3–21. Boston: Beacon, 1961.

Grewe, Cordula. "Aesthetic Religion, Religious Aesthetics, and the Romantic Quest for Epiphany." In *A Companion to Nineteenth-Century Art*, edited by Michelle Facos, 175–92. Hoboken, NJ: John Wiley and Sons, 2019.

Grey, Thomas S. "Wagner the Degenerate: Fin de Siècle Cultural 'Pathology' and the Anxiety of Modernism." *Nineteenth Century Studies* 16, no. 1 (2002): 73–92.

Grootenboer, Hanneke. *Treasuring the Gaze: Intimate Vision in Late Eighteenth-Century Eye Miniatures*. Chicago: University of Chicago Press, 2012.

Großpietsch, Christoph. "Pilgerreisen zu Mozart und ein Salzburger Bürgerhaus." In *Häuser der Erinnerung: Zur Geschichte der Personengedenkstätte in Deutschland*, edited by Anne Bohnenkamp, Constanze Breuer, Paul Kahl, and Stefan Rhein, 153–86. Leipzig: Evangelische Verlagsanstalt, 2015.

Gugitz, Gustav. *Österreichs Gnadenstätten in Kult und Brauch*. 5 vols. Vienna: Verlag Brüder Hollinek, 1958.

Gur, Golan. "Music and 'Weltanschauung': Franz Brendel and the Claims of Universal History." *Music and Letters* 93, no. 3 (2012): 350–73.

Guthrie, Kate. *The Art of Appreciation: Music and Middlebrow Culture in Modern Britain*. Berkeley: University of California Press, 2021.
Habermas, Jürgen. *The Structural Transformation of the Public Sphere*. Translated by Thomas Burger. Cambridge, MA: MIT Press, 1989.
Habisch, André. "Spiritual Capital." In *Handbook on the Economics of Reciprocity and Social Enterprise*, edited by Luigino Bruni, Stefano Zamagni, and Antonella Ferrucci, 336–43. Cheltenham: Edward Elgar, 2013.
Hagner, Michael. *Geniale Gehirne: Zur Geschichte der Elitegehirnforschung*. Göttingen: Wallstein Verlag, 2004.
———. "Skulls, Brains, and Memorial Culture: On Cerebral Biographies of Scientists in the Nineteenth Century." *Science in Context* 16, no. 1–2 (2003): 195–218.
Hahn, Cynthia. "Absent No Longer: The Saint and the Sign in Late Medieval Pictorial Hagiography." In *Hagiographie und Kunst: Der Heiligenkult in Schrift, Bild und Architektur*, edited by Gottfried Kerscher, 152–75. Berlin: Dietrich Reimer Verlag, 1993.
Hallam, Elizabeth, and Jenny Hockey. *Death, Memory, and Material Culture*. New York: Berg, 2001.
Halse, Sven. "The Literary Idyll in Germany, England, and Scandinavia, 1770–1848." In *Romantic Prose Fiction*, edited by Gerald Gillespie, Manfred Engel, and Bernard Dieterle, 383–411. Amsterdam: John Benjamins, 2008.
Hamburger, Jeffrey F. *The Visual and the Visionary: Art and Female Spirituality in Late Medieval Germany*. New York: Zone Books, 1998.
Hamilton, Kenneth. "Wagner and Liszt: Elective Affinities." In *Richard Wagner and His World*, edited by Thomas S. Grey, 27–64. Princeton, NJ: Princeton University Press, 2009.
Hansen, Miriam Bratu. "Benjamin's Aura." *Critical Inquiry* 34, no. 2 (2008): 336–75.
Hanslick, Eduard. "Beethoven in Wien." In *Zur Enthüllung des Beethoven-Denkmals in Wien am 1. Mai, 1880*, 1–11. Vienna: Beethoven-Denkmal-Comités, 1880.
———. "Das Monument." In *Aus dem Opernleben der Gegenwart: Neue Kritiken und Studien von Eduard Hanslick*, 4th ed., 368–72. Berlin: Allgemeiner Verein für Deutsche Literatur, 1901.
———. *Hanslick: Sämtliche Schriften*. Edited by Dietmar Strauß. 7 vols. Vienna: Böhlau Verlag, 1993.
———. *Musikalisches Skizzenbuch: Neue Kritiken und Schilderungen*. Berlin: Allgemeiner Verein für Deutsche Literatur, 1888.
———. *On the Musically Beautiful*. Translated by Lee Rothfarb and Christoph Landerer. New York: Oxford University Press, 2018.
———. "Wagner-Kultus." 1882. In *Aus dem Opernleben der Gegenwart: Neue Kritiken und Studien von Eduard Hanslick*, 4th ed., 338–52. Berlin: Allgemeiner Verein für Deutsche Literatur, 1901.
Hargraves, John A. *Music in the Works of Broch, Mann, and Kafka*. Rochester, NY: Camden House, 2001.
Hau, Michael. *The Cult of Health and Beauty in Germany: A Social History, 1890–1930*. Chicago: University of Chicago Press, 2003.
———. "The Holistic Gaze in German Medicine, 1890–1930." *Bulletin of the History of Medicine* 74, no. 3 (2000): 495–524.
Haydon, Benjamin Robert. *The Diary of Benjamin Robert Haydon*. Edited by Willard Bissell Pope. 5 vols. Cambridge, MA: Harvard University Press, 1963.

Head, Matthew. "C. P. E. Bach 'in Tormentis': Gout Pain and Body Language in the Fantasia in A Minor, H278 (1782)." *Eighteenth-Century Music* 13, no. 2 (2016): 211–34.

Henke, Burkhard, Susanne Kord, and Simon Richter, eds. *Unwrapping Goethe's Weimar: Essays in Cultural Studies and Local Knowledge*. Rochester, NY: Camden House, 2000.

Herdt, Jennifer A. *Forming Humanity: Redeeming the German Bildung Tradition*. Chicago: University of Chicago Press, 2019.

Heuss, Alfred. "Ein Tag in Sankt Ludwig." *Die Musik* 12, no. 9 (1912/1913): 131–45.

———. "Wie steht es heute um Bruckner? Allerlei Brucknerfragen." *Zeitschrift für Musik* 91, no. 9 (1924): 488–93.

Hiebl, Ewald. "German, Austrian, or 'Salzburger'? National Identities in Salzburg, c. 1830–70." In *Different Paths to the Nation: Regional and National Identities in Central Europe and Italy, 1830–70*, edited by Laurence Cole, 100–121. Basingstoke: Palgrave Macmillan, 2007.

Higgins, David. "Art, Genius, and Racial Theory in the Early Nineteenth Century: Benjamin Robert Haydon." *History Workshop Journal* 58 (2004): 17–40.

Hiller, Jonathan R. "Lombroso and the Science of Literature and Opera." In *The Cesare Lombroso Handbook*, edited by Paul Knepper and P. J. Ystehede, 226–52. London: Routledge, 2013.

His, Wilhelm. "Anatomische Forschungen über Johann Sebastian Bach's Gebeine und Antlitz." In *Abhandlungen der mathematisch-physischen Classe der Königlich Sächsischen Gesellschaft der Wissenschaften* 22:379–420. Leipzig: S. Hirzel, 1895.

———. *Johann Sebastian Bach: Forschungen über dessen Grabstätte, Gebeine und Antlitz; Bericht an den Rath der Stadt Leipzig*. Leipzig: F. C. W. Vogel, 1895.

Hobsbawm, Eric J. *The Age of Empire, 1875–1914*. New York: Vintage Books, 1989.

Hobsbawm, Eric J., and Terence Ranger, eds. *The Invention of Tradition*. Cambridge: Cambridge University Press, 1983.

Hoffmann, Robert. "Vom Mozartdenkmal zur Festspielgründung: Musik- und Vereinskultur im 19. Jahrhundert." In *Salzburger Musikgeschichte: Vom Mittelalter bis ins 21. Jahrhundert*, edited by Jürg Stenzl, Ernst Hintermaier, and Gerhard Walterskirchen, 401–23. Salzburg: Verlag Anton Pustet, 2005.

Hofmannsthal, Hugo von. *Festspiele in Salzburg*. Vienna: Bermann-Fischer, 1938.

———. *Gesammelte Werke in Einzelausgaben*. Edited by Herbert Steiner, Vol. 1, *Prosa*. Frankfurt: S. Fischer, 1956.

Holl, Oskar. "Dokumente zur Entstehung der Salzburger Festspiele: Unveröffentliches aus der Korrespondenz der Gründer." *Maske und Kothurn* 13, nos. 2–3 (1967): 148–79.

Horkheimer, Max, and Theodor W. Adorno. *Dialectic of Enlightenment: Philosophical Fragments*. Translated by Edmund Jephcott. Stanford, CA: Stanford University Press, 2002.

Horlacher, Rebekka. *The Educated Subject and the German Concept of Bildung: A Comparative Cultural History*. New York: Routledge, 2016.

Horner, Johann. *Die internationale Wallfahrt zum Mozart-Häuschen am Kapuzinerberge in Salzburg: Zur 124. Geburts-Feier W. A. Mozart's, 1880*. Salzburg: Internationale Mozart-Stiftung, 1880.

Huelsenbeck, Richard. *Dada Almanach*. Berlin: Erich Reiss, 1920.

Hummel, Walter. "Das Mozart-Album der Internationalen Mozartstiftung." In *Festschrift Otto Erich Deutsch zum 80. Geburtstag am 5. September, 1963*, edited by Walter Gerstenberg, Jan LaRue, and Wolfgang Rehm, 110–19. Kassel: Bärenreiter, 1963.

Huneker, James. *Chopin: The Man and His Music*. New York: Charles Scribner's Sons, 1900.

Hust, Christoph. *August Bungert: Ein Komponist im deutschen Kaiserreich.* Tutzing: H. Schneider, 2005.
Hutcheon, Linda, and Michael Hutcheon. *Four Last Songs: Aging and Creativity in Verdi, Strauss, Messiaen, and Britten.* Chicago: University of Chicago Press, 2015.
Hutschenruyter, Willem. *Das Beethovenhaus.* Stuttgart: Greiner und Pfeiffer, 1911.
Huyssen, Andreas. "Mass Culture as Woman: Modernism's Other." In *After the Great Divide: Modernism, Mass Culture, Postmodernism,* 44–64. Bloomington: Indiana University Press, 1986.
Irving, Washington. 1820. *The Sketch Book of Geoffrey Crayon, Esq.* Paris: Baudry's European Library, 1836.
Janetschek, Ottokar. *Mozart: Ein Künstlerleben.* Berlin: Richard Bong, 1924.
Jefferies, Matthew. *Imperial Culture in Germany, 1871–1918.* Houndmills: Palgrave Macmillan, 2003.
Jonsson, Stefan. *Crowds and Democracy: The Idea and Image of the Masses from Revolution to Fascism.* New York: Columbia University Press, 2013.
Jordan, Max. *Geselschap.* Bielefeld: Velhagen and Klasing, 1906.
Kagel, Mauricio. *Ludwig van.* 1969; Munich: Winter and Winter, 2007. DVD.
———. *Tamtam: Dialoge und Monologe zur Musik.* Zurich: Piper, 1975.
Kahler, Otto-Hans, ed. *Theodor Billroth als Musikkritiker.* Rockville, MD: Kabel Verlag, 1988.
Kallberg, Jeffrey. *Chopin at the Boundaries: Sex, History, and Musical Genre.* Cambridge, MA: Harvard University Press, 1996.
———. "Chopin's Last Style." *Journal of the American Musicological Society* 38, no. 2 (1985): 264–315.
Kam, Lap-Kwan. "Between Musicology and Mythology at the *Stunde Null*: Austria's 950th 'Birthday' and the 50th Anniversary of Bruckner's Death." In *Dreams of Germany: Musical Imaginaries from the Concert Hall to the Dance Floor,* edited by Neil Gregor and Thomas Irvine, 221–46. New York: Berghahn, 2020.
Kammel, Frank Matthias. "Der Gipsabguss: Vom Medium der ästhetischen Norm zur toten Konserve der Kunstgeschichte." In *Ästhetische Probleme der Plastik im 19. und 20. Jahrhundert,* edited by Andrea Kluxen, 47–72. Nuremberg: Aleph, 2001.
Kämpken, Nicole, and Michael Ladenburger. *Bewegte und bewegende Geschichte: 125 Jahre Beethoven-Haus; Sonderausstellung zum Jubiläum Beethoven-Haus Bonn; 24. Februar bis 17. August 2014.* Bonn: Beethoven-Haus, 2014.
Kane, Brian. *Sound Unseen: Acousmatic Sound in Theory and Practice.* New York: Oxford University Press, 2014.
Karhausen, Lucien R. "The Mozarteum's Skull: A Historical Saga." *Journal of Medical Biography* 9, no. 2 (2001): 109–17.
Karnes, Kevin. *A Kingdom Not of This World: Wagner, the Arts, and Utopian Visions in* Fin-de-Siècle *Vienna.* New York: Oxford University Press, 2013.
———. *Music, Criticism, and the Challenge of History: Shaping Modern Musical Thought in Late Nineteenth-Century Vienna.* Oxford: Oxford University Press, 2016.
Katzenstein, Peter J. *Disjoined Partners: Austria and Germany since 1815.* Berkeley: University of California Press, 1976.
Keefe, Simon P. *Mozart's Requiem: Reception, Work, Completion.* Cambridge: Cambridge University Press, 2012.
Kennaway, James. *Bad Vibrations: The History of the Idea of Music as a Cause of Disease.* London: Routledge, 2016.

Kerber, Erwin, ed. *Ewiges Theater: Salzburg und seine Festspiele*. Munich: R. Piper. Verlag, 1935.
Kerner, Dieter. *Krankheiten großer Musiker*. 4th ed. Stuttgart: Schattauer, 1986.
Kerst, Friedrich. *Die Erinnerungen an Beethoven*. Vol. 2. Stuttgart: J. Hoffmann, 1913.
Kessler, Herbert L. "Configuring the Invisible by Copying the Holy Face." In *The Holy Face and the Paradox of Representation: Papers from a Colloquium Held at the Bibliotheca Hertziana, Rome, and the Villa Spelman, Florence, 1996*, edited by Herbert L. Kessler and Gerhard Wolf, 6:125–51. Villa Spelman Colloquia. Bologna: Nuova Alfa Editoriale, 1998.
Kimber, Marian Wilson. "Never Perfectly Beautiful: Physiognomy, Jewishness, and Mendelssohn Portraiture." In *Mendelssohn Perspectives*, edited by Nicole Grimes and Angela R. Mace, 9–30. Burlington, VT: Ashgate, 2012.
Klages, Ludwig. *Die Grundlagen der Charakterkunde*. 5th ed. Leipzig: J. A. Barth, 1928.
Klebes, Martin. "Glauben und Wissen (um das Wissen um das Wissen): Krisenbekämpfung bei Broch." In *Hermann Broch und die Ökonomie*, edited by Jürgen Heizmann, Bernhard Fetz, and Paul Michael Lützeler, 151–68. Wuppertal: Arco, 2018.
Knittel, Kristin M. [K. M.] "'Late,' Last, and Least: On Being Beethoven's Quartet in F Major, op. 135." *Music and Letters* 87, no. 1 (2006): 16–51.
———. "Pilgrimages to Beethoven: Reminiscences by His Contemporaries." *Music and Letters* 84, no. 1 (2003): 19–54.
———. *Seeing Mahler: Music and the Language of Antisemitism in* Fin-de-Siècle *Vienna*. Burlington, VT: Ashgate, 2010.
———. "Wagner, Deafness, and the Reception of Beethoven's Late Style." *Journal of the American Musicological Society* 51, no. 1 (1998): 49–82.
Knickenberg, Fritz, and Ferdinand August Schmidt, eds. *Verein Beethoven-Haus in Bonn: Bericht über die ersten fünfzehn Jahre seines Bestehens (1889–1904)*. Bonn: Beethoven-Haus Verein, 1904.
Kobbé, Gustav. *The Loves of Great Composers*. New York: Thomas Y. Crowell, 1905.
Köberle, Adolf. *Bach, Beethoven, Bruckner als Symbolgestalten des Glaubens: Eine frömmigkeitsgeschichtliche Deutung*. Berlin: Im Furche-Verlag, 1936.
Koch, Ludwig. "Chronik des Denkmals." In *Zur Enthüllung des Mozart-Denkmals in Wien am 21. April 1896*, 16–25. Vienna: Reisser und Werthner; Mozart-Denkmal-Comité, 1896.
Köhne, Julia Barbara. *Geniekult in Geisteswissenschaften und Literaturen um 1900 und seine filmischen Adaptionen*. Vienna: Böhlau Verlag, 2014.
Kohrs, Klaus Heinrich. *Anton Bruckner: Angst vor der Unermeßlichkeit*. Frankfurt: Stroemfeld; Roter Stern, 2017.
Konrad, Ulrich, ed. *Wolfgang Amadeus Mozart (1756–1791): Neue Ausgabe sämtlicher Werke*. Ser. 10 (supp.), work group 30, vol. 4, *Fragmente*. Kassel: Bärenreiter, 2002.
Kopitz, Klaus Martin, and Rainer Cadenbach. *Beethoven aus der Sicht seiner Zeitgenossen in Tagebüchern, Briefen, Gedichten und Erinnerungen*. 2 vols. Munich: G. Henle Verlag, 2009.
Korsyn, Kevin. "J. W. N. Sullivan and the *Heiliger Dankgesang*: Questions of Meaning in Late Beethoven." In *Beethoven Forum 2*, edited by Christopher Reynolds, 133–74. Lincoln: University of Nebraska Press, 1993.
Koshar, Rudy. *German Travel Cultures*. Oxford: Berg, 2000.
Kramer, Elizabeth. "The Idea of *Kunstreligion* in German Musical Aesthetics of the Early Nineteenth Century." PhD diss., University of North Carolina, Chapel Hill, 2005.
———. "The Idea of Transfiguration in the Early German Reception of Mozart's Requiem." *Current Musicology* 81 (2006): 73–107.

Kramer, Richard. *Unfinished Music*. Oxford: Oxford University Press, 2008.
Kreczi, Hanns. *Das Bruckner-Stift St. Florian und das Linzer Reichs-Bruckner-Orchester (1942–1945)*. Graz: Akademische Druck- und Verlagsanstalt, 1986.
Kregor, Jonathan. *Liszt as Transcriber*. New York: Cambridge University Press, 2010.
Kreissle von Hellborn, Heinrich. *Franz Schubert*. Vienna: Carl Gerold's Sohn, 1865.
Kretschmer, Ernst. *Geniale Menschen*. 2nd ed. Berlin: Springer, 1931.
Kretzschmar, Hermann. "Beethoven als Märtyrer: Zum 17. Dezember (1901)." In *Gesammelte Aufsätze über Musik und anderes aus den Grenzboten*, edited by Alfred Heuss, 421–32. Leipzig: F. W. Grunow, 1910.
Kreuzer, Gundula. *Curtain, Gong, Steam: Wagnerian Technologies of Nineteenth-Century Opera*. Berkeley: University of California Press, 2018.
Kris, Ernst, and Otto Kurz. *Legend, Myth, and Magic in the Image of the Artist: A Historical Experiment*. New Haven, CT: Yale University Press, 1979.
Kristeva, Julia. *The Severed Head: Capital Visions*. New York: Columbia University Press, 2012.
Kundera, Milan. *The Unbearable Lightness of Being*. Translated by Michael Henry Heim. New York: Harper, 1999.
Kurlander, Eric. *Hitler's Monsters: A Supernatural History of the Third Reich*. New Haven, CT: Yale University Press, 2017.
Kuryluk, Ewa. *Veronica and Her Cloth: History, Symbolism, and Structure of a "True" Image*. Cambridge: Basil Blackwell, 1991.
Kutschke, Beate. "The Celebration of Beethoven's Bicentennial in 1970: The Antiauthoritarian Movement and Its Impact on Radical Avant-Garde and Postmodern Music in West Germany." *Musical Quarterly* 93, no. 3/4 (2010): 560–615.
Landau, Hermann Joseph, ed. *Erstes poetisches Beethoven-Album: Zur Erinnerung an den grossen Tondichter und an den Säcularfeier begangen den 17. December 1870*. Prague: self-publication, 1872.
Landon, H. C. Robbins, ed. *Beethoven: A Documentary Study*, translated by Richard Wadleigh and Eugene Hartzell. New York: Macmillan, 1970.
Lange, Gustav. *Musikgeschichtliches*. Berlin: R. Gaertners, 1900.
Lange-Eichbaum, Wilhelm. *Genie, Irrsinn und Ruhm: Eine Pathographie des Genies*. 1928. Munich: Ernst Reinhardt Verlag, 1956.
———. *The Problem of Genius*. Translated by Eden Paul and Cedar Paul. New York: Macmillan, 1932.
LaPorte, Charles. *The Victorian Cult of Shakespeare: Bardology in the Nineteenth Century*. Cambridge: Cambridge University Press, 2021.
Laqueur, Thomas W. *The Work of the Dead: A Cultural History of Mortal Remains*. Princeton, NJ: Princeton University Press, 2015.
Latour, Bruno. *Reassembling the Social: An Introduction to Actor-Network Theory*. Oxford: Oxford University Press, 2005.
Lavater, Johann Caspar. *Essays on Physiognomy, Designed to Promote the Knowledge and the Love of Mankind*. Translated by Thomas Holloway. 3 vols. London: John Murray, 1789.
Leistra-Jones, Karen. "Hans von Bülow and the Confessionalization of *Kunstreligion*." *Journal of Musicology* 35, no. 1 (2018): 42–75.
———. "Improvisational Idyll: Joachim's 'Presence' and Brahms's Violin Concerto, Op. 77." *Nineteenth-Century Music* 38, no. 3 (2015): 243–71.

———. "(Re-)Enchanting Performance: Joachim and the Spirit of Beethoven." In *The Creative Worlds of Joseph Joachim*, edited by Valerie Woodring Goertzen and Robert Whitehouse Eshbach, 86–103. Woodbridge: Boydell and Brewer, 2021.

———. "Staging Authenticity: Joachim, Brahms, and the Politics of *Werktreue* Performance." *Journal of the American Musicological Society* 66, no. 2 (2013): 397–436.

Leitzmann, Albert, ed. *Ludwig van Beethoven: Berichte der Zeitgenossen, Briefe und persönliche Aufzeichnungen*. 2 vols. Leipzig: Insel, 1921.

Lekan, Thomas M. *Imagining the Nation in Nature: Landscape Preservation and German Identity, 1885–1945*. Cambridge, MA: Harvard University Press, 2004.

———. "A 'Noble Prospect': Tourism, *Heimat*, and Conservation on the Rhine, 1880–1914." *Journal of Modern History* 81, no. 4 (2009): 824–58.

Leppert, Richard. "The Musician of the Imagination." In *The Musician as Entrepreneur, 1700–1914: Managers, Charlatans, and Idealists*, edited by William Weber, 25–58. Bloomington: Indiana University Press, 2004.

Leslie, Esther. "Souvenirs and Forgetting: Walter Benjamin's Memory Work." In *Material Memories: Design and Evocation*, edited by Marius Kwint, Christopher Breward, and Jeremy Aynsley, 107–22. Oxford: Berg, 1999.

Lewis, Wyndham. *Tarr*. London: Chatto and Windus, 1928.

Lindner, Andreas. "Die Posthume Reisen des Anton B." In *Bruckner-Tagung: Anton Bruckner auf Reisen; Ebrach, 29. und 30. Juli 2012; Bericht*, edited by Theophil Antinocek, Andreas Lindner, and Klaus Petermayr, 127–47. Vienna: Musikwissenschaftlicher Verlag, 2013.

Liszt, Franz. *Life of Chopin*. Translated by Martha Walter Cook. 4th ed. Boston: Ditson, 1863.

Lockwood, Lewis. *Beethoven's Lives: The Biographical Tradition*. Woodbridge: Boydell, 2020.

Lombroso, Cesare. *Genie und Irrsinn in ihren Beziehungen zum Gesetz, zur Kritik und zur Geschichte*. Leipzig: Philipp Reclam, 1887.

———. *The Man of Genius*. London: Walter Scott, 1891.

Loos, Adolf. "Ornament and Crime." In *Ornament and Crime: Selected Essays*, translated by Michael Mitchell, 167–76. Riverside, CA: Ariadne, 1998.

Loos, Helmut. "Die Texturierung Beethovenscher Instrumentalwerke: Ein Kapitel der Beethoven-Deutung." In *Beethoven und die Nachwelt: Materialien zur Wirkungsgeschichte Beethovens*, 117–37. Bonn: Beethoven-Haus, 1986.

Ludendorff, Mathilde. *Der ungesühnte Frevel an Luther, Lessing, Mozart, Schiller im Dienste des allmächtigen Baumeisters aller Welten*. Munich: Ludendorffs Verlag, 1929.

———. *Mozarts Leben und gewaltsamer Tod*. Munich: Ludendorffs Verlag, 1936.

Lutz, Deborah. "The Dead Still among Us: Victorian Secular Relics, Hair Jewelry, and Death Culture." *Victorian Literature and Culture* 39, no. 1 (2011): 127–42.

———. *Relics of Death in Victorian Literature and Culture*. Cambridge: Cambridge University Press, 2015.

Lux, Josef August. *Franz Schuberts Lebenslied: Ein Roman der Freundschaft*. 1915. Vienna: Österreichische Buchgemeinschaft, 1953.

Lynch, Deidre Shauna. *Loving Literature: A Cultural History*. Chicago: University of Chicago Press, 2015.

Maack, Ferdinand. *Wie Steht's mit dem Spiritismus*. Hamburg: Xenologischer Verlag, 1901.

Macdonald, Dwight. *Masscult and Midcult: Essays against the American Grain*. New York: New York Review of Books, 2011. First published as *Essays against the American Grain*, 1962.

———. "A Theory of Mass Culture." *Diogenes* 1, no. 3 (1953), 1–17.

MacLeod, Catriona. "Sweetmeats for the Eye: Porcelain Miniatures in Classical Weimar." In *The Enlightened Eye: Goethe and Visual Culture*, edited by Evelyn K. Moore and Patricia Anne Simpson, 41–72. Amsterdam: Brill, 2007.

Magendie, François. *Précis élémentaire de physiologie*. 1924. 2nd ed. Paris: Méquignon-Marvis, 1825.

Mai, François Martin. *Diagnosing Genius: The Life and Death of Beethoven*. Montreal: McGill-Queen's University Press, 2007.

Maier, Elisabeth. *Années de Pèlerinage: Neue Dokumente zu August Göllerichs Studienzeit bei Franz Liszt und Anton Bruckner*. Edited by Renate Grasberger. 2 vols. Wiener Bruckner-Studien 4. Vienna: Musikwissenschaftlicher Verlag, 2013.

Maier, Elisabeth, and Renate Grasberger, eds. *Die Bruckner-Bestände des Stiftes St. Florian: Katalog*. 3 vols. Wiener Bruckner-Studien 6. Vienna: Musikwissenschaftlicher Verlag, 2014.

Maier-Graefe, Julius. *Der Fall Böcklin und die Lehre von Einheiten*. Stuttgart: Julius Hoffmann, 1905.

Margry, Peter Jan, ed. *Shrines and Pilgrimage in the Modern World: New Itineraries into the Sacred*. Amsterdam: Amsterdam University Press, 2004.

Martin, Russell. *Beethoven's Hair: An Extraordinary Historical Odyssey and a Scientific Mystery Solved*. New York: Random House / Broadway Books, 2000.

Mathew, Nicholas. *The Haydn Economy: Music, Aesthetics, and Commerce in the Late Eighteenth Century*. Chicago: University of Chicago Press, 2022.

———. *Political Beethoven*. Cambridge: Cambridge University Press, 2013.

———. "Review of *The Beethoven Syndrome: Hearing Music as Autobiography*, by Mark Evan Bonds." *Journal of the American Musicological Society* 75, no. 3 (2022): 614–18.

Mathew, Nicholas, and Mary Ann Smart, eds. "Quirk Historicism." Special forum, *Representations* 132, no. 1 (2015).

McColl, Sandra. *Music Criticism in Vienna, 1896–1897: Critically Moving Forms*. Oxford: Clarendon, 1996.

McCormick, Lisa. "The Agency of Dead Musicians." *Contemporary Social Science* 10, no. 3 (2015): 323–35.

McMahon, Darrin M. *Divine Fury: A History of Genius*. New York: Basic Books, 2013.

McManus, Laurie. *Brahms in the Priesthood of Art: Gender and Art-Religion in the Nineteenth-Century German Musical Imagination*. Oxford: Oxford University Press, 2021.

McMullan, Gordon. *Shakespeare and the Idea of Late Writing: Authorship in the Proximity of Death*. Cambridge: Cambridge University Press, 2007.

McMullan, Gordon, and Sam Smiles, eds. *Late Style and Its Discontents: Essays in Art, Literature, and Music*. Oxford: Oxford University Press, 2016.

McVeigh, Simon. "The Concert Series." In *Oxford Handbook of Music and Intellectual Culture in the Nineteenth Century*, edited by Paul Watt, 293–316. Oxford: Oxford University Press, 2020.

Meier, Albert, Alessandro Costazza, and Gérard Laudin, eds. *Kunstreligion: Ein ästhetisches Konzept der Moderne in seiner historischen Entfaltung*. 3 vols. Berlin: De Gruyter, 2011.

Meredith, William. "The History of Beethoven's Skull Fragments: Part One." *Beethoven Journal* 20, no. 1/2 (2005): 3–46.

———. "The History of Beethoven's Skull Fragments: Part Two." *Beethoven Journal* 30, no. 1 (2015): 25–29.

———, ed. Special issue, *Beethoven Journal* 20, no. 1/2 (Summer 2005).

Messing, Scott. "The Vienna Beethoven Centennial Festival of 1870." *Beethoven Newsletter* 6, no. 3 (1991): 57–63.

Mielichhofer, Ludwig. *Das Mozart-Denkmal zu Salzburg und dessen Enthüllungs-Feier im September 1842*. Salzburg: Mayr'sche Buchhandlung, 1843.

Minor, Ryan. *Choral Fantasies: Music, Festivity, and Nationhood in Nineteenth-Century Germany*. Cambridge: Cambridge University Press, 2012.

———. "Prophet and Populace in Liszt's 'Beethoven' Cantatas." In *Franz Liszt and His World*, edited by Christopher H. Gibbs and Dana Gooley, 113–65. Princeton, NJ: Princeton University Press, 2006.

Mole, Tom. *Byron's Romantic Celebrity: Industrial Culture and the Hermeneutic of Intimacy*. New York: Palgrave Macmillan, 2007.

Morales, Helen. *Pilgrimage to Dollywood: A Country Music Road Trip through Tennessee*. Chicago: University of Chicago Press, 2014.

Morgenstern, Anja. "Constanze Nissen in Salzburg 1824–1842: Neue Aspekte zur Entstehung des Mozartkults." In *Salzburgs Musikgeschichte im Zeichen des Provinzialismus? Die ersten Jahrzehnte des 19. Jahrhunderts*, edited by Dominik Šedivý, 304–45. Vienna: Hollitzer Wissenschaftsverlag, 2014.

Morris, Mitchell. *The Persistence of Sentiment: Display and Feeling in Popular Music of the 1970s*. Berkeley: University of California Press, 2013.

Morris, Rosalind C., ed. *The Returns of Fetishism: Charles de Brosses and the Afterlives of an Idea*. Text of de Brosses introduced, edited, and translated by Daniel H. Leonard. Chicago: University of Chicago Press, 2017.

Morton, Marsha. *Max Klinger and Wilhelmine Culture: On the Threshold of German Modernism*. London: Routledge, 2014.

Mugglestone, Erica. "Guido Adler's 'The Scope, Method and Aim of Musicology' (1885): An English Translation with an Historico-analytical Commentary." *Yearbook for Traditional Music* 13 (1981): 1–21.

Mundy, Rachel. "Evolutionary Categories and Musical Style from Adler to America." *Journal of the American Musicological Society* 67, no. 3 (2014): 735–68.

Must, Gustav. "The Origin of the German Word *Ehre*, 'Honor.'" *PMLA* 76, no. 4 (1961): 326–29.

Nettl, Bruno. *Heartland Excursions: Ethnomusicological Reflections on Schools of Music*. Urbana: University of Illinois Press, 1995.

Ngai, Sianne. *Our Aesthetic Categories: Zany, Cute, Interesting*. Cambridge, MA: Harvard University Press, 2012.

Nicolai, Ernst Anton. *Die Verbindung der Musik mit der Artzneybelahrheit*. Halle: Carl Hermann Hemmerde, 1745.

Niemack, Dr. med. J. "Herzschlag und Rhythmus: Ein Versuch, dem Verständnis von Beethovens Werken durch das Studium seiner Ohren- und Herzkrankheit näher zu kommen." *Die Musik* 7, no. 27 (1907/1908): 19–25.

Noble, Jonathan. *That Jealous Demon, My Wretched Health: Disease, Death, and Composers*. Woodbridge: Boydell, 2018.

Nora, Pierre. "Between Memory and History: *Les lieux de mémoire*." 1984. Translated by Marc Roudebush. *Representations* 26, no. 1 (1989): 7–24.

———. *Realms of Memory: Rethinking the French Past*. 1984. 3 vols. Edited by Lawrence D. Kritzman. Translated by Arthur Goldhammer. New York: Columbia University Press, 1996–98.

Norden, Martin F. *The Cinema of Isolation: A History of Physical Disability in the Movies*. New Brunswick, NJ: Rutgers University Press, 1994.

North, Julian. *The Domestication of Genius: Biography and the Romantic Poet*. Oxford: Oxford University Press, 2009.
Notley, Margaret. *Lateness and Brahms: Music and Culture in the Twilight of Viennese Liberalism*. Oxford: Oxford University Press, 2006.
Nohl, Ludwig. *Beethoven, Liszt, Wagner: Ein Bild der Kunstbewegung unseres Jahrhunderts*. Vienna: Wilhelm Braumüller, 1874.
———. *Beethovens Brevier*. Edited by Paul Sakolowski. 2nd ed. Leipzig: Hermann Seemann Nachfolger, 1901.
———. *Beethovens Leben*. 4 vols. Vienna: Hermann Markgraf, 1864.
———. *Der Geist der Tonkunst*. Frankfurt: J. D. Sauerländer's Verlag, 1861.
———. "Die Entstehung der Zauberflöte und ihre innere Verbindung mit dem Ring des Nibelungen." In *Mosaik: Für Musikalisch-Gebildete*, 200–221. Leipzig: Gebrüder Senf, 1882.
———. *Eine Stille Liebe zu Beethoven: Nach dem Tagebuche einer Jungen Dame*. Leipzig: Ernst Julius Günter, 1875.
Nordau, Max. *Entartung*. 2 vols. Berlin: Carl Duncker, 1892.
Novello, Vincent. *A Mozart Pilgrimage: Being the Travel Diaries of Vincent and Mary Novello in the Year 1829*. Edited by Rosemary Hughes. London: Novello, 1955.
Nowak, Leopold, ed. *Anton Bruckner: Sämtliche Werke; Kritische Gesamtausgabe*. Vienna: Musikwissenschaftlicher Verlag, 1951.
Nußbaumer, Martina. *Musikstadt Wien: Die Konstruktion eines Images*. Freiburg: Rombach Verlag, 2007.
Oehler-Klein, Sigrid. *Die Schädellehre Franz Joseph Galls in Literatur und Kritik des 19. Jahrhunderts: Zur Rezeptionsgeschichte einer medizinisch-biologisch begründeten Theorie der Physiognomik und Psychologie*. Stuttgart: Gustav Fischer Verlag, 1990.
Olalquiaga, Celeste. *The Artificial Kingdom: On the Kitsch Experience*. Minneapolis: University of Minnesota Press, 1998.
Ostrower, Francie. *Why the Wealthy Give: The Culture of Elite Philanthropy*. Princeton, NJ: Princeton University Press, 1995.
Oulibischeff, Alexander. *Mozarts Leben und Werke*. Translated by Albert Schraishuon. Stuttgart: Ad. Becher's Verlag, 1847.
Overy, Paul. "'The Whole Bad Taste of Our Period': Josef Frank, Adolf Loos, and 'Gschnas.'" In "Redefining Kitsch: The Politics of Design," ed. Judy Attfield, special issue, *Home Cultures* 3, no. 3 (2006): 213–33.
Pachler, Faust. *Beethoven und Marie Pachler-Koschak*. Berlin: B. Behr's Buchhandlung, 1866.
Painter, Karen. "Mozart at Work: Biography and a Musical Aesthetic for the Emerging German Bourgeoisie." *Musical Quarterly* 86, no. 1 (2002): 186–235.
———. *Symphonic Aspirations: German Music and Politics, 1900–1945*. Cambridge, MA: Harvard University Press, 2007.
———. "W. A. Mozart's Beethovenian Afterlife: Biography and Musical Interpretation in the Twilight of Idealism." In *Late Thoughts: Reflections on Artists and Composers at Work*, edited by Karen Painter and Thomas Crow, 117–43. Los Angeles: Getty Institute, 2006.
Parsons, Nicholas. *Vienna: A Cultural History*. Oxford: Oxford University Press, 2009.
Pascoe, Judith. *The Hummingbird Cabinet: A Rare and Curious History of Romantic Collectors*. Ithaca, NY: Cornell University Press, 2006.
Paudler, Fritz. "Die Rasse Beethovens." *Der Auftakt, Prague* 7, no. 3 (1927): 57–61.

Pazaurek, Gustav. *Guter und schlechter Geschmack im Kunstgewerbe*. Stuttgart: Deutsche Verlags-Anstalt, 1912.

Perciaccante, Antonio, Philippe Charlier, Camille Negri, Alessia Coralli, Otto Appenzeller, and Raffaella Bianucci. "Did Frédéric Chopin Die from Heart Failure?" *Journal of Cardiac Failure* 24, no. 5 (2018): 342–44.

Perec, Georges. *Species of Spaces, and Other Pieces*. Edited and translated by John Sturrock. Harmondsworth: Penguin, 1997.

Peritz, Jessica Gabriel. "The Castrato Remains—or, Galvanizing the Corpse of Musical Style." *Journal of Musicology* 39, no. 3 (2022): 371–403.

Person, Jutta. *Der pathographische Blick: Physiognomik, Atavismustheorien, und Kulturkritik, 1870–1930*. Würzburg: Königshausen and Neumann, 2005.

Pesce, Dolores. *Liszt's Final Decade*. Rochester, NY: University of Rochester Press, 2014.

Phillips, Reuben. "Exhumations, Honorary Graves, and the Fashioning of Vienna's Self-Image as the 'City of Music.'" *Musical Quarterly* 102, nos. 2–3 (2019): 303–49.

Pietschmann, Klaus, and Melanie Wald-Fuhrmann, eds. *Der Kanon der Musik: Theorie und Geschichte*. Munich: Richard Boorberg Verlag, 2013.

Plath, Wolfgang. "Gefälschte Mozart-Autographen: Der Fall Nicotra." *Acta Mozartiana* 26 (1979): 2–10.

Pointon, Marcia. "Casts, Imprints, and the Deathliness of Things: Artifacts at the Edge." *Art Bulletin* 96, no. 2 (2014): 170–95.

Polanyi, Karl. *The Great Transformation*. New York: Farrar and Rinehart, 1944.

Pollock, Emily Richmond. *Opera after the Zero Hour: The Problem of Tradition and the Possibility of Renewal in Postwar West Germany*. New York: Oxford University Press, 2019.

Popper, Karl. *Conjectures and Refutations: The Growth of Scientific Knowledge*. London: Routledge and Kegan Paul, 1963.

Poskett, James. *Materials of the Mind: Phrenology, Race, and the Global History of Science, 1815–1920*. Chicago: University of Chicago Press, 2019.

Pötschner, Peter. *Das Schwarzspanierhaus: Beethovens letzte Wohnstätte*. Vienna: Paul Zsolnay Verlag, 1970.

Potter, Pamela M. *Art of Suppression: Confronting the Nazi Past in Histories of the Visual and Performing Arts*. Oakland: University of California Press, 2016.

Preis-Verzeichnis plastischer Bildwerke: Gebrüder Micheli Bildhauer-Werkstätten, gegründet 1824. Berlin, ca. 1900.

Rabinbach, Anson. "Unclaimed Heritage: Ernst Bloch's *Heritage of Our Times* and the Theory of Fascism." *New German Critique* 11 (1977): 5–21.

Raffa, Guy P. *Dante's Bones: How a Poet Invented Italy*. Cambridge, MA: Belknap Press of Harvard University Press, 2020.

Rank, Otto. *Psychology and the Soul: A Study of the Origin, Conceptual Evolution, and Nature of the Soul*. Baltimore: Johns Hopkins University Press, 1998.

Rau, Heribert. *Beethoven: Historischer Roman*. 4 vols. Frankfurt: Verlag Meidinger Sohn, 1859.

Rauschenberger, Walther. *Erb- und Rassenpsychologie schöpferischer Persönlichkeiten*. Jena: Gustav Fischer, 1942.

———. "Rassenmerkmale Beethovens und seiner nächsten Verwandten." *Volk und Rasse* 7 (1934): 194–203.

———. "Über die Rassichen Grundlagen der deutschen Tonkunst." *Zeitschrift für Rassenkunde* 12 (1941): 1–9.

Reece, Frederick. *Forgery in Musical Composition: Aesthetics, History, and the Canon.* Oxford University Press, 2025.

Rehding, Alexander. *Music and Monumentality: Commemoration and Wonderment in Nineteenth-Century Germany.* Oxford: Oxford University Press, 2009.

Reiter, Christian. "The Causes of Beethoven's Death and His Locks of Hair: A Forensic-Toxicological Investigation." *Beethoven Journal* 22, no. 1 (2007): 2–5.

———. "On the Authenticity of Beethoven's Skull Fragments from the Estate of Prof. Dr. Romeo Seligmann." *Wiener medizinische Wochenschrift*, November 2022. http://doi.org/10.1007/s10354-022-00985-4.

Rév, István. "Parallel Autopsies." *Representations* 49, no. 1 (1995): 15–39.

Richards, Annette. *The Temple of Fame and Friendship: Portraits, Music, and History in the C. P. E. Bach Circle.* Chicago: University of Chicago Press, 2022.

Ricœur, Paul. *Memory, History, Forgetting.* Chicago: University of Chicago Press, 2004.

Riegl, Alois. "The Modern Cult of Monuments: Its Character and Its Origin." Translated by Kurt W. Forster and Diane Ghirardo. *Oppositions* 25 (1982): 21–51.

Rifai, Nader, William Meredith, Kevin Brown, Sarah A. Ehrdahl, and Paul J. Jannetto. "Letter to the Editor: High Lead Levels in 2 Independent and Authenticated Locks of Beethoven's Hair." *Clinical Chemistry*, May 6, 2024. https://doi.org/10.1093/clinchem/hvae054.

Riggs, Christina. *Photographing Tutankhamun: Archaeology, Ancient Egypt, and the Archive.* London: Bloomsbury Visual Arts, 2019.

Rilke, Rainer Maria. *Die Aufzeichnungen des Malte Laurids Brigge.* Frankfurt: Suhrkamp Verlag, 1948.

Ringer, Fritz. *The Decline of the German Mandarins: The German Academic Community, 1890–1933.* 2nd ed. Hanover, NH: University Press of New England, 1990.

Roach, Joseph. *It.* Ann Arbor: University of Michigan Press, 2007.

Rojek, Chris. *Celebrity.* London: Reaktion Books, 2001.

———. *Presumed Intimacy: Para-social Relationships in Media, Society, and Celebrity Culture.* Cambridge: Polity, 2016.

Rost, Henrike. *Musik-Stammbücher: Erinnerung, Unterhaltung und Kommunikation im Europa des 19. Jahrhunderts.* Vienna: Böhlau Verlag, 2020.

Sacks, Oliver. *Musicophilia: Tales of Music and the Brain.* New York: Random House, 2008.

Sadie, Julie Anne, and Stanley Sadie. *Calling on the Composer: A Guide to European Composer Houses and Museums.* New Haven, CT: Yale University Press, 2005.

Saffle, Michael, and Jeffrey R. Saffle. "Medical Histories of Prominent Composers: Recent Research and Discoveries." *Acta Musicologica* 65, no. 2 (1993): 77–101.

Said, Edward. *On Late Style: Music and Literature against the Grain.* New York: Vintage Books, 2006.

Saliot, Anne-Gaëlle. *The Drowned Muse: Casting the Unknown Woman of the Seine across the Tides of Modernity.* Oxford: Oxford University Press, 2015.

Salten, Felix. *Das österreichische Antlitz.* Berlin: S. Fischer, 1910.

Sayler, Oliver. *Max Reinhardt and His Salzburg.* New York: Brentano's, 1923.

Schaaffhausen, Hermann. "Einige Reliquien berühmter Männer." *Correspondenz-Blatt der deutschen Gesellschaft für Anthropologie* 16 (1885): 147–50.

Schaffer, Simon. "How Disciplines Look." In *Interdisciplinarity: Reconfigurations of the Social and Natural Sciences*, edited by Andrew Barry and Georgina Born, 57–81. New York: Routledge, 2013.

Schindler, Anton Felix. *Beethoven as I Knew Him*. Edited by Donald W. MacArdle. Translated by Constance S. Jolly. Mineola: Dover, 1996.

Schnapp, Friedrich, ed. and trans. *Liszts Testament*. Weimar: Hermann Böhlaus Nachfolger, 1931.

Schneller, Julius Franz Borgias. *Julius Schneller's Lebensumriss und vertraute Briefe an seine Gattin und seine Freunde*. Edited by Ernst Münch. Leipzig: J. Scheible, 1834.

Schorske, Carl E. *Fin-de-Siècle Vienna: Politics and Culture*. 1961. New York: Knopf Doubleday, 1980.

Schulz, L. F. *Edle Charakterzüge, schöne und große Handlungen, wichtige Anekdoten, Scenen, witzige Einfälle und letzte Worte berühmter Menschen der ältern und neuern Zeit*. Vienna: Doll, 1804.

Schurich, Hans. *Das Zauberflötenhäuschen*. Salzburg: Internationale Stiftung Mozarteum, 1950.

Schweisheimer, Waldemar. *Beethovens Leiden: Ihr Einfluss auf sein Leben und Schaffen*. Munich: Georg Müller Verlag, 1922.

Scobey, Katherine Lois, and Olive Brown Horne. *Stories of Great Musicians*. New York: American Book Company, 1905.

Segel, Harold B., ed. and trans. *The Vienna Coffeehouse Wits, 1890–1938*. West Lafayette, IN: Purdue University Press, 1993.

Semmel, Stuart. "Reading the Tangible Past: British Tourism, Collecting, and Memory after Waterloo." *Representations* 69 (2000): 9–37.

Senfelder, Leopold. *Die Katakomben bei St. Stefan in Sage und Geschichte*. Vienna: Hölder-Pichler-Tempsky, 1924.

Senior, Kathryn. "Did Beethoven Die of Lead Poisoning?" *Lancet* 356, no. 9240 (October 28, 2000): 1498.

Shadle, Douglas W. *Orchestrating the Nation: The Nineteenth-Century American Symphonic Enterprise*. Oxford: Oxford University Press, 2016.

Silverman, Lisa. *Becoming Austrians: Jews and Culture between the World Wars*. Oxford: Oxford University Press, 2012.

Simmons, Sherwin. "Chaplin Smiles on the Wall: Berlin Dada and Wish-Images of Popular Culture." *New German Critique* 84 (2001): 3–34.

Slonimsky, Nicolas. *Slonimsky's Book of Musical Anecdotes*. 1948. London: Routledge, 2013.

Solie, Ruth. *Music in Other Words: Victorian Conversations*. Berkeley: University of California Press, 2004.

Sontag, Susan. "Notes on Camp." 1964. In *Against Interpretation, and Other Essays*, 275–92. London: Eyre and Spottiswood, 1967.

Spitzer, Daniel. *Verliebte Wagnerianer*. 2nd ed. Vienna: Julius Kilkhardt, 1880.

Spitzer, Michael. *Music as Philosophy: Adorno and Beethoven's Late Style*. Bloomington: Indiana University Press, 2006.

Springer, Robert. *Die klassische Stätten von Jena und Ilmenau: Ein Beitrag zur Goethe-Literatur*. Berlin: Julius Springer, 1869.

Stachel, Peter. "'Das Krönungsjuwel der österreichischen Freiheit': Die Wiedereröffnung der Wiener Staatsoper 1955 als Akt österreichischer Identitätspolitik." In *Bühnen der Politik: Die Oper in europäischen Gesellschaften im 19. und 20. Jahrhundert*, edited by Sven Oliver Müller and Jutta Toelle, 90–107. Vienna: Oldenbourg, 2008.

Stadlbauer, Christina, Christian Reiter, Beatrix Patzak, Gerhard Stingeder, and Thomas Prohaska. "History of Individuals of the 18th/19th Centuries Stored in Bones, Teeth, and Hair Analyzed by LA-ICP-MS—a Step in Attempts to Confirm the Authenticity of Mozart's Skull." *Analytical and Bioanalytical Chemistry* 388, no. 3 (2007): 593–602.

Stadler, Margarete. "Ein musikalisches Märchen: Zum Geburtstage Beethovens." *Beilage zur Neuen Musik-Zeitung* 5 (1912): 115.
Stafford, William. *The Mozart Myths: A Critical Reassessment.* Stanford, CA: Stanford University Press, 1991.
———. *Mozart's Death: A Corrective Survey of the Legends.* London: Macmillan, 1991.
Stanyek, Jason, and Benjamin Piekut. "Deadness: Technologies of the Intermundane." *Drama Review* 54, no. 1 (2010): 14–38.
Stavlas, Nikos. "Reconstructing Beethoven: Mauricio Kagel's *Ludwig van*." PhD thesis, Goldsmiths, University of London, 2012.
Steger, Florian, and Joachim Thiery. "Theodor Puschmann: *Richard Wagner: Eine psychiatrische Studie*, 1873; Näherungen." *Wagnerspectrum* 11, no. 1 (2015): 105–40.
Stein, Erwin. "Eine heilige Stätte: Die Grabkirche in Jerusalem," *Illustrierte Welt* 14, no. 22 (1866): 294–95.
Steinberg, Michael P. *The Meaning of the Salzburg Festival: Austria as Theater and Ideology, 1890–1938.* Ithaca, NY: Cornell University Press, 1990.
Sterling, Brett E. *Hermann Broch and Mass Hysteria: Theory and Representation in the Age of Extremes.* Rochester, NY: Camden House, 2022.
Stockert-Meynert, Dora von. *Theodor Meynert und seine Zeit: Zur Geistesgeschichte Österreichs in der 2. Hälfte des 19. Jahrhunderts.* Vienna: Österreichischer Bundesverlag, 1930.
Straus, Joseph N. *Broken Beauty: Musical Modernism and the Representation of Disability.* New York: Oxford University Press, 2018.
Sullivan, J. W. N. *Beethoven: His Spiritual Development.* 1927. New York: Knopf, 1944.
Sumner Lott, Marie. *The Social Worlds of Nineteenth-Century Chamber Music: Composers, Consumers, Communities.* Urbana: University of Illinois Press, 2015.
Swenson, Astrid. *The Rise of Heritage: Preserving the Past in France, Germany, and England, 1789–1914.* Cambridge: Cambridge University Press, 2013.
Tappert, Wilhelm. *Ein Wagner-Lexicon: Wörterbuch der Unhöflichkeit enthaltend grobe, höhnende, gehässige und verläumderische Ausdrücke, welche gegen den Meister Richard Wagner, seine Werke und seine Anhänger von den Feinden und Spöttern gebraucht worden sind.* Leipzig: E. W. Fritzsch, 1877.
Telesko, Werner. *Kulturraum Österreich: Die Identität der Regionen in der bildenden Kunst des 19. Jahrhunderts.* Vienna: Böhlau, 2008.
Thorau, Christian. "Werk, Wissen, und touristisches Hören: Popularisierende Kanonbildung in Programmheften und Konzertführern." In *Der Kanon der Musik: Theorie und Geschichte,* edited by Klaus Pietschmann and Melanie Wald-Fuhrmann, 540–66. Munich: Richard Boorberg Verlag, 2013.
Tibbetts, John C., ed. *Schumann: A Chorus of Voices.* Milwaukee: Amadeus Press, 2010.
Tichy, Gottfried. "Zur Anthropologie des Genies: Mozarts Schädel." *Jahrbuch der Universität Salzburg,* 1989, 251–65.
Timms, Edward. *Karl Kraus, Apocalyptic Satirist.* Vol. 1, *Culture and Catastrophe in Habsburg Vienna.* New Haven, CT: Yale University Press, 1986. Vol. 2, *The Postwar Crisis and the Rise of the Swastikas.* New Haven, CT: Yale University Press, 2005.
Treitel, Corinna. *A Science for the Soul: Occultism and the Genesis of the German Modern.* Baltimore: Johns Hopkins University Press, 2004.
Tümmers, Horst-Johs. *Rheinromantik: Romantik und Reisen am Rhein.* Cologne: Greven Verlag, 1968.

Türck, Hermann. *Der geniale Mensch*. 1896. 4th ed. Berlin: Dümmlers Verlagsbuchhandlung, 1899.

Tzschirner, Heinrich Gottlieb. *Tzschirner's letzte Worte an heiliger Stätte gesprochen*. Leipzig: G. Fleischer, 1828.

Ulmer, Renate. "Zwischen Geniekult und Existenzmaskerade." In *Masken: Metamorphosen des Gesichts von Rodin bis Picasso*, edited by Ralf Beil, 149–59. Ostfildern: Hatje Cantz, 2009.

Unowsky, Daniel L. *The Pomp and Politics of Patriotism: Imperial Celebrations in Habsburg Austria, 1848–1916*. West Lafayette, IN: Purdue University Press, 2005.

Unseld, Melanie. *Biographie und Musikgeschichte: Wandlungen biographischer Konzepte in Musikkultur und Musikhistoriographie*. Cologne: Böhlau Verlag, 2014.

Varwig, Bettina. *Music in the Flesh: An Early Modern Musical Physiology*. Chicago: University of Chicago Press, 2023.

Vazsonyi, Nicholas. "Beethoven Instrumentalized: Richard Wagner's Self-Marketing and Media Image." *Music and Letters* 89, no. 2 (2008): 195–211.

———. *Richard Wagner: Self-Promotion and the Making of a Brand*. Cambridge: Cambridge University Press, 2010.

Verdery, Katherine. *The Political Lives of Dead Bodies: Reburial and Postsocialist Change*. New York: Columbia University Press, 1999.

Vetter, Isolde. "Wagner in the History of Psychology." In *Wagner Handbook*, edited by Ulrich Müller and Peter Wapnewski, 118–55. Cambridge, MA: Harvard University Press, 1992.

Vincent, Esther H. "The Doctors Look at Music." *Quarterly Bulletin of the Northwestern University Medical School*, September 25, 1945, 240–46.

Waddell, Nathan. *Moonlighting: Beethoven and Literary Modernism*. Oxford: Oxford University Press, 2019.

Wade, Rebecca. *Domenico Brucciani and the Formatori of Nineteenth-Century Britain*. New York: Bloomsbury, 2019.

Wagner, Karl. *Das Mozarteum: Geschichte und Entwicklung einer kulturellen Institution*. Innsbruck: Helbling, 1993.

Wagner, Richard. "Das Publikum in Zeit und Raum." 1878. In *Religion and Art*. Translated by William Ashton Ellis, 85–94. Lincoln: University of Nebraska Press, 1994.

———. *Gesammelte Schriften und Dichtungen*. 10 vols. Leipzig: E. W. Fritzsch, 1871.

———. *Richard Wagner's Prose Works*. Translated by William Ashton Ellis. 8 vols. 1892–99. New York: Broude Brothers, 1966.

———. *Sämtliche Briefe*. Edited by Gertrud Strobel, Werner Wolf, Werner Breig, Klaus Burmeister, Johannes Forner, Andreas Mielke, and Martin Dürrer. Vol. 4. Leipzig: Deutscher Verlag für Musik, 1967.

Waldvogel, Richard. *Auf der Fährte des Genius: Biologie Beethovens, Goethes, Rembrandts*. Hannover: Hansche Buchhandlung, 1925.

Walker, Alan. *Franz Liszt*. 3 vols. Ithaca, NY: Cornell University Press, 1983–96.

Walton, Benjamin. "Quirk Shame." In "Quirk Historicism," ed. Nicholas Mathew and Mary Ann Smart, special forum, *Representations* 132, no. 1 (2015): 121–29.

Watkins, Holly. "Schoenberg's Interior Designs." *Journal of the American Musicological Society* 61, no. 1 (2008): 123–206.

Watson, Nicola J. *The Author's Effects: On Writer's House Museums*. Oxford: Oxford University Press, 2020.

———. *The Literary Tourist: Readers and Places in Romantic and Victorian Britain*. New York: Palgrave Macmillan, 2007.

Weber, William. *The Great Transformation of Musical Taste: Concert Programming from Haydn to Brahms*. Cambridge: Cambridge University Press, 2008.

Wedekind, Frank. *Kitsch*. In *Gesammelte Werke*, edited by Artur Kutscher and Joachim Friedenthal, 2:207–74. Munich: Georg Müller Verlag, 1924.

Weissmann, Adolf. *Die Entgötterung der Musik*. Berlin: Max Hesse, 1926.

Westover, Paul. *Necromanticism: Traveling to Meet the Dead, 1750–1860*. New York: Palgrave Macmillan, 2012.

Wild, Irene [Margarete Koelman]. "Dschang und Dschau: Nach dem Leben von Irene Wild." In *Deutsche Roman-Bibliothek*, 316–22. Stuttgart: Deutsche Verlag-Anstalt, 1910.

———. *Ein Liebesschicksal in Liedern*. Dresden: E. Pierson, 1904.

Williamson, George S. *The Longing for Myth in Germany: Religion and Aesthetic Culture from Romanticism to Nietzsche*. Chicago: University of Chicago Press, 2004.

Witt, Michał, Artur Szklener, Jerzy Kawecki, Witold Rużyłło, Marta Negrusz-Kawecka, Michał Jeleń, Renata Langfort, Wojciech Marcwica, and Tadeusz Dobosz. "A Closer Look at Frédéric Chopin's Cause of Death." *American Journal of Medicine* 131, no. 2 (2018): 211–12.

Worthen, John. *Robert Schumann: Life and Death of a Musician*. New Haven, CT: Yale University Press, 2007.

Yang, Mina. *Planet Beethoven: Classical Music at the Turn of the Millennium*. Middletown, CT: Wesleyan University Press, 2014.

Yearsley, David. *Bach and the Meanings of Counterpoint*. Cambridge: Cambridge University Press, 2002.

Youens, Susan. "Hugo Wolf and the Operatic Grail: The Search for a Libretto." *Cambridge Opera Journal* 1, no. 3 (1989): 277–98.

Zilsel, Edgar. *Die Geniereligion: Ein kritischer Versuch über das moderne Persönlichkeitsideal mit einer historischen Begründung*. 1918. Edited by Johann Dvořák. Frankfurt: Suhrkamp, 1990.

———. *Die sozialen Ursprünge der neuzeitlichen Wissenschaft*. Frankfurt: Suhrkamp, 1976.

———. "Mozart und die Zeit: Eine didaktische Phantasie." *Der Brenner* (December 1912): 268–71.

Zur Enthüllung des Mozart-Denkmals in Wien am 21. April 1896. Vienna: Verlag des Mozart-Denkmal-Comités, 1896.

Zweig, Stefan. *The World of Yesterday*. 1942. London: University of Nebraska Press, 1964.

Index

Page numbers in italics refer to illustrations.

actor-network theory (ANT), 9
Adler, Guido, 98, 203n97
Adorno, Theodor W., 83, 118, 119, 135, 137–38, 141, 156, 207n47
Aktualitätskitsch, 142
Alboin (king of the Lombards), 38
Alleaume, Ludovic, 130
anachronism, 130, 131–33, 139–40, 143, 155, 162
anonymity, 126, 132, 136
anthropometry, 82, 88, 102. *See also* phrenology/ cranioscopy
antisemitism, 7, 79, 89, 139
Aragon, Louis, *Aurélien*, 208n69
archival fever, 5
Arendt, Hannah, 171
Ariès, Philippe, 111, 113, 205n18
art-religion, as concept, 2–3. *See also* composer devotion
Atterbohm, Daniel Amadeus, 201n53
Attfield, Nicholas, 158
Auden, W. H., 77
Auer, Max, 167, 168–69, 201n59, 217n138, 218n150
Austrian identity, and pan-Germanism, 73–74

Bach, Johann Sebastian: exhumation of, 97; portraits/images of, 72–73, *74*; relics of, 10; skull of/ in phrenology rhetoric, 98, 101–2, *101*
Bach family, 85–86
Bach Museum (Leipzig), 10
Bach Society (Cologne), 27
Bahr, Hermann, 75, 76–77
Balestrieri, Lionello, *Beethoven*, 105, *106*
Barach, Moritz (pseud. Dr. Märzroth), 66, 71
Barrias, Felix Joseph, 114

Bartók, Béla, 161
Baruda, J., 44
Bary, Alfred von, 199n21
Baudrillard, Jean, 153
Bayreuth: as model for Beethoven heritage in Bonn, 26; as model for Mozart heritage in Salzburg, 55, 56
beautiful death paradigm, 106, 110–15, 162
Beethoven, Ludwig van: admiration for Haydn, 22; in biofiction, 6, 7, 111; birth house (*see* Beethoven, Ludwig van, birth house [Bonn]); death house (Schwarzspanierhaus), 22, 45–50, *47*, 126, 192n119, 208n73; exhumations of, 46, 63, 91, 92, 94, 95, 97; festivals in honor of, 23–24, 28, 146, 159, 187–88n43; Heiligenstadt Testament, 200n36; humanism of, 135; instruments of, 29–30, 188n48, 188n52; and Karlskirche monument project, 184n49; in kitsch commodities, 142, 143; late and last works of, 117–20, 206n41; in medical culture (*see* Beethoven, Ludwig van, in medical culture); monuments of, 24–25, 33, 132, *133*, 146; Mozart compared to, 53, 74; and occult movement, 11; portraits/ images of, 41, *41*, 73, *74* (*see also* Beethoven, Ludwig van, masks); relics from belongings of, 1, 14, 29–30, 96, 106, 143, 184–85nn62–63, 188n48, 188n52; relics from body of, 1, 14, 106, 112, 113–14, 143, 185n64, 212n27; relics from buildings associated with, 15, 46–48, 49, 185n70; temples dedicated to, 159
Beethoven, Ludwig van, birth house (Bonn), *32*; birth room reimagined as Nativity, 30–33, *33*; chamber festivals in, 28, 187–88n43; confusion over location of, 18, 25; death imagery, 33–34;

Beethoven, Ludwig van, birth house (*cont.*)
Ehrenpflicht rhetoric, 22, 25–26; mission statements, 26, 29; piety rhetoric, 26, 28, 29; relic acquisition efforts, 14, 29–30; relics from pieces of, 15, 185n70; visitors' books, 31, 33, 34–45, *35*, *37*, *38*
Beethoven, Ludwig van, in medical culture: genome testing, 95–96; hearing organ examinations, 84, 92; heart examinations, 81; lead-level testing, 80, 198n2; in pathographies, 81, 86, 87; skull examinations/in phrenology rhetoric, 89, 90, 94–95, 96, 98, *99*, *100*, 101, 103–4, 163
Beethoven, Ludwig van, masks: and admiration for stony-faced visage, 41, *41*; as anachronism, 130, 131–33, 139–40; context for creation of, 89–90; deathbed accounts, 110–11, 112–14; feminine admiration for, 121–22, 128–30; in feuilletons, 133–34; and *Inconnue de la Seine*, 109, 123–26, 130; as kitsch, 128–29, 130, 139–40; life vs. death masks, 105–6, *108*; and masks made from Schwarzspanierhaus pieces, 49, 126, 192n119, 208n73; as object of memory, 106–8; in paintings/etchings/photographs, 105, *106*, *107*, 123, *124*, *125*, 126–28, *126*, *127*, 130, 131–32, *131*, 209n90; replicas of, 105, 106, 121, 122–23, 126; as *vera icon*, 120–22
Beethoven, Ludwig van, works: Adelaïde, op. 46, 50; Bagatelle in B flat Major, WoO 60, 117, 206n41; Fifth Symphony, 11, 37, 190n73; *Missa Solemnis*, 118, 135; "Moonlight" Sonata, 130; Ninth Symphony (Ode to Joy), 23, 33, 37, 40, 146, 159; Piano Sonata, op. 26, xi; Piano Sonata, op. 109, ix; String Quartet, op. 127, 207n54; String Quartet, op. 130, 11, 28, 30, 135; String Quartet, op. 135, 46, 48, 49, 192n119, 207n56; String Quintet in C Major, WoO 62, 206n41; *Tempest* Sonata, xi; Tenth Symphony, 40; *Waldstein* Sonata, xi
Beethovenhalle (Bonn), 23, 29
Bellini, Vincenzo, 117
Benjamin, Walter, 125, 130, 135, 141, 142, 152, 171, 172, 209n89, 212nn24–25
Benkard, Ernst, *The Undying Face*, 136
Bennett, Jane, 9
Berg, Carl, 38–39, *38*, 190–91n86
Bergmüller, C. W., *Beethoven-Sonate*, 123, *123*
Berlin, claim to Beethoven's instruments, 29–30, 188n48
Berlioz, Hector, 85, 86, 200n31; *Evenings with the Orchestra*, 157
Bertuch, Friedrich Justin, 6
Berwald, Franz, 199n21
Besitzbürgertum, defined, 3
Biber, Heinrich, 54
Biedermeier domesticity, 142
Bildung: as concept, 2–3; crisis of, 137–39, 141 (*see also* kitsch); and Western canon formation, 4
Bildungsbürgertum, defined, 3

Billroth, Theodor, 83, 84–85, 199n19; *Who Is Musical?*, 85
Binhold, Hubert, 189n59
Binkley, Sam, 142, 211n20
biofiction, 6, 8, 111, 116–17, 202n71
birth houses: Haydn's, 22; Mozart's, 54–55; Schubert's, 10. *See also* Beethoven, Ludwig van, birth house (Bonn)
Bismarck, Otto von, 26
Bloch, Ernst, "Dust," 142, 212n25
Böcklin, Arnold, 133, 152
Bonanza Kings, 60
Bonds, Mark Evan, 8, 109
Bonn: Beethovenhalle, 23, 29; Beethoven monument in, 24–25, 33, 146; as *Beethoven-Stadt*, 22. *See also* Beethoven, Ludwig van, birth house (Bonn)
Bösendorfer Saal (Vienna), 49, 192n119
Bourdieu, Pierre, 26, 145
bourgeoisie, as label, 3
Bovet, Hermine, 7, 34
Brahms, Johannes, 6, 24, 26, 83, 84
brain-fold studies, 100, 204n109
Bransen, Walter, 199n21
Brendel, Franz, 74
Brentano, Antonie, 201n53
Breuning, Gerhard von, 22, 94, 95, 98, 103–4
Broca, Paul, 88
Broch, Hermann, 138, 141, 154, 157–58; *A Theory of Mass Mania*, 157
Brodbeck, David Lee, 73–74
brotherhood rhetoric, 54, 65
Brown, Bill, 9
Browning, Robert and Elizabeth Barrett, 115
Brucciani Atelier, 122
Bruckner, Anton, 7, 139, 161–70, *164*, 216n120, 217n138, 218n143; *The Musician of God*, 182n21
Brüser Atelier, 126
Büchner, Emil, 190n75
Budapest: Bartók reburied in, 161; Liszt's jubilee celebrated in, 147–48, 213n48
Bülow, Hans von, 119
Bungert, August, 139, 150–52, 214n62; *The Homeric World*, 151
burials: *Ehrengräber* (honorable graves), 46, 97; embalmed bodies, 161–70, 216n116; exhumations, 46, 63, 91, 92, 94, 95, 97, 149, 161; pilgrimage to tombs, 46, 97, 162, 165, 169, 218n151
Bursy, Karl, 201n53
Byron, Lord, 13

camp sensibility, 172
Canetti, Elias, 215n91
canon, defined, 4
canon formation, challenging traditional narrative of, 4–8

INDEX

canonization process (Catholic), 4–5
Carchen, Gerolamo, 94
Carl Friedrich von Sachsen-Weimar-Eisenach, 216n116
Carus, Carl Gustav, 97
Carus, Ernst August, 83
Casals, Pablo, 192n119
Catholicism, 2, 4–5, 10–11, 23, 31, 41, 127, 131, 162, 166, 169, 218n143
celebrity culture, and parasocial relationships, 13. *See also* parasociality
Celenza, Anna Harwell, 132
Chaplin, Charlie, 129
character heads (*Charakterköpfe*), 122–23
characterology/graphology, 97, 203n93
Chop, Max, 151, 152, 214n69
Chopin, Frédéric: death of, 114; heart of, 80, 115; late and last works of, 117; in pathographies, 86, 200n34; portraits/images of, 6, 114–15, 183n25, 205–6n27; relics of, 114, 115
choral societies, men's, 64–66, *64*, 74, 75, 165, 195n47
Christ: bride of, 44; death/suffering of, 111, 117–18, 119–20, 131–32, *131*; face of, 113, 121, 132; humanity of, 132–33, 135; *imitatio Christi*, 144, 145, 146, 149, 212n31; Nativity imagery, 30–33, *33*, 56; and parasocial longing, 40, 41
Chua, Daniel K. L., 135
Clésinger, Auguste, 114
Coleridge, Samuel Taylor, 13
collective memory, 5, 46
Cologne, Bach Society in, 27
Comini, Alessandra, 109, 122, 126, 132, 208n69
composer devotion, 1–3; and canon formation, challenging traditional narrative of, 4–8; guilty pleasure in, 17–19; in satire, ix–xi, *x*, 63, 143–44, 148–50, *150*, 153–57, 159–61. See also *Ehrenpflicht* (obligation to honor); kitsch; medicine and medical culture; piety; pilgrimage; relics and relic culture
Cook, Thomas, 55
Corbineau-Hoffmann, Angelika, 109
cranioscopy/phrenology, 82, 85, 87–88, 89, 90, 91, 94, 97–102, 103
crown imagery, 31, 33, *33*, 93, 121, 123, *125*, *126*, 189n62
curiosities, collections of, 82, 123

Dadaism, 129
Danhauser, Carl, 106
Danhauser, Josef, 105–6; *Liszt Fantasizing at the Piano*, 106, 204n4
Dante Alighieri, 103
Darwinism, cultural, 98
Davies, James Q., 80
deafness, 36–37, 81

death: aesthetization of white, 103; and anonymity, 126, 132, 136; beautiful death paradigm, 106, 110–15, 162; embalmed bodies, 161–70, 216n116; exhumations, 46, 63, 91, 92, 94, 95, 97, 149, 161; juxtaposed with birth imagery, 31, 33–34; late and last works before, 115–20; memento mori, 106, 111, 115, 126, 128, 135–36, 168; variety in material culture of, 115
death house, Beethoven's (Schwarzspanierhaus), 22, 45–50, *47*, 126, 192n119, 208n73
death masks, as practice, 105–6, *108*. *See also* Beethoven, Ludwig van, masks
de Brosses, Charles, 9–10
degeneration, and creativity, 85–86
Delacroix, Eugène, 72
Demel, Karl, 55
Denkmäler deutscher Tonkunst (*Monuments of German Musical Art*), 27, 74
Denkmalwut (monument fever), 12, 23, 97
Derrida, Jacques, 5, 182n19
Deutsch, Otto Erich, 81
Deutscher Bund Heimatschutz (German Association for Homeland Protection), 27
Diabelli, Anton, 22, 117, 206n41
disability, and pathography, 81–83, 84–87, 102–3
divine inspiration, 72
DNA testing, 95–96
Döblin, Alfred, 136, 210n111
doctors. *See* medicine and medical culture
Dom-Musik-Verein, 54, 55
Donizetti, Gaetano, 94, 117
Dorfles, Gillo, 141
Dotter, Anton, 92
Dović, Marijan, 4
Droste-Hülshoff, Annette von, *The Spiritual Year*, 127–28
Dumba, Nikolaus, 24, 25, 51
Dvořák, Max, 49
Dyer, Richard, 103

Eckhardt, Leopold, 92
Eckstein, Frederick, 166, 217n132
Edenberg, Carl Langer von, 48
Egger-Lienz, Albin, *Ninth Symphony*, 41
Egyptian imagery, 57, 60, *61*
Ehrengräber (honorable graves), 46, 97
Ehrenpflicht (obligation to honor): and Beethoven birth house, 22, 25–26; and burials, 52, 91, 94; competitive, 23–25, 52, 78, 147, 148, 149; defined, 21–22; in historic preservation rhetoric, 21–22, 25–26, 49, 53, 78–79, 197n88; and scientific inquiry, 83, 88, 91, 94, 100–101
Elben, Otto, 195n47
Elias, Norbert, 158, 215n99
embalmed bodies, 161–70, 216n116
embeddedness, 142, 211n20

Engelmann, Theodor Wilhelm, 199n19
Engl, Johann Evangelist, 71
England: collecting culture in, 14–15; contribution to German heritage formation, 14, 15; literary culture in, 13, 17–18, 184n56
Enlightenment ideals, 54, 65
environmental preservation rhetoric, 27
eroticism, 40, 44, 129–30
exhibitionary complex, 10
exhumations, 46, 63, 91–92, 94, 95, 97, 149, 161

Falkoff, Rebecca R., 199n14
Feis, Oswald, 85
Felix, Orlando, 88
femininity: and cult of the face, 121–22, 129–30; and kitsch, 129, 140–41
festival cantatas, 12, 146, 183–84n48, 195–96n52
festivals: in honor of Beethoven, 23–24, 28, 146, 159, 187–88n43; in honor of Liszt, 147–48; in honor of Mozart, 51, 54, 55; and men's choral societies, 64–66, 64, 74, 75, 165, 195n47
fetishism, 9–10
Feuerstein, Johann Heinrich, 199n21
feuilletons, 133–34, 139
Franco-Prussian War, 23
Frankl, Ludwig August, 7, 24, 51–52, 92
freemasons, 64, 65
Freilichtmuseum (Salzburg), 63
Freud, Sigmund, 5, 10
friendship/keepsake albums (*Stammbücher*), 6, 8, 34, 36, 147
Frimmel, Theodor von, 98
Frisch, Walter, 133
funeral culture: ceremony for demolition of Beethoven's death house, 46–48, 47; funeral photography, 126

Gall, Franz Joseph: followers of, 92, 97, 98, 101; phrenology theories, 82, 87–88, 89; plaster mask collection, 89–90; skull capacity of, 200n41; skull collection, 87
Garas, François, 159
Gehmacher, Friedrich, 194n19
Gehring, Franz, 186n16
genius/"geniology," 82, 88–89, 90
genome testing, 95–96
German Association for Homeland Protection (Deutscher Bund Heimatschutz), 27
Germanic superiority rhetoric, 85, 90, 101
Geselschap, Friedrich, *Beethoven's Birth*, 31–33, 33
Gluck, Christoph Willibald, 72, 74, 184n49
Gobbi, Heinrich, 148
Goebbels, Joseph, 168, 170, 217n135
Goebels, Johannes, 194n24
Goethe, August von, 91

Goethe, Johann Wolfgang von: face mask of, 122; interest in phrenology and physiognomy, 90–91; last words of, 206n42; as model for Salzburg's marketing of Mozart, 6, 58, 71; "On Contemplating Schiller's Skull," 91; in pathographies, 86; "Wand'rers Nachtlied II," 53, 71
Goldscheider ceramics, 122–23
Göllerich, August, 164, 217n132
Goodhill, Victor, 202n77
Goßler, Gustav von, 30
Gouk, Penelope, 83
Graf, Conrad, 204n4
Gräflinger, Franz, 165
graphology/characterology, 97, 203n93
grave sites. *See* burials
Gray, Richard T., 90–91
Green, Emily H., 145
Greenberg, Clement, 141, 210n2
Grieg, Edvard, 158
Griesinger, Georg August von, 89
Grillparzer, Franz, 48, 118
Grootenboer, Hanneke, 111
Gschnas, 154
Gustav Mahler Society (Vienna), 78

Habermas, Jürgen, 34
Haiger, Ernst, 159
hair, locks of, 1, 14, 95–96, 97, 106, 112, 113–14, 115, 143, 184n61, 185n64, 203n87, 212n27
Halbbildung (pseudo-culture), 138, 156
Hallam, Elizabeth, 115
Handel, George Frideric, 73, 74
Hanslick, Eduard, 18–19, 24–25, 76, 83, 84–85; "The Cult of Wagner," 144; *On the Musically Beautiful*, 18
Hargraves, John A., 158
Hart, Mary, 15
Hartmann, Franz, 113–14
Hau, Michael, 83
Hauptmann, Gerhart, 76
Haydn, Franz Joseph: birth house, 22; death of, false rumor, 149–50; and Karlskirche monument project, 184n49; portraits/images of, 72, 74; skull of/in phrenological studies, 89, 91, 92, 93, 93, 98, 100, 161
Haydn, Michael, 54
Haydon, Benjamin Robert, 14
Haynald, Lajos, 213n48
hearing trumpets, Beethoven's, 96
Heidl, Mrs. Alfred, 57–58
Heiligenstadt, Beethoven monument in, 24
Heimat rhetoric, 45
Helgason, Jón Karl, 4
Henke, Burkhard, 211n22
Henle, Jacob, 83
heritage. *See* historic preservation

INDEX

Hermitage Museum (St. Petersburg), 31
Herzfelde, Wieland, 129
Heuss, Alfred, 158–59, 166; "A Day at the Monastery of St. Ludwig," 153–54, 159–60
Hindemith, Paul, 158–59
His, Wilhelm, 101–2, *101*
historic preservation: *Ehrenpflicht* rhetoric, 21–22, 25–26, 49, 53, 78–79, 197n88; vs. modernization and growth, 46, 48; piety rhetoric, 26, 28–29, 49, 57–58, 60, 197n88. *See also* Beethoven, Ludwig van, birth house (Bonn); Mozart, Wolfgang Amadeus, *Magic Flute* cottage (Salzburg)
Hitler, Adolf, 168, 170, 217n135
Hoch, Joseph, 200n29
Hockey, Jenny, 115
Höfer, Werner, ix–x
Hoffmann, E. T. A., "Ritter Gluck," 134
Hoffmann, J. (publisher), 206n41
Hoffmann, Josef (architect), 60, *61*
Hofmannsthal, Hugo von, 74, 75–76
Holbein, Hans, *The Body of the Dead Christ in the Tomb*, 132
holy site rhetoric, 30–31, 36, 39
Holz, Karl, 184n61
honor. *See Ehrenpflicht* (obligation to honor)
Höppener, Hugo ("Fidus"), 130, 209n86
Horner, Johann, 57, 60–62
Hosmer, Harriet, 115
Hugo, Victor, *Le roi s'amuse*, 116
humility topos, 145
Hungary. *See* Budapest
Hust, Christoph, 150–51
Hutcheon, Linda, 102–3
Hutcheon, Michael, 103
Hutschenruyter, Willem, *Das Beethovenhaus*, 159
Hüttenbrenner, Anselm, 112–13, 117, 205n22
Huyssen, Andreas, 129, 140–41
hypophora, 66
Hyrtl, Jacob, 92
Hyrtl, Josef, 48, 84, 92–94, 163

iconicity, 120–22, 132
imitatio Christi, 144, 145, 146, 149, 212n31
Inconnue de la Seine, 109, 123–26, 130
industrial revolution, 45
International Mozart Foundation (IMS), 52, 55, 56, 58, 60, 71, 72, 75
Irving, Washington, 18
Italy, Mahler heritage in, 77–79

Janetschek, Ottokar, 202n71
Jelinek, Franz Xavier, 15
Jews: admiration for Beethoven, 39, 41; antisemitism, 7, 79, 89, 139; financing of monument projects, 24

Joachim, Joseph, 11, 24, 25–26, 27–28, *29*, 30, 147, 187n29
Joachim Albrecht (prince of Prussia), 30
Jonsson, Stefan, 156
Jugendstil, 122, 130, 141

Kagel, Mauricio, *Ludwig van*, ix–xi, *x*, 134, 159, 171
Kalbeck, Max von, 196n60
Kallberg, Jeffrey, 117
Kallman, Chester, 77
Kapuzinerberg, 53. *See also* Mozart, Wolfgang Amadeus, *Magic Flute* cottage (Salzburg)
Karl Bernhard von Sachsen-Weimar-Eisenach, 216n116
Karlskirche (Vienna), failed musical monument in, 12, 184n49
Karnes, Kevin, 132
Kaufmann, Paul, 96
keepsake/friendship albums (*Stammbücher*), 6, 8, 34, 36, 147
Kerber, Erwin, 77
Kerner, Dieter, 81, 198n10
kitsch: and anachronism, 130, 139–40, 143, 155; Beethoven masks as, 128–29, 130, 139–40; and camp sensibility, 172; cheap imitations as, 150–53; cult of the living as, 143–50; and Dadaism, 129; discourses on, overview, 141–43; and femininity, 129, 140–41; Klinger blamed for hastening Beethoven's *Verkitschung*, 133; and mass culture criticisms, 137–39, 153–61; as term, 17, 138, 140–41
Kitsch: The World of Bad Taste (Dorfles), 141
Klages, Ludwig, 97
Klein, Franz, 90, 105
Kleinbürgertum, defined, 3
Klimeš, František, 183n25
Klinger, Max: Beethoven monument, 132, 133; *Brahms Fantasy* series, 133; *The Crucifixion of Christ*, 209n92; *Dead Mother*, 132; *Death*, 132; *Death as Savior*, 132; *Pietà: Maria and John Mourning at the Body of Christ*, 131–32, *131*, 140
Klotz, Ernst, *101*, 102
Knittel, K. M., 35–36, 39
Köberle, Adolf, 218n143
Koelman, Margarete (pseud. Irene Wild), 39–44, 129–30, 172
Kohrs, Klaus Heinrich, 163
Kolb, Alois, 130
Kolisch family, 199n21
Körner, Theodor, 14
Korpal, Tadeusz, 183n25
Kosztolányi, Desider, 134
Kraus, Hedwig, 93
Kraus, Karl, 154
Kreissle von Hellborn, Heinrich, 97–98
Kristeva, Julia, 121

Kunstreligion, as concept, 2–3. See also composer devotion
Kuppe, Wilhelm, 26
Kurlander, Eric, 218n153
Kürnberger, Ferdinand, 24
Kwiatkowski, Teofil, *The Last Moments of Frédéric Chopin*, 205–6n27

Landau, Hermann Josef, 23
landscape preservation movement (*Naturdenkmalschutz*), 27
landscape symphonies, 66
Lang, Fritz, *Metropolis*, 156
Lange-Eichbaum, Wilhelm, 86–87, 200n39
Langer von Edenberg, Carl, 84, 98
LaPorte, Charles, 19, 186n82
late style discourse, 109–10, 118, 207n47
Latour, Bruno, 9
laurel wreaths, 31, *33*, 93, 121, 123, *125*, *126*, 148
Lavater, Johann Caspar, 88–89, 90, 201n51
Le Beau, Louise Adolpha, 195n40
Le Bon, Gustave, 153, 214n73
Leipzig, Bach Museum in, 10
Leistra-Jones, Karen, 25, 119
Lekan, Thomas, 27
Léon, Viktor, *The Musician of God*, 182n21
Leppert, Richard, 109
Lergetporer, Louisa, 56–57, 71–72, 196n63
Levi, Hermann, 144
Lewis, Wyndham, *Tarr*, 128–29
life reform movements, 11, 159
Lindenschmit, Wilhelm (the Younger), *Hall of Heroes of German Music*, 72–74, *74*
Lindner, Andreas, 164
Liszt, Franz: burial of, 149, *150*, 213–14n59; cult of, 7, 147–49, 213n45; face mask of, 122; as heir of Beethoven, 119, 145; Joachim compared to, 25; library of, 116; and New German School controversy, 24; plaster copy of Beethoven's death mask, 106; portraits/images of, 73, *74*; relics of, 143, 212n27; ritual consecration of home, 213n58; self-promotion strategies, 146–47
Liszt, Franz, works: "At Wagner's Graveside," 146, 212n43; *Beethoven-Cantate*, 146, 184n48; *The Bells of Strasbourg*, 146; *Cantate zur Inauguration des Beethoven-Monuments*, 146, 148, 184n48, 195–96n52; *Christus*, 148; *Dante Sonata*, 147; "Licht, mehr Licht!," 206n42; *Life of Chopin*, 114
Loewenthal, Richard, 188n55
Loewi, Otto, 199n19
Lombroso, Cesare, 85–86
Loos, Adolf, 138, 154
Loos, Helmut, 207n54
Lorenzi Atelier, 122
Lotheissen, Georg, 199n21

Ludendorff, Mathilde, 81
Ludwig, Carl Friedrich Wilhelm, 199n19
Lueger, Karl, 185n76
Lutz, Deborah, 6, 115, 206n29
Lyser, Johann Peter, 31

Macdonald, Dwight, 210–11n2
Maelzel, Johann Nepomuk, 96
Magendie, François, 82, 198n11
Magic Flute cottage. See Mozart, Wolfgang Amadeus, *Magic Flute* cottage (Salzburg)
Mahler, Gustav, 53, 77–79, 197n88; *Das Lied von der Erde*, 77, 197n87; Symphony No. 9, 77; Symphony No. 10, 77
Mahler-Werfel, Alma, 84
Maier, Elisabeth, 216n120
Maier-Graefe, Julius, 133
Mandylion of Edessa, 121
Manski, Dorothee, 199n21
Marleni, Emma, 194n23
Marsop, Paul, 215n105
martyrology, 86–87
Marx, Karl, 10
mass culture, criticism of, 137–39, 153–61
material studies, and musicology, 9, 19, 183n34
Mathew, Nicholas, 9
Matthisson, Friedrich von, 50
Maximilian (archduke of Austria and emperor of Mexico), 163, 164, *164*
McCormick, Lisa, 183n37
McMahon, Darrin, 82
McManus, Laurie, 6, 25, 146–47
McMullan, Gordon, 109–10
Męcina-Krzesz, Józef Feliks, 183n25
medicine and medical culture: brain-fold studies, 100, 204n109; characterology/graphology, 97, 203n93; exhumations, 91–92, 94, 95, 97; genius/"geniology," 82, 88–89, 90; musical networks, 83–84, 89–90, 199n21; pathography, 81–83, 84–87, 102–3; phrenology/cranioscopy, 82, 85, 87–88, 89, 90, 97–102, 103; physiognomy, 82, 88, 90–91; racialism in, 85, 90, 101, 200n39, 204n109; skulls valued as relics, 80, 82, 87, 88, 91, 92, 94–95, 96
Meinert, Carl, 181n1, 212n27
memory: Beethoven's masks as object of, 106–8; collective, 5, 46; kitsch as medium of, 142–43; relics as medium of, 115
Mendelssohn, Felix, 73, *74*, 83; *Lobgesang*, 184n48
men's choral societies, 64–66, *64*, *74*, 75, 165, 195n47
Menzler, Wilhelm, *Beethoven or Polyhymnia*, *126*
Meredith, William, 95
Metternich, Klemens von, 64
Meynert, Theodor, 84, 98, 100
Micheli Brothers, 106, 122, 128

INDEX

middlebrow, 138, 157
Mielichhofer, Ludwig, 54
Minor, Ryan, 12, 146, 195n52, 213n58
Mittelstand, defined, 3
mizpah, 44–45, 173
modernism, 129
Mole, Tom, 13
monument fever (*Denkmalwut*), 12, 23, 97
monuments: of Beethoven, 24–25, 33, 132, 133, 146; and *Ehrenpflicht* rhetoric, 24–25, 52; Karlskirche proposal, 12, 184n49; of Mozart, 51, 52, 52; and piety, 12; of Wagner, 51
Monuments of German Musical Art (*Denkmäler deutscher Tonkunst*), 27, 74
Morris, Rosalind C., 9
Mortier de Fontaine, Henri-Louis-Stanislaus, 207n56
Mozart, Constanze, 15, 54
Mozart, Wolfgang Amadeus: in biofiction, 6, 7, 111, 116–17, 202n71; birth house, 54–55; festivals in honor of, 51, 54, 55; Goethe as model for Salzburg's marketing of, 6, 58, 71; and Karlskirche monument project, 184n49; in kitsch commodities, 142; monuments of, 51, 52, 52; in pathographies, 81, 86, 198n10; portraits/images of, 72–73, 73, 74, 116–17; reimagined as serious artist, 53–54, 71–72, 74–75; relics of, 1, 15, 58 (see also Mozart, Wolfgang Amadeus, *Magic Flute* cottage [Salzburg]); and Salzburg Festival ideology, 76–77; skull of/in phrenology rhetoric, 89, 91, 92, 93–94
Mozart, Wolfgang Amadeus, *Magic Flute* cottage (Salzburg), 52–54, 52; in historic preservation rhetoric, 57–58, 60, 62; location changes, 53, 58, 63; men's choral performances at, 64, 64, 65–66, 75; protective cover proposals for, 60, 61, 62–63, 62; and reimagining of Mozart as serious artist, 53–54, 71–72, 75; relic/reliquary dynamics, 58, 59; unveiling ceremony, 53, 56–57, 65; visitors' books, 194n23
Mozart, Wolfgang Amadeus, works: *Der Schauspieldirektor*, 75; *Don Giovanni*, 1, 52, 55; *The Magic Flute*, 53, 54, 57, 64, 65, 74–75, 77; *Requiem*, 116–17
Mozarteum, 54, 55, 63
Mozart und Schikaneder (operetta), 75
Müller, Franz Xaver, 169
Munkácsy, Mihály, 116
Musealisierung (museal impulse), x, 10
musicology, and material studies, 9, 19, 183n34

Nagel, Willibald, 152
Nägeli, Hans Georg, 65
Nagy, Imre, 161
Napoleonic imagery, 106, 112, 117–18
National Socialists, 90, 159, 161, 166, 168–69, 170, 218n153

Nativity imagery, 30–33, 33, 56
nature: landscape preservation movement, 27; landscape symphonies, 66; Mozart reimagined as lover of, 71–72
Naunyn, Bernard, 199n19
Nazi ideology, 90, 159, 161, 166, 168–69, 170, 218n153
Nettl, Bruno, *Heartland Excursions*, 172
neuroscience, 88, 100
New German School, 6–7, 24, 25–26, 75, 119, 196n66
new materialism, 9
Ney, Elly, x
Nicolai, Ernst Anton, 199n26
Niemack, J., 81
Nietzsche, Friedrich, 38, 214n62
Nissen, Georg Nikolaus von, 54, 199n21
Noble, Jonathan, 102
Nohl, Ludwig, 6–7, 75, 119–20, 148–49, 207n54
Nora, Pierre, 5, 46, 63, 182n17
Nordau, Max, *Degeneracy*, 144
Novello, Sabilla, 62
Novello, Vincent and Mary, 15
Nußbaumer, Martina, 186n19

occult rhetoric, 11, 28, 133–34, 153, 218n153
Olalquiaga, Celeste, 143

Pachler-Koschak, Marie, 185n63
Painter, Karen, 166
pan-Germanism, 73–74, 75
parasociality, 13, 39, 41, 44, 144
Parry, Patrick, 213n45
pathography, 81–83, 84–87, 102–3
Paumgartner, Bernhard, 76
Paunzen, Arthur, 216n107
Pazaurek, Gustav, 141, 142, 211n17
Penn, Hermann, 49–50
Perec, Georges, 172
Pestalozzi, Johann Heinrich, 65
Peter, Johann, 92
Pfitzner, Hans, *Palestrina*, 182n21
philanthropy: and cycle of economic transformation, 26–27; *Ehrenpflicht* rhetoric, 25–26; fund-raising initiatives, 26, 60–62; piety rhetoric, 26
Phillips, Reuben, 46
photography: funeral, 126; as medium, 125, 135
phrenology/cranioscopy, 82, 85, 87–88, 89, 90, 91, 94, 97–102, 103
physicians. *See* medicine and medical culture
physiognomy, 82, 88, 90–91
piety: contexts for, 12–13; and *Ehrenpflicht* rhetoric, 22; in historic preservation rhetoric, 26, 28–29, 49, 57–58, 60, 197n88; and *imitatio* rhetoric, 145; in landscape preservation rhetoric, 27; vs.

piety (*cont.*)
    modernity, 48; and relic culture, 11; and skull collecting, 88, 94
pilgrimage: and cycle of economic transformation, 26–27; and *imitatio* rhetoric, 145; narrative tropes, 35–36; rhetoric of, overview, 20; Salzburg as popular destination for, 54, 56; in satire, 156, 159–60; to tombs, 46, 97, 162, 165, 169, 218n151; true devotion rhetoric, 28, 36, 39, 40–41, 114, 145; visitors' books, 31, 33, 34–45, *35, 37, 38*, 194n23. *See also* Beethoven, Ludwig van, birth house (Bonn); Mozart, Wolfgang Amadeus, *Magic Flute* cottage (Salzburg); relics and relic culture
Plattensteiner, Richard, *Beethoven: The Great Musician in God's Honor*, 182n21
Pöck, Gregor, 46, 48
Pohl, Ferdinand, 151
Pointon, Marcia, 111
Polanyi, Karl, 211n20
Polgar, Alfred, 128, 129, 156–57, 209n79
Popper, Karl, 198n11
Poskett, James, 88
postcards, 6, 31, *32*, 41, *41*, 182–83n25
postmodernism, 211n19
Potter, Pamela, 166
Prechtl, Michael Mathias, *Beethovens Erotica*, 130
preservation rhetoric. *See* historic preservation
priesthood rhetoric, 6, 26, 28, 30, 64, 147
proto-surrealism, 133
pseudoscience, as term, 82, 198n11
psychopathography, 86
Puccini, Giacomo, 158
Pushee, Elsbeth, 189n73
Pushkin, Alexander, 116

"quirk shame," 19

Raaf, Anton August, 48
race: and character heads, 122–23; in medical culture, 85, 90, 101, 200n39, 204n109
Rákóczy Ferenc II, 149, *150*
Ramann, Lina, *150*, 213n50
Rank, Otto, 183n43
Rath, Walther, "My songs will live on when I myself am gone," 127–28, *127*
Rauschenberger, Walther, 90, 201n59
Reece, Frederick, 96
Reger, Max, 26
Rehding, Alexander, 12, 138, 146
Reinhardt, Max, 75
Reißiger, Carl, 117, 206n41
relics and relic culture, 1–2, 9; as anachronism, 143; buildings anthropomorphized as human remains, 46–49, 58, 63, 192n119; cloth, 121; and composer devotion, overview, 10–11; death masks as, 106 (*see also* Beethoven, Ludwig van, masks); defined, 9; and exhumations, 46, 63, 91–92, 94, 95, 97, 149, 161; hair, locks of, 1, 14, 95–96, 97, 106, 112, 113–14, 115, 143, 184n61, 185n64, 203n87, 212n27; and iconicity, 120–22, 132; late and last works as, 116, 117, 119–20; politicization of, 161, 168–69, 170; in satire, 144; scholarly paradigms, 9–10; skulls valued as, 80, 82, 87, 88, 91, 92, 94–95, 96; as technology for remembering, 115. *See also* pilgrimage
relic snatching (*Reliquienhascherei*), 88, 94, 201n48
Rheinberger, Joseph Gabriel, *Festchor*, 183n48
Rhine valley: *Heimat* rhetoric, 45; landscape preservation movement in, 27
Richarz, Franz, 203n87
Richter, Hans, 55, 148
Ricœur, Paul, 182n17
Riegl, Alois, 48–49
Ries, Franz, 188n52
Rilke, Rainer Maria, 124
Roach, Joseph, 13, 150
Rochlitz, Friedrich, 53, 71
Rohrau, Haydn's birth house in, 22
Rokitansky, Carl von, 83–84
Romako, Anton, *Mozart at the Spinet*, 72, *73*, 196n63
Romanticism, 2, 3, 13, 27, 54, 117–18, 158, 215n99
Rosenbaum, Carl, 92
Rossini, Gioachino, 1, 97, 181n2
Roth, Dieter, ix, x
Rothe, Anna, 153
Rubinstein, Anton, 29
Rudorff, Ernst, 7, 27–28, 187n38
Ruttmann, Walter, *Berlin: Symphony of a Metropolis*, 156

Sacks, Oliver, *Musicophilia*, 85
Saget, Hans Maria, 37
Said, Edward, 118
"saintly sage" archetype, 82, 86–87, 114, 199n13
Saint-Saëns, Camille, 158
Salieri, Antonio, 89, 116
Saliot, Anne-Gaëlle, 109, 125, 126, 130
Salzburg: concert hall proposals, 55–56, *56*; Goethe as model for marketing of Mozart, 6, 58, 71; Mozart birth house in, 54–55; Mozart festivals in, 51, 54, 55, 56; Mozart monument in, 51; as *Mozart-Stadt*, 22, 51, 54–55, 57; self-fashioning after World War I, 75–77. *See also* Mozart, Wolfgang Amadeus, *Magic Flute* cottage (Salzburg)
Salzburg Festival, 51, 75–77
Santer, Saskia, 197n89
satire, ix–xi, *x*, 63, 143–44, 148–50, *150*, 153–57, 159–61
Sauter, Ferdinand, 113

INDEX

Schaaffhausen, Hermann, 97, 98–100, 203n87
Schikaneder, Emanuel, 53, 75
Schiller, Friedrich, 90, 91, 147, 187n38
Schindler, Anton, 14, 160, 181n1, 185n63, 188n48, 200n41, 212n27
Schleiffer, Betti, 7
Schlemmer, Oskar, 128, 140
Schlesinger, Adolf Martin, 117, 206n41
Schmalhausen, Otto, 129
Schmidt, Carl, 188n58
Schmidt, Friedrich, 207n54
Schneller, Julius, 1, 2
Schoenberg, Arnold, 84, 199n21
Schorske, Carl, 76, 138
Schrattenholz, Josef, 151
Schubert, Franz: in biofiction, 6, 111; birth house, 10; and Bruckner, 163; criticism of, 158; exhumations of, 46, 63, 91, 94, 97; and Karlskirche monument project, 184n49; in kitsch commodities, 142, 211–12n22; in pathographies, 86; in phrenology/physiognomy rhetoric, 89, 97–98, 99, 100; portraits/images of, 73, 74; *Reliquie*, 117; in Salzburg, 54
Schumann, Clara: and medicine-music networks, 83; and New German School controversy, 24; portraits/images of, 73, 74; and priesthood of art, 6, 26, 147; and Robert Schumann's exhumation, 97, 203n87; and Rudorff, 27
Schumann, Robert: and medicine-music networks, 83; mental illness, 86, 100, 203n87; portraits/images of, 73, 74; and priesthood of art, 147; skull of, 97, 203n87
Schütz, Gustav, 192n119, 208n73
Schwammerl (novella), 6
Schwarzspanierhaus, 22, 45–50, 47, 126, 192n119, 208n73
séances, 11, 28, 133–34
Selbstinszenierung (self-staging), 145–47, 150–51
Seligmann, Franz Romeo, 94–95
Setkowicz, Adam, 183n25
sexual deviance, 86, 200n34
sexual eroticism, 40, 44, 129–30
Shaffer, Peter, *Amadeus*, 116
Shakespeare, William, 15, 18, 122
Shroud of Turin, 121
Simmel, Georg, 153, 214n73
Simmons, Sherwin, 129
Simrock, Karl, 186n16
skulls: anatomist collections of, 87, 91–94; phrenology, 82, 85, 87–88, 89, 90, 91, 94, 97–102, 103; theft of, 91, 92, 94, 200n41
Society for Friends of Music, 1, 24, 93, 96, 97, 98
Soemmerring, Samuel Thomas von, 88
Solie, Ruth, 23
Sontag, Susan, 172
Sound of Music, The (musical), 76

253

specimens. *See* skulls
spirit mediums, 11, 28, 29, 133–34, 153
spiritual presence rhetoric, 29, 37, 48, 52
Spitzer, Daniel, *Wagnerians in Love*, 143–44, 145–46, 160
Stadler, Margarete, 33
stage mediums, 11, 133–34, 153
Stammbücher (keepsake/friendship albums), 6, 8, 34, 36, 147
Starhemberg, Camillo von, 58
Stassen, Franz, 28, 29, 130
statues. *See* monuments
Steinacker, Edmund, 214n59
Steinberg, Michael P., 75–76
Steiner, Rudolph, 166, 217n132
Stern, Wolf, 7
Sterneck, Carl von, 53, 57, 58, 60
Stift Heiligenkreuz (Vienna), 46–48
Strauss, Richard, *Also Sprach Zarathustra*, 39
Streicher, Nannette and Andreas, 89–90
Struve, Gustav, 97
Stuck, Franz von, 131, 209n90; *Beethoven Mask with Laurel Wreath*, 123, 125
Sullivan, J. W. N., 118
surrealist art, 125, 129, 130
Swedenborg, Emanuel, 201n49
Sylva, Carmen, 151
symbolist movement, 133
syphilis, 86

Tandler, Julius, 84, 199n20
Tappert, Wilhelm, 144
Tchaikovsky, Pyotr Ilyich, 1, 158, 181n2
Teltscher, Josef, 112
Teutonic superiority rhetoric, 85, 90, 101
Thaly, Kálmán von, 213n59
thing theory, 9
Third Reich, Nazi ideology, 90, 159, 161, 166, 168–69, 170, 218n153
Thomas à Kempis, *De Imitatione Christi*, 212n31
Thoreau, Henry David, 53
Tilgner, Viktor, 52
timelessness, 154–55, 162, 166
Timms, Edward, 153–54
Tisza, István, 214n59
Toblach, Italy, 77–78
tombs, pilgrimage to, 46, 97, 162, 165, 169, 218n151
Torggler, Hermann, *Beethoven-Phantasieportrait*, 209n90
tourism industry, 14, 18, 27–28. *See also* pilgrimage
Treitel, Corinna, 218n153
Trendelenburg, Friedrich, 199n19
tuberculosis, 80, 86
Türck, Hermann, 117–18

Ulißen, Regina, 190n78

Vazsonyi, Nicholas, 26, 145
vera icon (true image), 120–22
Verdery, Katherine, 104, 161
Verdi, Giuseppe, 26
Verein Beethoven-Haus, 26–27, 28, 29, 30
verkitscht. *See* kitsch
Veronica, Saint, 113, 121
Viardot, Louis, 181n2
Viardot, Pauline, 1
Vienna: Beethoven death house in (Schwarzspanierhaus), 22, 45–50, *47*, 126, 192n119, 208n73; Beethoven monument in, 24–25; Beethoven's one-hundredth birth year celebrated in (1870), 23–24; biofiction popularity in, 6; as city of music (*Musikstadt Wien*), 24, 25, 46, 166; Karlskirche monument proposal, 12, 184n49; kitsch culture in, 154; and Mahler heritage, 78; modernization and growth in, 45, 49; Mozart monument in, 51, 52; Mozart's *Magic Flute* cottage in, 57–58
Vienna City Museum, 48
Vienna State Opera, 55, 60, 155
visitors' books, 31, 33, 34–45, *35*, *37*, *38*, 194n23

Waddell, Nathan, 109, 128–29
Wagner, Richard: cult of, 18, 143–46, 151, 158; face mask of, 122; as heir of Beethoven, 119, 145; influence on Carl Berg, 38; and men's choirs, 65, 195n50; monuments of, 51; and New German School controversy, 24; in pathographies, 86; portraits/images of, 73, *74*; self-promotion strategies, 26, 55, 145
Wagner, Richard, works: *Parsifal*, 146; "A Pilgrimage to Beethoven," 15, 145, 185n72; "The Public in Time and Space," 75; *Ring* cycle, 55, 60, 75, 148, 151; *Tannhäuser*, 65
Währing, Beethoven's grave in, 24
Waldvogel, Richard, 86
Walhalla, 97, 134, 159, 169, 185n71

Walton, Benjamin, 19
Watson, Nicola, 13, 184n56
Wawruch, Andreas, 112, 117
Weber, Carl Maria von, 117, 206n41
Weber, William, 182n12
Wedekind, Frank, *Kitsch*, 157
Weihekuss (kiss of consecration), 145
Weimar: festivals in, 146, 147; as *Goethe-Stadt*, 22
Weinberg, Richard, 204n109
Weinzierl, Max von, "Des Künstlers Genius," 65–66, *67–70*
Weissmann, Adolf, 138, 211n4
Weltner, Albert, 62, 195n40
Wenusch, Adolf, 167
Wermonty, Alfons, 116
Westover, Paul, 13
white supremacy, 103
Wiener Werkstätte, 60, 141
Wild, Irene (Margarete Koelman), 39–44, 129–30, 172
Wilhelm II (kaiser), 45
Williams, Frank Ernest, 20
Williamson, George S, 181–82n2
Wimmer, Georg, *Head of Beethoven over Nocturnal Waters*, 123, *124*
Wittmann, Hugo, 147–49, 160
Wolf, Hugo, 38
Wunderkammer tradition (curiosity cabinets), 82, 123

Yang, Mina, 81
Yearsley, David, 101

Zauberflötenhäuschen (*Magic Flute* cottage). *See* Mozart, Wolfgang Amadeus, *Magic Flute* cottage (Salzburg)
Zilsel, Edgar: *Die Geniereligion*, 156; "Mozart and Time," 153–56
Zweig, Stefan, 46; *The World of Yesterday*, 49

www.ingramcontent.com/pod-product-compliance
Lightning Source LLC
Chambersburg PA
CBHW022045290426
44109CB00014B/993